Subcellular Biochemistry

Volume 21
Endoplasmic Reticulum

SUBCELLULAR BIOCHEMISTRY

SERIES EDITOR

J. R. HARRIS, Institute of Zoology, University of Mainz, Mainz, Germany

ASSISTANT EDITORS

H. J. HILDERSON, University of Antwerp, Antwerp, Belgium
D. A. WALL, SmithKline Beecham Pharmaceuticals, King of Prussia, Pennsylvania, U.S.A.

A Continuation Order Plan is available for this series. A continuation order will bring delivery of each new volume immediately upon publication. Volumes are billed only upon actual shipment. For further information please contact the publisher.

Subcellular Biochemistry

Volume 21
Endoplasmic Reticulum

Edited by

N. Borgese
CNR Center for Cytopharmacology
University of Milan
Milan, Italy

and

J. R. Harris
Institute of Zoology
University of Mainz
Mainz, Germany

SPRINGER SCIENCE+BUSINESS MEDIA, LLC

The Library of Congress cataloged the first volume of this title as follows:

Sub-cellular biochemistry.
   London, New York, Plenum Press.
   v. illus. 23 cm. quarterly.
   Began with Sept. 1971 issue. Cf. New serial titles.
   1. Cytochemistry—Periodicals. 2. Cell organelles—Periodicals.
QH611.S84   574.8'76   73-643479

ISBN 978-0-306-44450-0   ISBN 978-1-4615-2912-5 (eBook)
DOI 10.1007/978-1-4615-2912-5

This series is a continuation of the journal *Sub-Cellular Biochemistry*,
Volumes 1 to 4 of which were published quarterly from 1972 to 1975

Contributors

Günter Blobel   Laboratory of Cell Biology, Rockefeller University, New York, New York 10021

Stefano Bonatti   Department of Biochemistry and Medical Biotechnology, University of Naples "Federico II," Naples, Italy

Nica Borgese   CNR Center for Cytopharmacology and Department of Pharmacology, University of Milan, Milan, Italy

Vincenzo Cerundolo   Institute of Molecular Medicine, John Radcliffe Hospital, Headington, Oxford OX3-9DU, United Kingdom

Lan Bo Chen   Dana-Farber Cancer Institute and Harvard Medical School, Boston, Massachusetts 02115

Gustav Dallner   Department of Biochemistry, University of Stockholm, S-106 91 Stockholm, and Clinical Research Center at Huddinge Hospital, Karolinska Institute, S-141 86 Huddinge, Sweden

Antonello D'Arrigo   CNR Center for Cytopharmacology and Department of Pharmacology, University of Milan, Milan, Italy

Roger A. Davis   Department of Biology, San Diego State University, San Diego, California 92182

Marcella De Silvestris   CNR Center for Cytopharmacology and Department of Pharmacology, University of Milan, Milan, Italy

Philippe F. Devaux   Institut de Biologie Physico-Chimique, F-75005 Paris, France

Thomas Dierks   Zentrum Biochemie/Abteilung Biochemie II der Universität, D-37073 Göttingen, Germany

Johan Ericsson   Department of Biochemistry, University of Stockholm, S-106 91 Stockholm, and Clinical Research Center at Huddinge Hospital, Karolinska Institute, S-141 86 Huddinge, Sweden

AnnaMaria Fra   San Raffaele Scientific Institute, Milan, Italy

Robert B. Freedman   Biological Laboratory, University of Kent, Canterbury, Kent CT2 7NJ, United Kingdom

Peter Klappa   Zentrum Biochemie/Abteilung Biochemie II der Universität, D-37073 Göttingen, Germany

Christopher Lee   Dana-Farber Cancer Institute and Harvard Medical School, Boston, Massachusetts 02115. *Present address:* Department of Cell Biology, Stanford University Medical Center, Stanford, California 94305

Jennifer Lippincott-Schwartz   Cell Biology and Metabolism Branch, National Institute of Child Health and Human Development, National Institutes of Health, Bethesda, Maryland 20892

Ryuichi Masaki   Department of Physiology, Kansai Medical University, Moriguchi, Osaka 570, Japan

Jacopo Meldolesi   Department of Pharmacology, CNR Cytopharmacology and B. Ceccarelli Centers, and San Raffaele Scientific Institute, University of Milan, Milan, Italy

Grazia Pietrini   CNR Center for Cytopharmacology and Department of Pharmacology, University of Milan, Milan, Italy

Pamela J. E. Rowling   Biological Laboratory, University of Kent, Canterbury, Kent CT2 7NJ, United Kingdom

Sanford M. Simon   Laboratory of Cellular Biophysics, Rockefeller University, New York, New York 10021

Roberto Sitia   San Raffaele Scientific Institute, Milan, and National Institute for Cancer Research, Genoa, Italy

Alan M. Tartakoff   Institute of Pathology, Case Western Reserve University, Cleveland, Ohio 44106

Yutaka Tashiro   Department of Physiology, Kansai Medical University, Moriguchi, Osaka 570, Japan

Maria Rosaria Torrisi   Department of Experimental Medicine, University of Rome "La Sapienza," Rome, Italy

Antonello Villa   Department of Pharmacology, CNR Cytopharmacology and B. Ceccarelli Centers, and San Raffaele Scientific Institute, University of Milan, Milan, Italy

Akitsugu Yamamoto   Department of Physiology, Kansai Medical University, Moriguchi, Osaka 570, Japan

Maria Zimmermann   Zentrum Biochemie/Abteilung Biochemie II der Universität, D-37073 Göttingen, Germany

Richard Zimmermann   Zentrum Biochemie/Abteilung Biochemie II der Universität, D-37073 Göttingen, Germany

Preface

For cell biologists interested in protein secretion and membrane biogenesis in the eukaryotic cell, the endoplasmic reticulum (ER) has always been a central research theme. In the past, however, attention was focused mainly on the translocation machinery permitting the passage of nascent proteins from the cytosol to the lumen of the ER. The importance and complexity of the events occurring *within* the lumen have begun to be appreciated only recently. Moreover, novel ER functions have been discovered, and investigators not traditionally involved with the ER—e.g., immunologists and students of signal transduction—have turned their attention to this ubiquitous organelle. Indeed, the processes in which the ER is involved appear at this point so numerous and varied that it is not easy to combine them into a global picture of this organelle.

Given the recent surge of interest in the ER, we considered the time ripe to put together in a single volume recent information and ideas on this subcellular compartment. In this book, we have collected 15 articles contributed by experts working on different aspects of the ER. Thus, the volume covers events occurring in the lumen as well as on the cytoplasmic side of the ER membrane, and deals with such diverse functions as protein translocation and export, lipid metabolism, Ca^{2+} homeostasis, and antigen presentation.

The first eight chapters are concerned with the translocation and processing of proteins destined for transport along the secretory pathway. In Chapter 1, Simon and Blobel review their electrophysiological evidence in favor of the existence of protein-conducting channels in translocation-competent membranes, and present a general conceptual framework within which to consider the problem of protein translocation. Klappa *et al.*, in Chapter 2, give a detailed account of the components so far identified in this process in mammals and yeast. The following two chapters are dedicated to the modifications that translocated proteins undergo within the ER lumen. In Chapter 3, Rowling and Freedman give a comprehensive review of folding, assembly, and posttranslational modifications, and in Chapter 4, Tartakoff focuses on one important and interesting modification: the attachment of glycolipid anchors to certain membrane proteins. Chapter 5 and 6, by Lippincott-Schwartz and by Bonatti and Torrisi, respectively, deal with the exit of newly synthesized proteins from the ER. The problem of quality control, the mechanisms of specific retention of ER resident proteins, and the routes of anterograde ER–Golgi and retrograde Golgi–ER membrane traffic are dealt with in these two chapters. The existence of quality control mechanisms in the ER lumen, which prevent the exit of improperly folded or assembled proteins, raises the problem of the fate of those proteins that do not meet the requirements for transport. This problem is extensively covered by Fra and Sitia in Chapter 7. Finally, Chapter 8, by Davis, focuses on the special case of lipoprotein assembly and secretion, because of the particular problems it poses and because of its physiopathological implications.

Chapters 9 and 10 cover two functions of the ER in which there has been a great deal of recent interest, i.e., Ca^{2+} signaling and antigen presentation. Chapter 9, by Meldolesi and Villa, first describes the requirements a subcellular compartment must meet to qualify as a rapidly exchanging Ca^{2+} store, and then illustrates the complexity and heterogeneity of the cytology of these compartments. Chapter 10, by Cerundolo, deals with peptide transport across the ER membrane, as well as with the role of peptides in MHC class I molecule assembly and transport.

Chapters 11–14 deal with ER lipid metabolism and with enzymes that function on the cytosolic face of the membrane. The complex ramifications of the mevalonate pathway and the regulation of the ER enzymes involved are comprehensively reviewed by Ericsson and Dallner in Chapter 11. Chapter 12, by Devaux, deals with a unique attribute of the ER membrane, central to its role in membrane biogenesis, i.e., its capacity to rapidly relocate phospholipids. Chapters 13 and 14, by Tashiro *et al.* and Borgese *et al.* respectively, review the biochemistry, from the cell biologist's point of view, of three ER enzymes with cytosolically oriented active domains involved in drug and lipid metabolism: the cytochrome P-450 family (Chapter 13) and NADH-cytochrome b_5 reductase and cytochrome b_5 (Chapter 14). The two chapters illustrate how newly synthesized ER enzymes use different targeting pathways to reach the ER membrane.

The book closes with the contribution of Lee and Chen (Chapter 15), which looks at the fascinating problem of the genesis of ER architecture, a field that has only recently become amenable to investigation, thanks to new technological developments.

Although we realize that a subject as vast as the ER cannot be fully covered in one volume, we nonetheless hope that this book will be useful to those who would like to take an integrated look at this multifaceted organelle.

We thank Stefano Bonatti, Kathryn Howell, and Roberto Sitia for help and advice in putting together this book.

N. Borgese
J. R. Harris

Milan, Italy
Mainz, Germany

Contents

Chapter 3
**Folding, Assembly, and Posttranslational Modification of Proteins
within the Lumen of the Endoplasmic Reticulum**
Pamela J. E. Rowling and Robert B. Freedman

Chapter 4
Biological Functions and Biosynthesis of Glycolipid-Anchored Membrane Proteins
Alan M. Tartakoff

Chapter 5
Membrane Cycling between the ER and Golgi Apparatus and Its Role in the Regulation of Biosynthetic Transport
Jennifer Lippincott-Schwartz

Chapter 6

**The Intermediate Compartment between Endoplasmic Reticulum
and Golgi Complex in Mammalian Cells**

Stefano Bonatti and Maria Rosaria Torrisi

Chapter 7

The Endoplasmic Reticulum as a Site of Protein Degradation

AnnaMaria Fra and Roberto Sitia

Chapter 8

**The Endoplasmic Reticulum Is the Site of Lipoprotein Assembly
and Regulation of Secretion**

Roger A. Davis

Chapter 9

Endoplasmic Reticulum and the Control of Ca²⁺ Homeostasis

Jacopo Meldolesi and Antonello Villa

Chapter 10

Antigen Processing and Presentation: The Role of the
Endoplasmic Reticulum

Vincenzo Cerundolo

Chapter 11

**Distribution, Biosynthesis, and Function of Mevalonate
Pathway Lipids**

Johan Ericsson and Gustav Dallner

Chapter 15
**Motility and Construction of the Endoplasmic Reticulum
in Living Cells**
Christopher Lee and Lan Bo Chen

Chapter 1

Mechanisms of Translocation of Proteins across Membranes

Sanford M. Simon and Günter Blobel

1. INTRODUCTION

The movement of proteins from their site of synthesis in the cytosol to their final destination is fundamental for organelle and membrane biogenesis, secretion, and cellular organization. During the past 20 years, many advances have been made in our understanding of protein topogenesis—the process by which a protein achieves its topological distribution within a cell and, for an integral membrane protein, how it achieves its particular transmembrane topography (Blobel, 1980).

Many of these advances have been shaped by the signal hypothesis (Blobel and Sabatini, 1971; Blobel and Dobberstein, 1975). The signal hypothesis was initially proposed to explain how proteins could be targeted to the secretary pathways (Blobel and Sabatini, 1971). It was known at the time that as proteins are synthesized they translocate (cross) the membrane of the endoplasmic reticulum (ER) (Redman *et al.*, 1966; Redman and Sabatini, 1966). This hypothesis

Abbreviations used in this chapter: ER, endoplasmic reticulum; RM, rough microsomes; SRP, signal recognition particle.

Sanford M. Simon   Laboratory of Cellular Biophysics, Rockefeller University, New York, New York 10021.   **Günter Blobel**   Laboratory of Cell Biology, Rockefeller University, New York, New York 10021.
Subcellular Biochemistry, Volume 21: Endoplasmic Reticulum, edited by N. Borgese and J. R. Harris. Plenum Press, New York, 1993.

posited that a signal in the initial segment of a protein targets it to the ER membrane. The hypothesis initially was a conceptual framework for formulating experimental questions such as: Is there a role for the amino-terminal extension that is synthesized on secretory proteins? Is this (signal) sequence necessary for targeting? Is it sufficient? However, it has since been elaborated into a general theory for protein topogenesis, one that suggests that signals exist both to initiate as well as to terminate translocation (Blobel, 1980). The model proposes that the transmembrane topography of a protein is determined by a sequential reading of the start/stop signals in its amino acid sequence. This elaboration of the signal hypothesis implies not only that signal sequences must function in targeting translocation but that they must also directly interface with the translocation machinery.

The role of signal sequences in the targeting process has been clearly established for eukaryotic ER (Walter et al., 1984), chloroplasts (Dobberstein et al., 1977), mitochondria (Maccecchini et al., 1979), nuclei (Kalderon et al., 1984), peroxisomes (Gould et al., 1988; Swinkels et al., 1991), and prokaryotic plasma membrane (Chang et al., 1978). However, the role of signal sequences in the translocation process has not been established. Indeed, little is known of the mechanism of translocation or the transmembrane translocation machinery. Further, there has been considerable disagreement over whether proteins partition into and translocate directly through the lipid bilayer (Engelman and Steitz, 1981; Von Heijne and Blomberg, 1979) or whether they translocate through an aqueous pore through the membrane (Blobel and Dobberstein, 1975; Blobel, 1980; Singer et al., 1987; Simon and Blobel, 1991). Membranes are barriers that segregate and, thus, compartmentalize cellular functions. The ability of a molecule to permeate a membrane is directly proportional to its solubility in lipid. The only exception to this rule are those substrates for which the membrane has specific transporters. Many proteins that vary in their charge and hydrophobicity are able to cross the membrane without destroying the permeability barrier. This requires a specific translocation machinery in the membrane. Further, the observation that the different proteins compete with each other for translocation implies that they use a common biochemical machinery.

The primary structure of signal sequences targeted to the ER have little in common other than positive charges at their amino-terminals and a stretch of hydrophobic amino acids in their middles. This ubiquitous hydrophobic stretch has prompted the notion that signal sequences partition directly into the lipid bilayer and that the rest of the protein then follows suit, that is, partitioning into the lipid bilayer on one side and exiting on the other (Engelman and Steitz, 1981; Von Heijne and Blomberg, 1979). It has been argued that although it would cost considerable energy to partition these proteins, which are often hydrophilic, into the membrane, the energy would be recovered when the protein returns to the aqueous milieu at the opposite membrane surface (Engelman and Steitz, 1981).

However, nature is ruled by kinetics rather than equilibria. Although a protein may be able to *eventually* partition across a bilayer, it is important to establish whether it can occur in the lifetime of the cell. Thus, we would argue that proteins must be shielded from the bilayer during the process of translocation: they must move across the membrane through protein-conducting channels.

2. EVIDENCE FOR THE EXISTENCE OF PROTEIN-CONDUCTING CHANNELS IN MAMMALIAN ER AND IN PROKARYOTIC PLASMA MEMBRANE

If protein-conducting channels exist, they should be considerably larger than conventional ion-conducting channels—and accessible using the arsenal of tools employed to study ion-conducting channels. What follows is a description of some of our efforts to use such tools to probe for the existence of protein-conducting channels. The general technique we used was to reconstitute translocation-competent membranes into planar lipid bilayers (Mueller *et al.*, 1962; Mueller and Rudin, 1969; Zimmerberg *et al.*, 1980; Miller, 1986). Then a voltage clamp was used to study the conductance properties of the reconstituted membrane (Simon *et al.*, 1989).

Initially, vesicles of pancreatic ER [rough microsomes (RM)] were fused to a planar lipid bilayer. We observed ion-conducting channels that were large in terms of their conductance (the ease with which ions could flow through the channel) and large in terms of the kinds of ions that could flow through them (such as glutamate and Hepes).

With the fusion of each RM to the bilayer, 5 to 15 ion channels were observed (Simon *et al.*, 1989; Simon and Blobel, 1991). However, with an electron microscope, hundreds of ribosomes could be seen on the surface of each RM. Each of these ribosomes is potentially synthesizing proteins that are translocating the bilayer. If the channels are the paths through which proteins translocate, we wondered, why were so few observed? One possible answer was that when occupied, these channels might not be freely permeable to ions. We tested this possibility by using puromycin to release nascent chains from their membrane-bound ribosomes (Simon and Blobel, 1991). [The peptidyltransferase of the ribosome adds puromycin to the carboxy-terminus of the nascent chain (Traut and Monro, 1964).] RMs were fused to a planar lipid bilayer and then puromycin was added to a concentration of 100 μM (a concentration usually used to block protein synthesis). This elicited a 100-fold increase in the conductance of the pancreatic membranes, a result consistent with an "unplugging" of larger numbers of protein-conducting channels (Simon and Blobel, 1991).

This conductance increase was dependent on the presence of RMs. Puromycin had no effect when added to the solutions bathing only a lipid bilayer.

Even after fusion of RMs to the bilayer, puromycin had no effect when added to the solutions bathing what is topologically equivalent to the lumenal side of the ER (when a vesicle fuses to the bilayer its lipids and proteins maintain their *in situ* topological orientation). When puromycin is subsequently added to the solutions bathing the cytosolic side of the very same bilayer, it evokes a large increase in the conductance of the membrane.

After adding puromycin (100 μM final concentration), there was a consistent large increase in the conductance of the membranes. To observe the release of individual translocating chains, we lowered the concentration of puromycin to 0.3 μM. After adding puromycin, we observed discrete increases of conductance in steps of 220 pS (in 50 mM KCl) (Simon and Blobel, 1991). We concluded that these conductance changes represent the unplugging of individual protein-conducting channels.

It is difficult and, more often than not, misleading to use the conductance to determine a channel's radius. However, these channels were large enough to let glutamate (180 pS in 50 mM K-glutamate) and Hepes (360 pS in 50 mM KCl and 50 mM K-Hepes) pass through. Thus, they could be of sufficient size to let unfolding polypeptide chains pass through.

The preceding studies were done at low salt concentrations (50–100 mM) where the ribosomes remain associated with the ER after puromycin treatment (Adelman *et al.*, 1973). What would happen to these channels at higher salt concentrations (≥ 150 mM) when the ribosomes dissociate? After fusing RM to a bilayer, and adding puromycin, we raised the concentration of salt bathing the membranes in steps of 50 mM. At first there was a gradual increase of conductance upon addition of salt to the solutions bathing the membrane. This is the consequence of increasing the number of charge carriers available to flow through the open channels. This increase was gradual, not instantaneous: it took time for the ions to diffuse to the surface of the membrane. The conductance continued to increase with subsequent increases of salt. However, as the salt concentration was raised to approximately 300 mM, the large channels closed. Not every channel closed at 300 mM. In general, they closed over the range of 150 mM to 600 mM (Simon and Blobel, 1991). This is the same range over which ribosomes dissociate from the membrane (Adelman *et al.*, 1973). Not all channels closed under these circumstances and we do not know if they all would have closed if we waited longer periods of time. Consequently, one conclusion is that the protein-conducting channels remain open if the ribosomes remain associated with the membrane. The treatment used to probe for these channels—puromycin—is nonphysiological. However, the observation makes sense in a physiological context. If these channels were to be open when they were not translocating polypeptides, then they would provide a pathway for equilibration of ions and metabolites between the lumen of the ER and the cytoplasm. It is important to note that in physiological conditions, upon the termination of protein translocation, the ribosomes dissociate from the membrane. Thus, in a physiological milieu, the channels would never be freely conductive to metabolites. They would be either open and translocating a chain (thus blocked) or closed.

If the channels close at the end of protein translocation, they must open at its initiation. Many of the steps involved in targeting a nascent peptide to the membrane to initiate translocation have been defined in mammalian systems. For example, after synthesis of the signal sequence on free ribosomes, a cytosolic factor [signal recognition particle (SRP)] binds to both the signal sequence and the ribosome (Walter *et al.*, 1981). Binding to the ribosome arrests protein synthesis (Walter and Blobel, 1981) until SRP, as part of an SRP–signal sequence–ribosome complex, binds to a receptor on the ER (Gilmore *et al.*, 1982). Then, in a series of steps requiring GTP hydrolysis (Connolly and Gilmore, 1989), the signal sequence dissociates from SRP and translocation is initiated.

The simplest hypothesis (and the easiest to test) holds that the signal sequence alone is both necessary and sufficient to initiate the opening of the protein-conducting channels. More complex possibilities require both SRP and the signal sequence, or both and the ribosome, or all of the above and nucleotide hydrolysis. We started by testing the simplest hypothesis. The experimental question was: are chemically synthesized signal peptides sufficient for opening large channels in the plasma membrane of *E. coli?* There were a few clear advantages to this experimental approach. First, in *E. coli* translocation across the plasma membrane can be dissociated from protein synthesis. Thus, it might be possible to uncouple requirements for a signal sequence or cytosolic factors from requirements for ribosomes. The results obtained with puromycin suggest that opening of the protein-conducting channel is tightly coupled to protein translocation. We were concerned that if we tried to open the channels with a complete protein it was possible that, as the channels opened, the protein would insert and block ion flow—our only measure of channel opening. Our hope was that by probing with only a signal sequence we could open the channel without occluding it with a translocating chain.

Plasma membrane vesicles from *E. coli* were fused to a planar lipid bilayer. In general, very few ion channels were observed under our experimental conditions [50 mM KCl, 5 mM K-Hepes (pH 7.5), 3 mM MgCl$_2$]: few, very small potassium channels; a large 60 pS channel (that was only seen when the membrane potential was positive on the cytoplasmic side), and an occasional channel of 115 pS. When right-side-out plasma membrane vesicles are made, there are fragments of the outer membrane attached to the outside of the vesicles. When these preparations were assayed for protein content on SDS-PAGE, some porins could be observed. However, these porins were rarely seen in our electrophysiological recordings. When the inner membrane vesicles fuse, the outer membrane fragments do not become integrated into the planar lipid bilayer and, thus, they are not seen by the electrophysiological voltage clamp setup.

Our probe for potential effects on the signal sequence was the amino-terminal 25-amino-acid fragment of the preLamB protein (Emr and Silhavy, 1983): MMITLRKLPLAVAVAAGVMSAQAMA. After fusion of plasma membrane vesicles to the planar bilayer, the signal peptide was added to a concentration of 200 nM. This corresponds to 120 signal peptides per volume of *E. coli* (Roberts *et al.*, 1963). This elicited a 500-fold increase in the conductance of the membrane (Simon and Blobel, 1992). This increase was observed only when the voltage across the membrane was kept positive on the cytosolic side (the physiological membrane potential is negative on the cytosolic side relative to the outside). To examine the microscopic nature of this large conductance increase, we added a substantially smaller concentration of signal peptide: 0.2 nM. This is roughly one signal peptide for every eight *E. coli* volumes. After addition of the signal peptide, there was a new channel of 200 pS. This channel was predominantly seen when the membrane potential was positive and tended to be closed when the cytosolic side of the membrane was negative.

These channels were further characterized by studying how voltage and salt concentration affected two parameters: (1) the amount of time the channel spent open and (2) the ease with which ions flowed through an open channel. The conductance of the channels fluctuated open and closed (at positive voltages) in low salt ($\sim$ 50 mM) and remained closed at negative membrane potentials. At higher salt concentrations ($>$ 500 mM) the channels still opened only at positive voltages, but then they remained open. They did not close, even at negative membrane potentials. There are a number of possible explanations for the effects of voltage and salt. The opening and closing of the channels could be directly dependent on the transmembrane voltage. Like the sodium or calcium channels, they would be closed at negative membrane potentials, and their probability of opening increases as the membrane potential becomes more positive. However, this could not explain the observations at high salt where the channels do not close at negative voltages. An alternative explanation is suggested by the observation that there are two features in common to signal sequences: (1) they have a positive charge at one end and (2) they have a stretch of hydrophobic amino acids. The positive charge could explain the voltage dependence. As the membrane potential is made more positive, the signal sequence would be driven farther into the channel. Similar observations have been made for many other ion channels. For example, insertion of the positively charged tetramethylammonium into potassium channels is strongly dependent on the transmembrane voltage. Thus, the voltage is affecting the channel indirectly by affecting the probability of a signal peptide being in the channel in the proper orientation for binding. The hydrophobic stretch could explain the salt effect. Hydrophobic–hydrophobic interactions are strengthened at higher salt concentrations. It is possible that the hydrophobic stretch of amino acids is important for binding to a hydrophobic site in the channel. One would expect that at low salt concentrations the signal peptide would bind and dissociate, causing the channel to open and close. However, at higher salt concentrations the peptide would bind and remain bound, stabilized by the hydrophobic interactions. Thus, the channel would open at positive voltages when the signal bound, but then remain open since the signal peptide does not dissociate.

In conclusion we have observed large aqueous channels in membranes that can translocate proteins (the mammalian ER and prokaryotic plasma membrane) under conditions that disturb translocation. Under similar salt conditions (50 mM KCl) the two channels are of similar conductance: 220 pS. This does not mean that they are the same channel. However, together they begin to make a compelling case for the existence of transmembrane aqueous channels for protein translocation.

These are the first electrophysiological characterizations of channels for the transport of proteins across membranes. These observations raise a number of questions. How are these channels regulated? Do they exist in other organelles? Are transmembrane aqueous pores a general mechanism for moving macromolecules across membranes? If macromolecules are in transmembrane pores, what moves them across?

3. REGULATION OF PROTEIN-CONDUCTING CHANNELS

There are three questions that are raised about regulation of these channels. (1) What are the physiological signals to open them at the initiation of translocation and to (2) close them at the end of translocation? (3) How do these channels allow potential transmembrane segments of translocating proteins to move laterally out of the lumen of the channel into the lipid bilayer? Our results suggest that the signal sequence is a sufficient ligand for opening the channel. This is consistent with a number of other observations. It has been demonstrated that the SRP, while extremely important for targeting nascent chains to the ER, is not required for translocation. Short peptides (< 80 amino acids) can translocate across mammalian ER without SRP (Zimmermann et al., 1990; see Klappa et al., this volume). This suggests that SRP's role is more significant for longer proteins. This increases the significance of its role in delaying protein synthesis until a protein has engaged the translocation appartus of the ER. Deletion of SRP in yeast is not lethal—although the yeast grow *extremely* poorly (Hann and Walter, 1991). Thus, SRP, as a targeting agent, kinetically accelerates translocation. However, it may not be essential for translocation across the membrane.

How could these channels allow for transmembrane segments to be integrated into the bilayer? Proteins that will be fully secreted across the bilayer as well as those that will remain resident both use the same translocation machinery (Lingappa *et al.*, 1978). This raises a fundamental topological problem: How can the translocating chain move out of the lumen of the protein-conducing channel and into the bilayer? Two broad classes of models exist. First, the channel could be made of multiple subunits that dissociate when a transmembrane segment is reached. This would leave the transmembrane segment embedded in the membrane. Alternatively, the channel could remain largely intact but open laterally to allow the chain to move parallel to the plane of the bilayer. Either way, a mechanism is needed for the translocation machinery to recognize when a latent transmembrane domain has been reached. Clearly, hydrophobicity alone is not a sufficient signal to indicate a latent transmembrane segment. Many proteins that fully translocate across the membrane have long stretches of hydrophobic amino acids. If any of these hydrophobic stretches were allowed to partition out of the channel into the bilayer, then they would most likely be trapped in the membrane. During protein synthesis the hydrophobic amino acids are assembled into a chain in the aqueous milieu of the channel inside the ribosome. To move them along within a hydrophilic environment takes little additional energy. However, once the hydrophobic amino acids embed in the bilayer, it would take considerable energy to pull them back out. It is not known how these segments are recognized. They may be detected by the ribosome, the translocation machinery, a cytosolic SRP-like factor, or any combination of the above.

4.   OTHER ORGANELLES

4.1.   Mitochondria and Chloroplasts

Direct evidence for protein-conducting channels does not exist in the mitochondrion and chloroplast. However, the results from a number of different experiments suggest that they share with the ER a common mechanism for translocation. First, hydrophilic molecules can be translocated into mitochondria. Double-stranded DNA has been coupled to a polypeptide containing a mitochondrial targeted signal sequence. This construct was translocated into the mitochondria (Vestweber and Schatz, 1989). The energetics of embedding such a construct in a lipid milieu is quite unfavorable. Second, it has been demonstrated that extremely hydrophobic sequences can translocate across mitochondrial and chloroplast membranes and out at the other side. The vesicular stomatitis virus G-protein has a single hydrophobic transmembrane domain near its carboxy-terminus. It is usually synthesized into the ER. When its ER-targeted signal sequence is replaced with a mitochondrial sequence, the protein translocates across the outer membrane and is found as a transmembrane protein of the inner membrane (Nguyen and Shore, 1987). When it is made with a chloroplast targeted signal sequence, it fully translocates across both chloroplast membranes (Lubben *et al.*,

1987). Two conclusions are suggested from these observations. First, the protein must not be translocating through the hydrocarbon core of the bilayer. The transmembrane domain of VSV G-protein is extremely hydrophobic. Once it partitioned into the membrane, it would not partition back into the aqueous phase on the opposite side—especially since it would mean simultaneously pulling subsequent hydrophilic segments into the membrane. The results also suggest that there is a specific machinery for decoding the sequences that indicate where to start and where to stop translocating a protein across the membrane. The topogenic sequence for the transmembrane segment of VSV G-protein that tells the ER "stop sending me across" apparently cannot be read by the machinery of the chloroplast or mitochondrion. Just as the signal sequences that target proteins are unique to each organelle, so are the stop-transfer sequences unique to each organelle.

4.2. Nucleus

Proteins and ribonucleoproteins are constantly traversing the nuclear pores. These are large structures that are held suspended in 100-nm circular pore complexes in the nuclear envelope. Molecules of up to 9 nm in diameter are able to freely diffuse in and out of these pores (Paine *et al.*, 1975; Jiang and Schindler, 1986). Larger molecules are able to enter only if they have a signal sequence that targets them for the nucleus. Like the ER and mitochondrial signal sequences, they are necessary (proteins do not enter the nucleus if the sequences are deleted) and sufficient (they can target cytosolic proteins to the nucleus) for nuclear transport. Thus, the nuclear pores are protein- and ribonucleoprotein-conducting channels. The only important differences are that they are significantly larger and they transport bidirectionally. Proteins can translocate across the ER, *E. coli* plasma membrane, mitochondria, and chloroplast only when they are in an unfolded conformation. This suggests that the pores have a maximum size of a few nanometers. In contrast, a gold sphere of 20 nm will be imported into the nucleus if it has a nuclear signal sequence. Little is known of the regulation of the nuclear pores. Are they like sphincters whose size is gated by the nuclear localization signal? Do they fully open, or only open large enough to allow passage? What is the role of ATP in transport through the pore?

5. DO TRANSMEMBRANE AQUEOUS PORES PROVIDE A GENERAL MECHANISM FOR MACROMOLECULAR TRANSPORT?

5.1. Toxin Transport

Most proteins secreted from bacteria use the same secretory pathway. Their signal sequences are interchangeable and their secretion is blocked by mutations in the "sec" pathway. However, a number of bacterial proteins use their own private transport pathway with its own signal sequences. This pathway is not blocked by "sec" mutations. Many bacterial toxins such as colicin V and hemolysin are members of this family. The genes required to secrete these proteins include TolC (which appears to be required by many different proteins) and specific proteins (such as HlyB and HlyD) that are encoded immediately adjacent to the structural gene for the toxin. There is no direct evidence that addresses the questions of whether these toxins are transported through the bilayer or through a transmembrane protein-conducting channel. Some of the toxins transported by these systems are extremely large (hemolysin is 1024 amino acids), suggesting that they may also be transported across the membrane in a linear, unfolded state. However, the only evidence supporting this point is the observation that the addition of LacZ to the amino-terminal of hemolysin blocks its export. The addition of LacZ (which folds into the cytosol) to the maltose-binding protein blocks its export through the "sec"-mediated transport pathway. The parallel observation with hemolysin suggests that a folded domain is not consistent with transport. A distinct signal sequence in the transported toxins has been identified that is both necessary and sufficient for transporting these toxins out of the cell. Thus, there is nothing specific about the structure of the toxin that allows it to cross the membrane. Instead, the competence for transport rests with the signal sequence targeting the protein to the right transporter.

5.2. DNA Transport

The transport of DNA across membranes has been studied in two bacterial systems: in phage infection of *E. coli* and in pseudomonas. When a number of different phages (T1, T4, T5, and T7) infect *E. coli*, there is a substantial increase in the permeability of the bacterial membrane (Boulanger and Letellier, 1988). The permeability of the membrane has been quantified during infection with T4 and with T5 phage. In both cases, the permeability of the membrane to potassium increases linearly with the number of infecting phage. The time course of the permeability increase correlates with the time course of DNA injection (Boulanger and Letellier, 1992). This correlation is particularly striking and clearly demonstrated in the case of T5 DNA injection, which occurs in two temporally discrete steps (Boulanger and Letellier, 1992). These results suggest that channels are the site of DNA transport.

Streptococcus pneumoniae naturally transfects itself with DNA from its environment. The transfection involves a specific increase in permeability to the DNA—there are no gross changes in the permeability of the bacterial membrane (e.g., no loss of proteins). A number of mutations have been found that block the ability to take up DNA (Morrison *et al.*, 1984). These mutations have been mapped to a protein that is strongly homologous to HlyB (see above), a key protein for transport of hemolysin out of bacteria (Alloing *et al.*, 1990). Mutation in these proteins also result in a depolarization of the membrane potential (although a depolarization, by itself, is not sufficient to block DNA uptake) (Trombe *et al.*, 1984). These results suggest that aqueous channels are being used for DNA uptake in *Pseudomonas*.

5.3. Metabolite Transport

Many metabolites such as maltose and histidine are transported across cellular membranes. Their transport has been particularly well characterized in *E. coli*. Numerous proteins have been identified that are either needed for binding the metabolite in the periplasm (Kang *et al.*, 1991; Davidson *et al.*, 1992) between the two cellular membranes, or required for transport across the inner cytoplasmic membrane (Kerppola and Ames, 1992; Davidson *et al.*, 1992). It has been assumed that the transport across the plasma membrane is via a transmembrane ATP-driven pump. However, many of the criteria used for establishing the presence of an ATP-driven pump have not been met. A phosphorylated intermediary has not been identified. Nor has a substrate-bound intermediary been identified. Alternatively, the transport may occur through a transmembrane aqueous pore. The periplasmic binding protein, upon binding substrate, could subsequently bind to, and open, the transmembrane channel. The released substrate would then be free to diffuse through the channel. Blockage of the channel by the binding protein on the periplasmic side would ensure that the substrate could only diffuse into the cell—thus transport would be vectorial.

6. MECHANISM OF TRANSPORT

If all of these molecules are in transmembrane aqueous pores, what is moving them across? Ions move through transmembrane pores by diffusion. There is a net movement across the channels because of transmembrane electrochemical gradient. No such gradient exists for the movement of macromolecules. If we assume that the macromolecules can diffuse back and forth, they should equilibrate on both sides of the membrane. We can explain transport if we add a second assumption: that mechanisms exist to modify macromolecules such that they get "trapped" or "ratcheted" on one side of the membrane. For example, if upon translocating even partially across the membrane a nascent chain is modified by glycosylation, that larger glycosylated moiety is less likely to diffuse back out across the pore. Other modifications such as binding of chaperones or differential folding (resulting from ΔpH or a change in the ionic milieu) would all ratchet the protein on one side of the membrane. The protein would then fluctuate back and forth from thermal energy until fully transported across the membrane. It has already been demonstrated that proteins translocate across the membrane through an aqueous channel (Simon and Blobel, 1991, 1992). It is also known that nascent proteins are modified as they translocate: by glycosylation, chaperones, cleavage of the signal sequence (which affects folding of the protein; see Rowling and Freedman, this volume). Any of these modifications could be sufficient to anchor the protein to one side of the membrane. Thus, this mechanism would result in the chain translocating across the membrane. However, would it be sufficient to move the protein across in time periods of physiological significance? This hypothesis has been tested by simulating the translocating chain using standard polymer dynamics simulation techniques (Simon *et al.*, 1992). Using both numerical and analytical solutions, it was demonstrated that such a "Brownian ratchet" was sufficient to move the chain across the membrane in milliseconds to seconds. The time for proteins to translocate cotranslationally is limited by the rate of protein synthesis. Consequently, a Brownian ratchet would move proteins across as quickly as they are synthesized. The time for a protein to translocate posttranslationally is faster than the rate we can measure (seconds). A Brownian ratchet mechanism is fast enough, and thus sufficient, to account for observed rates of co- and posttranslational translocation across membranes. Hence, it is not necessary to invoke a motor or pump to move the translocating chain.

7. CONCLUDING REMARKS

A growing body of evidence provides a compelling case for the existence of transmembrane aqueous channels for the translocation of proteins. These results raise many more questions than they answer: How are transmembrane proteins integrated into the bilayer? What are the components that form these protein-conducting channels? Are transmembrane aqueous channels responsible for other forms of macromolecular transport? We fully expect that various macromolecular transport systems will not be homologous. However, we hope the paradigms established here for the study of protein-conducting channels will be applicable to other systems.

8. REFERENCES

Adelman, M. R., Sabatini, D. D., and Blobel, G., 1973, Ribosome–membrane interaction. Non-destructive disassembly of rat liver rough microsomes into ribosomal and membranous components, *J. Cell Biol.* **56:**206–229.

Alloing, G., Trombe, M. C., and Claverys, J. P., 1990, The ami locus of the gram-positive bacterium Streptococcus pneumoniae is similar to binding protein-dependent transport operons of gram-negative bacteria, *Mol. Microbiol.* **4:**633–644.

Blobel, G., 1980, Intracellular protein topogenesis, *Proc. Natl. Acad. Sci. USA* **77**:1496–1500.

Blobel, G., and Dobberstein, B., 1975, Transfer of proteins across membranes. II. Reconstitution of functional rough microsomes from heterologous components, *J. Cell Biol.* **67**:852–862.

Blobel, G., and Sabatini, D. D., 1971, Ribosome–membrane interaction in eukaryotic cells, in: *Biomembranes 2* (L. A. Manson, ed.), pp. 193–195, Plenum Press, New York.

Boulanger, P., and Letellier, L., 1988, Characterization of ion channels involved in the penetration of phage T4 DNA into Escherichia coli cells, *J. Biol. Chem.* **263**:9767–9775.

Boulanger, P., and Letellier, L., 1992, Ion channels are likely to be involved in the two steps of phage T5 DNA penetration into *Escherichia coli* cells, *J. Biol. Chem.* **267**:3168–3172.

Chang, C. N., Blobel, G., and Model, P., 1978, Detection of prokaryotic signal peptidase in an Escherichia coli membrane fraction: Endoproteolytic cleavage of nascent f1 pre-coat protein, *Proc. Natl., Acad. Sci. USA* **75**:361–365.

Connolly, T., and Gilmore, R., 1989, The signal recognition particle receptor mediates the GTP-dependent displacement of SRP from the signal sequence of the nascent polypeptide, *Cell* **57**:599–610.

Davidson, A. L., Shuman, H. A., and Nikaido, H., 1992, Mechanism of maltose transport in Escherichia coli: Transmembrane signaling by periplasmic binding proteins, *Proc. Natl. Acad. Sci. USA* **89**:2360–2364.

Dobberstein, B., Blobel, G., and Chua, N. H., 1977, *In vitro* synthesis and processing of a putative precursor for the small subunit of ribulose-1,5-bisphosphate carboxylase of *Chlamydomonas reinhardtii*, *Proc. Natl. Acad. Sci. USA* **74**:1082–1085.

Emr, S. D., and Silhavy, T. J., 1983, Importance of secondary structure in the signal sequence for protein secretion, *Proc. Natl. Acad. Sci. USA* **80**:4599–4603.

Engelman, D. M., and Steitz, T. A., 1981, The spontaneous insertion of proteins into and across membranes: The helical hairpin hypothesis, *Cell* **23**:411–422.

Gilmore, R., Blobel, G., and Walter, P., 1982, Protein translocation across the endoplasmic reticulum. I. Detection in the microsomal membrane of a receptor for the signal recognition particle, *J. Cell Biol.* **95**:463–469.

Gould, S. J., Keller, G. -A., and Subramani, S., 1988, Identification of peroxisomal targeting signals located at the carboxy terminus of four peroxisomal proteins, *J. Cell Biol.* **107**:897–905.

Hann, B. C., and Walter, P., 1991, The signal recognition particle in S. cerevisiae, *Cell* **67**:131–144.

Jiang, L. W., and Schindler, M., 1986, Chemical factors that influence nucleocytoplasmic transport: A fluorescence photobleaching study, *J. Cell Biol.* **102**:853–858.

Kalderon, D., Roberts, B. L., Richardson, W. D., and Smith, A. E., 1984, A short amino acid sequence able to specify nuclear localization, *Cell* **39**:499–509.

Kang, C. H., Shin, W. C., Yamagata, Y., Gokcen, S., Ames, G. F., and Kim, S. H., 1991, Crystal structure of the lysine-, arginine-, ornithine-binding protein (LAO) from Salmonella typhimurium at 2.7-A resolution, *J. Biol. Chem.* **266**:23893–23899.

Kerppola, R. E., and Ames, G. F., 1992, Topology of the hydrophobic membrane-bound components of the histidine periplasmic permease. Comparison with other members of the family, *J. Biol. Chem.* **267**:2329–2336.

Lingappa, V. R., Katz, F. N., Lodish, H. F., and Blobel, G., 1978, A signal sequence for the insertion of a transmembrane glycoprotein. Similarities to the signals of secretory proteins in primary structure and function, *J. Biol. Chem.* **253**:8667–8670.

Lubben, T. H., Bansberg, J., and Keegstra, K., 1987, Stop-transfer regions do not halt translocation of proteins into chloroplasts, *Science* **238**:1112–1114.

Maccecchini, M. L., Rudin, Y., Blobel, G., and Schatz, G., 1979, Import of proteins into mitochondria: Precursor forms of the extramitochondrially made F1-ATPase subunits in yeast, *Proc. Natl. Acad. Sci. USA* **76**:343–347.

Miller, C., 1986, *Ion Channel Reconstitution*, Plenum Press, New York.

Morrison, D. A., Trombe, M. C., Hayden, M. K., Waszak, G. A., and Chen, J. D., 1984, Isolationof transformation-deficient Streptococcus pneumoniae mutants defective in control of competence, using insertion-duplication mutagenesis with the erythromycin resistance determinant of pAM beta 1, *J. Bacteriol.* **159:**870–876.

Mueller, P., and Rudin, D. O., 1969, Bimolecular lipid membranes: Techniques of formation, study of electrical properties, and induction of ionic gating phenomena, in: *Laboratory Techniques in Membrane Biophysics* (H. Passow and R. Stampfil, eds.), pp. 141–156, Springer-Verlag, Berlin.

Mueller, P., Rudin, D. O., Ti Tien, H., and Wescott, W. C., 1962, Reconstitution of cell membrane structure *in vitro* and its transformation into an excitable system, *Nature* **194:**979–980.

Nguyen, M., and Shore, G. C., 1987, Import of hybrid vesicular stomatitis G protein to the mitochondrial inner membrane, *J. Biol. Chem.* **262:**3929–3931.

Paine, P. L., Moore, L. C., and Horowitz, S. B., 1975, Nuclear envelope permeability, *Nature* **254:**109–114.

Redman, C. M., and Sabatini, D. D., 1966, Vectorial discharge of peptides released by puromycin from attached ribosomes, *Proc. Natl. Acad. Sci. USA* **56:**608–615.

Redman, C. M., Siekevitz, P., and Palade, G. E., 1966, Synthesis and transfer of amylase in pigeon pancreatic microsomes, *J. Biol. Chem.* **241:**1150–1158.

Roberts, R. B., Abelson, P. H., Cowie, D. B., Bolton, E. T., and Britten, R. J., 1963, *Studies of Biosynthesis in Escherichia coli*, Carnegie Institution of Washington, Washington, D.C.

Simon, S. M., and Blobel, G., 1991, A protein-conducting channel in the endoplasmic reticulum, *Cell* **65:**1–10.

Simon, S. M., and Blobel, G., 1992, Signal peptides open protein-conducting channels in E. coli, *Cell* **69:**677–684.

Simon, S. M., Blobel, G., and Zimmerberg, J., 1989, Large aqueous channels in membrane vesicles derived from the rough endoplasmic reticulum of canine pancreas or the plasma membrane of *Escherichia coli*, *Proc. Natl. Acad. Sci. USA* **86:**6176–6180.

Simon, S. M., Peskin, C. S., and Oster, G. F., 1992, What drives the translocation of proteins, *Proc. Natl. Acad. Sci. USA* **89:**3770–3774.

Singer, S. J., Maher, P. A., and Yaffe, M. P., 1987, On the translocation of proteins across membranes, *Proc. Natl. Acad. Sci. USA* **84:**1015–1019.

Swinkels, B. W., Gould, S. J., Bodnar, A. G., Rachubinski, R. A., and Subramani, S., 1991, A novel, cleavable peroxisomal targeting signal at the amino-terminus of the rat 3-ketoacyl-CoA thiolase, *EMBO J.* **10:**3255–3262.

Traut, R. R., and Monro, R. E., 1964, The puromycin reaction and its relation to protein synthesis, *J. Mol. Biol.* **10:**63–72.

Trombe, M. C., Laneelle, G., and Sicard, A. M., 1984, Characterization of a Streptococcus pneumoniae mutant with altered electric transmembrane potential, *J. Bacteriol.* **158:**1109–1114.

Vestweber, D., and Schatz, G., 1989, DNA–protein conjugates can enter mitochondria via the protein import pathway, *Nature* **338:**170–172.

Von Heijne, G., and Blomberg, C., 1979, Transmembrane translocation of protein, *Eur. J. Biochem.* **97:**175–181.

Walter, P., and Blobel, G., 1981, Translocation of proteins across the endoplasmic reticulum III. Signal recognition protein (SRP) causes signal sequence-dependent and site-specific arrest of chain elongation that is released by microsomal membranes, *J. Cell Biol.* **91:**557–561.

Walter, P., Ibrahimi, I., and Blobel, G., 1981, Translocation of proteins across the endoplasmic reticulum. I. Signal recognition protein (SRP) binds to in-vitro-assembled polysomes synthesizing secretory protein, *J. Cell Biol.* **91:**545–550.

Walter, P., Gilmore, R., and Blobel, G., 1984, Protein translocation across the endoplasmic reticulum, *Cell* **38:**5–8.

Zimmerberg, J., Cohen, F. S., and Finkelstein, A., 1980, Micromolar Ca^{2+} stimulates fusion of lipid vesicles with planar bilayers containing a calcium-binding protein, *Science* **210:**906–908.

Zimmermann, R., Zimmermann, M., Wiech, H., Schlenstedt, G., Müller, G., Morel, F., Klappa, P., Jung, C., and Cobet, W. W. E., 1990, Ribonucleoparticle-independent transport of proteins into mammalian microsomes, *J. Bioenerg. Biomembr.* **22:**711–723.

Chapter 2

Components and Mechanisms Involved in Transport of Proteins into the Endoplasmic Reticulum

Peter Klappa, Maria Zimmermann, Thomas Dierks, and Richard Zimmermann

1. MECHANISMS INVOLVED IN PROTEIN TRANSPORT INTO THE ENDOPLASMIC RETICULUM

Every polypeptide has a unique intra- or extracellular location where it fulfills its function. In all eukaryotes, proteins are synthesized in a single compartment, i.e., the cytosol (excluding protein synthesis in mitochondria and chloroplasts). Noncytosolic proteins, therefore, have to be directed to different subcellular locations. In these cases, the sites of synthesis and functional location are separated by at least one biological membrane. Thus, mechanisms that ensure the specific transport of proteins across membranes must exist.

In protein export, one can distinguish between signal peptide-independent

Abbreviations used in this chapter: BiP, heavy-chain binding protein; ER, endoplasmic reticulum; hps, heat shock protein; NEM, *N*-ethylmaleimide; ppcec, preprocecropin; ppl, preprolactin; SRP, signal recognition particle; SSR, signal sequence receptor; WBP1, wheat germ agglutinin binding protein 1.

Peter Klappa, Maria Zimmermann, Thomas Dierks, and Richard Zimmermann
Zentrum Biochemie/Abteilung Biochemie II der Universität, D 37073, Göttingen, Germany.

Subcellular Biochemistry, Volume 21: Endoplasmic Reticulum, edited by N. Borgese and J. R. Harris. Plenum Press, New York, 1993.

mechanisms and those in which a signal sequence is a prerequisite for specific transport. In mammals and yeasts, signal peptide-independent mechanisms (operating for export of certain newly synthesized proteins) take place at the plasma membrane where they involve transport components related to the multidrug resistance proteins, i.e., a family of ATP-dependent membrane proteins (Kuchler *et al.*, 1989; McGrath and Varshavsky, 1989; Rubartelli *et al.*, 1990). Another signal peptide-independent mechanism has been described recently for mammalian cells that is involved in antigen transport into the lumen of the endoplasmic reticulum (ER) where antigenic peptides are linked to MHC I molecules (Spies and DeMars, 1991; Parham, 1991, 1992; Levy *et al.*, 1991; Koppelman *et al.*, 1992; Cerundolo, this volume). Two peptide transporters, tap1 and tap2, with ATP-binding sites have been identified in this respect.

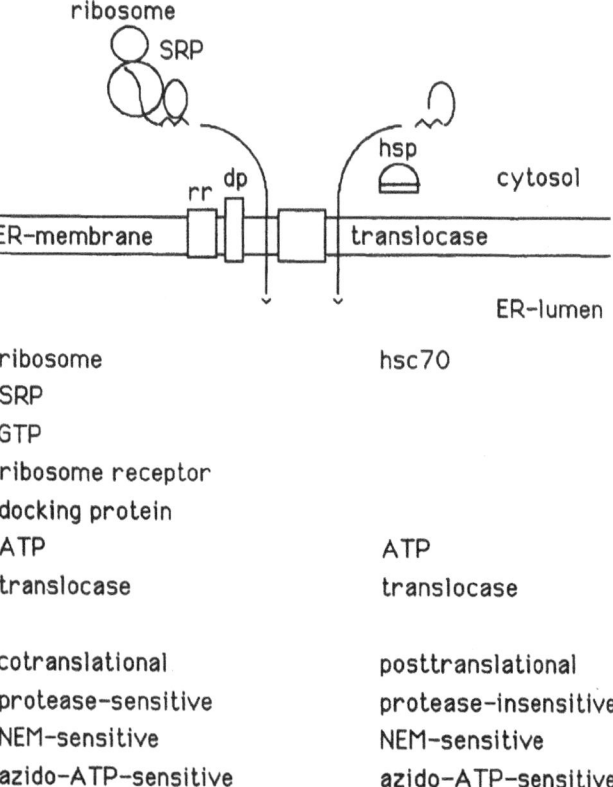

FIGURE 1. Components and mechanisms involved in protein transport into the endoplasmic reticulum. The cartoon summarizes the various components involved in ribonucleoparticle-dependent and -independent transport of proteins into microsomes. Furthermore, experimental details are indicated that point to differences and common features for the two mechanisms. See text for further details.

The signal peptide-dependent mechanism (operating for export of the vast majority of newly synthesized proteins) also is initiated at the level of the membrane of the ER (Figure 1). It includes various steps such as association of the precursor protein with the ER membrane, membrane insertion, and completion of translocation (in the case of soluble proteins). It is well established that precursor proteins are not transported in their native (i.e., folded) state and that signal peptides in the precursor proteins are involved in preserving the transport-competent (i.e., nonnative) state as well as in facilitating membrane specificity. Furthermore, there are two alternatively acting mechanisms involved in preserving transport competence in the cytosol (Zimmermann and Meyer, 1986; Wiech *et al.*, 1991). In the first mechanism, protein synthesis is slowed down. This mechanism involves two ribonucleoparticles, i.e., the ribosome itself and signal recognition particle (SRP), and their receptors on the microsomal surface, i.e., receptors for ribosome and SRP (docking protein). This mechanism requires hydrolysis of GTP. In the second mechanism, protein folding and/or aggregation is/are slowed down. This mechanism does not involve ribonucleoparticles and their receptors but depends on hydrolysis of ATP and on molecular chaperones, like hsc70. In both mechanisms a translocase in the microsomal membrane mediates protein translocation. It includes a signal peptide receptor on the *cis* side of the microsomal membrane and a component that also depends on hydrolysis of ATP.

2. COMPONENTS INVOLVED IN PROTEIN TRANSPORT INTO YEAST MICROSOMES

In a yeast cell-free system, the yeast presecretory protein prepro-α-factor is transported into yeast microsomes posttranslationally, i.e., as the fully synthesized precursor protein (Hansen *et al.*, 1986; Waters and Blobel, 1986; Rothblatt and Meyer, 1986). Genetic and biochemical evidence suggested the involvement of the *cis*-acting chaperone, hsc70, and at least a second cytosolic (NEM-sensitive) protein, which has not been identified as yet (Figure 1, Table I) (Waters *et al.*, 1986; Chirico *et al.*, 1988; Deshaies *et al.*, 1988). However, there also is ribonucleoparticle-dependent protein transport in yeast (Poritz *et al.*, 1988; Ribes *et al.*, 1988; Hann *et al.*, 1989; Amaya *et al.*, 1990; Hann and Walter, 1991). The SRP54 protein was first identified as the homologue of mammalian SRP54 protein and has since been shown to be essential for cell growth. The sec65 protein was identified genetically as a transport component and according to the sequence analysis contains a domain that has striking similarity to mammalian SRP19 protein (Stirling and Hewitt, 1992; Hann *et al.*, 1992). Recent genetic evidence suggests that the proteins sec70, sec71, and sec72 are involved in transport (Green *et al.*, 1992). However, no intracellular location or specific function has as yet been assigned to any of these proteins.

Table I
Proteinaceous Components and Complexes Involved in Transport of Precursor
Proteins into the ER or in Covalent Modifications[a]

Complex/protein	Yeast	Mammals
RNP[b]	7S RNA	7S RNA
	scR1 RNA	7SL RNA
		72kDa-p
		SRP72-su
		68kDa-p
		SRP68-su
	54kDa-p	54kDa-p
	SRP54-su.GTP-bp	SRP54-su.GTP-bp
	30kDa-p	19kDa-p
	sec65-p	SRP19-su
		14kDa-p
		SRP14-su
		9kDa-p
		SRP9-su
RNP receptor		69kDa-p
		DPα-su.GTP-bp
		30kDa-p
		DPβ-su.GTP-bp
Ribosome receptor		180kDa-p
		34kDa-p
cis-acting molecular chaperones	70kDa-p	70kDa-p
	Ssa1-p.ATPase	hsc70.ATPase
	40kDa-p	
	Sis1-p	
	90kDa-p	90kDa-p
	hsc82	hsp90
Translocase	41kDa-p	45kDa-p
	sec61-p	
	30kDa-p	37kDa-p
	sec62-p	
	31.5kDa-gp	36kDa-gp
		TRAM-p (mp39)
	23kDa-p	34kDa-p
		imp34
	73kDa-p	
	sec63-p	
trans-acting molecular chaperones	78kDa-p	78kDa-p
	Kar2-p.ATPase	BiP.ATPase
		94kDa-p
		grp94

(*continued*)

Table I   (*Continued*)

Complex/protein	Yeast	Mammals
Signal peptidase	19kDa-p	18/21kDa-p
	sec11-p	**SPC18/21**-su
	25kDa-gp	23kDa-gp
		SPC23-su
Oligosaccharyl transferase		65kDa-gp
		ribophorinI
		63kDa-gp
		ribophorinII
	45kDa-p	48kDa-p
	WBP1-p	

[a]Within table, roman type indicates molecular mass of the component; boldface type is the abbreviated name of the component.
[b]Abbreviations: bp, binding protein; DP, docking protein; gp, glycoprotein; grp, glucose-regulated protein; hsp, heat shock protein; imp, integral membrane protein; mp, membrane protein; p, protein; RNP, ribonucleoparticle; SPC, signal peptidase complex; SRP, signal recognition particle; su, subunit; TRAM, translocating chain-associated membrane.

We assume that the two pathways converge at the level of a putative signal peptide receptor, which has only been functionally characterized (Sanz and Meyer, 1989). Genetic evidence suggested that the membrane proteins sec61, sec62, and sec63 (also termed ptl1 or npl1) are part of the membrane translocase (Table I) (Deshaies and Schekman, 1987, 1989; Sadler *et al.,* 1989; Toyn *et al.,* 1988). sec63 contains a lumenal domain that has striking similarity to bacterial dnaJ protein, a protein that is known to functionally interact with dnaK, the bacterial hsp70 homologue (Sadler *et al.,* 1989). Biochemical evidence suggested that the sec61, sec62, and sec63 proteins transiently form complexes with a 31.5-kDa glycoprotein and a 23-kDa protein (Deshaies *et al.,* 1991), which could be related to the canine proteins mp39 and imp34, respectively. From the behavior of sec61 with respect to complex formation with the other proteins, one could speculate that it is related to either the 45-kDa signal peptide receptor that has been described for mammalian microsomes or to the azido-ATP-sensitive component that we have functionally defined for canine microsomes (see below). Precursor proteins in transit can be cross-linked to the sec61 and sec62 proteins as well as to BiP (Müsch *et al.,* 1992; Sanders *et al.,* 1992). Furthermore, the *trans*-acting chaperone BiP (KAR2-gene product) has been shown to have a direct role in transport (Vogel *et al.,* 1990; Nguyen *et al.,* 1991; Sanders *et al.,* 1992; see below).

Yeast signal peptidase contains more than one subunit and is not directly involved in protein transport (Böhni *et al.,* 1988; YaDeau and Blobel, 1989; YaDeau *et al.,* 1991). In addition, a wheat germ agglutinin binding protein (WBP1) has been identified as an essential component of oligosaccharyl transferase (te Heesen *et al.,* 1991, 1992).

A

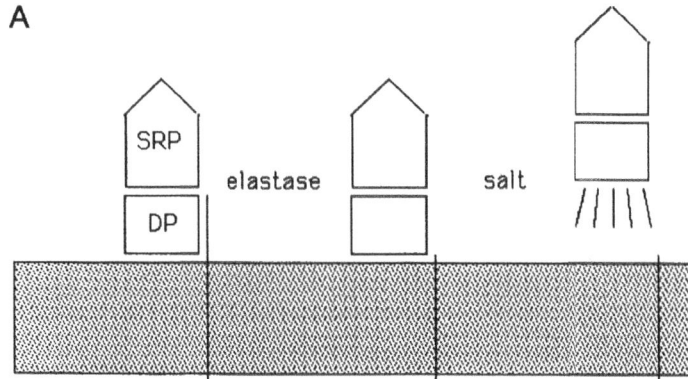

FIGURE 2. Photoaffinity labeling of dog pancreas microsomes with 8-azido-ATP leads to inactivation of the microsomes with respect to membrane insertion of ribonucleoparticle-dependent and -independent precursor proteins. Pretreatment of dog pancreas microsomes with limiting concentrations of trypsin or elastase leads to inactivation of the docking protein (A) and, therefore, inactivation of the microsomes with respect to membrane insertion of preprolactin (ribonucleoparticle-dependent) but not of preprocecropin (ribonucleoparticle-independent) (Meyer and Dobberstein, 1980a) (B). When the elastase fragment of docking protein α subunit (DP_f) is added back to pretrypsinized microsomes, the microsomes are reconstituted in their activity with respect to preprolactin (B). Photoaffinity labeling of dog pancreas microsomes with azido-ATP leads to inactivation of the microsomes with respect to membrane insertion of preprolactin and preprocecropin (B). The elastase fragment of docking protein as well as the pretrypsinized microsomes are inactivated by photoaffinity labeling with azido-ATP (C). The (in)ability of microsomes to facilitate membrane insertion of preprolactin (ppl) or preprocecropin (ppcec) after a certain pretreatment (assayed as conversion of the precursor to the mature form) is indicated as yes (no).

3. COMPONENTS INVOLVED IN PROTEIN TRANSPORT INTO MAMMALIAN MICROSOMES

In the mammalian system the ribonucleoparticle-dependent pathway has been analyzed in great detail (Figure 1, Table I). It involves SRP (Walter *et al.*, 1981; Walter and Blobel, 1981a,b, 1982; Krieg *et al.*, 1986; Kurzchalia *et al.*, 1986; Wolin and Walter, 1988, 1989; Zopf *et al.*, 1990; Römisch *et al.*, 1990; Lütcke *et al.*, 1992) and its receptor in the microsomal membrane (docking protein) (Meyer and Dobberstein, 1980a,b; Meyer *et al.*, 1982; Lauffer *et al.*, 1985; Tajima *et al.*, 1986) and the ribosome (Adelman *et al.*, 1973; Borgese *et al.*, 1974; Perara *et al.*, 1986) and its receptor (Savitz and Meyer, 1990; Tazawa *et al.*, 1991; Collins and Gilmore, 1991; Nunnari *et al.*, 1991). With respect to the nature of the ribosome receptor we assume that it may contain two different subunits (i.e., as long as data are missing that link the observed ribosome binding to protein transport). There is a GTP requirement in the transport of

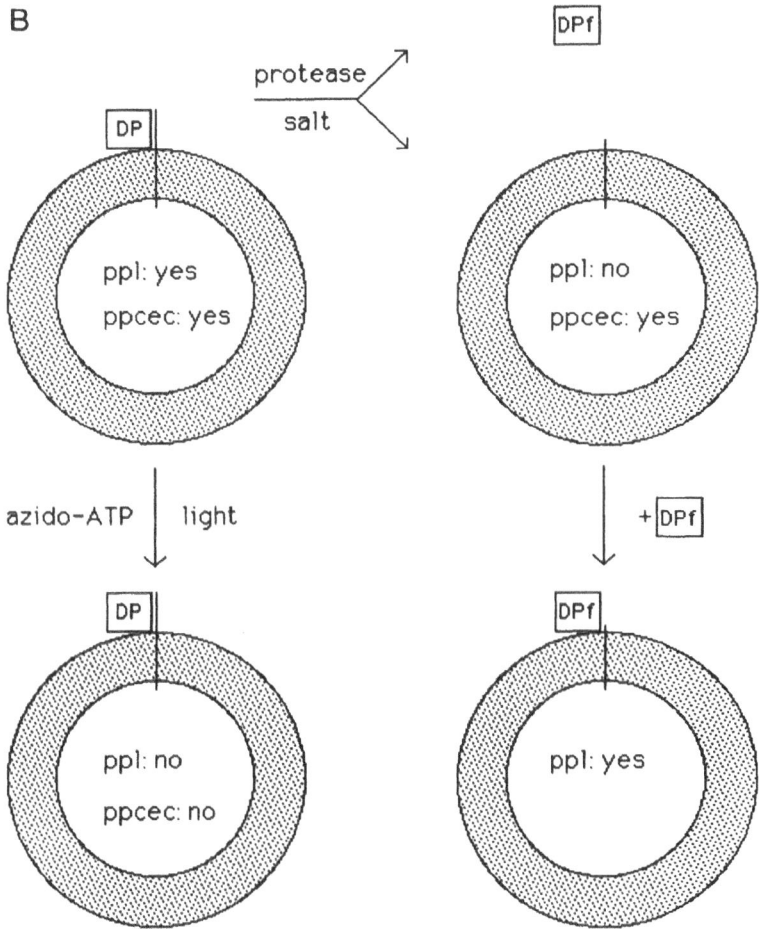

FIGURE 2. (*Continued*)

ribonucleoparticle-dependent precursor proteins, such as preprolactin (ppl) (Connolly and Gilmore, 1986, 1989; Connolly *et al.*, 1991). This GTP effect is related to the GTP-binding proteins SRP and docking protein (Connolly and Gilmore, 1989; Rapiejko and Gilmore, 1992). In the ribonucleoparticle-independent pathway, ribonucleoparticles and their receptors are not involved. This was demonstrated by our observation that small precursor proteins, like preprocecropin (ppcec), are efficiently transported into salt-washed or trypsinized microsomes, i.e., in the absence of SRP or docking protein (ppcec versus ppl in Figure 2B) (Zimmermann and Mollay, 1986; Müller and Zimmermann,

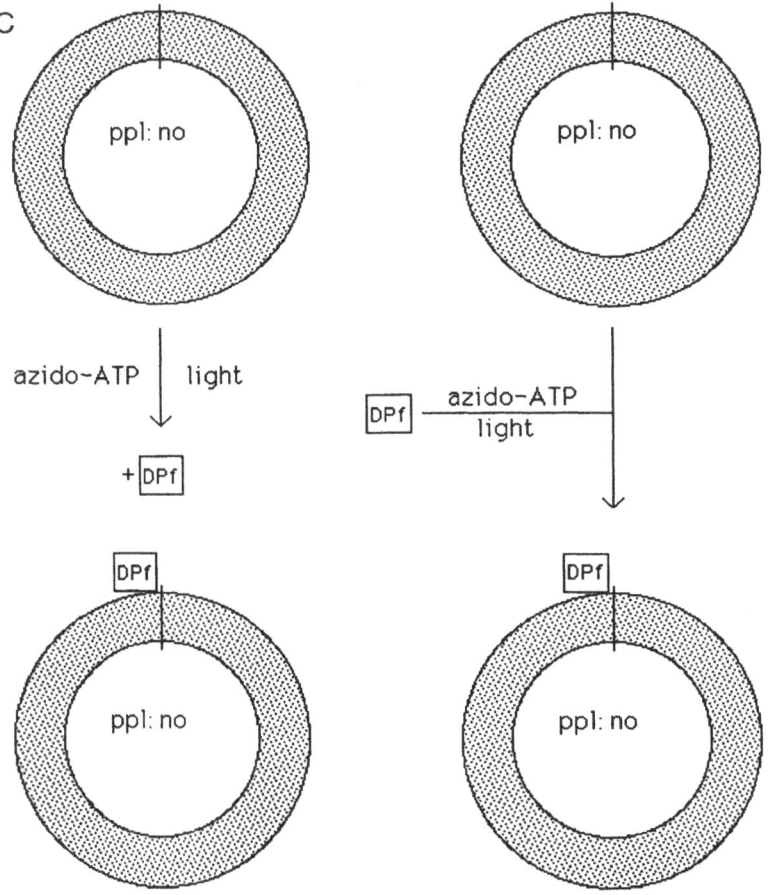

FIGURE 2. *(Continued)*

1987, 1988; Schlenstedt and Zimmermann, 1987; Schlenstedt *et al.*, 1990). Molecular chaperones can function as an alternative to this elaborate system. The *cis*-acting chaperone has been characterized as hsp70 and seems to collaborate with a second cytosolic (NEM-sensitive) protein (Wiech *et al.*, 1987; Zimmermann *et al.*, 1988; Wiech *et al.*, 1993). Our current working model proposes that hsp90 and dnaJ may be the proteins of interest (see below).

The signal peptidase of higher eukaryotic organisms contains at least two different subunits and is not directly involved in protein transport (Baker and Lively, 1987; Greenburg *et al.*, 1989; Shelness *et al.*, 1988; Shelness and Blobel,

1990).One of the two subunits has been shown to occur as a pair of homologues and to be highly similar to the yeast sec11 protein. In addition, ribophorins I and II together with a third protein are involved in oligosaccharyl transfer (Kelleher *et al.*, 1992; Crimaudo *et al.*, 1987). The third protein may be identical to the yeast protein WBP1.

We assume that the ribonucleoparticle-dependent and -independent pathways converge at the level of a putative signal peptide receptor which may be identical to the 45-kDa-protein signal sequence-binding protein in microsomal membranes (Austen *et al.*, 1984; Robinson *et al.*, 1987). Besides this protein, additional membrane proteins are parts of a general translocase (Table I). Ribonucleoparticle-independent transport of presecretory proteins involves a membrane component that is sensitive to chemical alkylation with NEM, i.e., that has an essential sulfhydryl group (Zimmermann *et al.*, 1990). The sulfhydryl is cytoplasmically exposed and the component is involved in membrane insertion but not in membrane association of the precursor proteins. This component may be identical to an NEM-sensitive component that acts past docking protein and ribosome receptor in ribonucleoparticle-dependent transport (Hortsch *et al.*, 1986; Nicchitta and Blobel, 1989). The so-called TRAM protein (mp39) appears to be part of the translocase (Wiedmann *et al.*, 1987b; Krieg *et al.*, 1989; Thrift *et al.*, 1991; Görlich *et al.*, 1992). Furthermore, interactions between the precursor proteins in transit and a 34-kDa and a 37-kDa protein have been shown (Kellaris *et al.*, 1991; High *et al.*, 1991). The so-called SSR (signal sequence receptor) complex (Hartmann *et al.*, 1989; Görlich *et al.*, 1990; Prehn *et al.*, 1990) no longer appears to be part of the translocation machinery (Görlich *et al.*, 1992; Migliaccio *et al.*, 1992). There exists an ATP-requiring subunit of the translocase that is involved in both mechanisms (Klappa *et al.*, 1991; Zimmermann *et al.*, 1991; see below).

3.1. Transport of Preprocecropin into Dog Pancreas Microsomes Depends on the Hydrolysis of ATP

We studied the transport of the chemically synthesized and purified precursor protein preprocecropin (a ribonucleoparticle-independent presecretory protein) into dog pancreas microsomes. After solubilization in dimethyl sulfoxide and subsequent dilution into an aqueous buffer, transport of preprocecropin occurred in the absence of molecular chaperones. Nevertheless, even under these conditions, transport depended on the hydrolysis of ATP. The basal transport level that occurred in the absence of added ATP was attributed to endogenous ATP. The concentration of ATP that led to half-maximal stimulation was on the order of 10 μM. At this concentration, other nucleotides (CTP, GTP, UTP) as well as AMP could not substitute for ATP (Table II). Furthermore, nonhydrolyz-

Table II

The Hydrolysis of ATP Is Involved in Transport of Purified Preprocecropin into Dog Pancreas Microsomes[a]

Addition	Preprocecropin processing (%)				
	10 μM	25 μM	100 μM	100 μM + 10 μM ATP	25 μM + 25 μM ADP
ATP	24	46	40		
ADP		30	35		
AMP		8	9	18	
8-Azido-ATP	8	34	40		
AMP-PCP	5	10	11	17	
AMP-PNP	4	11	11	12	
ATPγS	8	18	23	25	
AP$_5$A		8		25[b]	9
GTP	6	17	28		
GDP		11			
GMP-PCP		11	14	33	
GMP-PNP		11	12	30	
GTPγS		14	14	33	
CTP	6	14	24		
UTP	12	20	28		
CP	7				
CP + CK	41				
No addition	6				

[a]Purified and ^{14}C-labeled preprocecropin was added to dog pancreas microsomes as described (Klappa *et al.*, 1991). The mixture was divided into aliquots and the aliquots were supplemented with various amounts of different nucleotides. After an incubation for 30 min at 37°C, the samples were analyzed by gel electrophoresis and fluorography. The amounts of preprocecropin as well as of procecropin were quantified by laser densitometry of the fluorograph, and the efficiencies of processing of preprocecropin were determined (given as procecropin in percent of preprocecropin plus procecropin). CP, creatine phosphate; CK, creatine kinase (final concentration: 50 μg/ml).
[b]Addition: 25 μM + 10 μM ATP.

able ATP analogues, such as AMP-PCP or AMP-PNP, were not effective. Since these analogues competed with ATP when they were added together with ATP (i.e., in contrast to GMP-PCP and GMP-PNP), we concluded that the hydrolysis of ATP by a microsomal protein is required for transport of preprocecropin into mammalian microsomes (Klappa *et al.*, 1991). Membrane insertion assayed as removal of the signal peptide by lumenal signal peptidase was affected. At higher concentrations, other nucleotides (CTP, GTP, UTP) as well as 8-azido-ATP were able to substitute for ATP. The fact that ADP was able to substitute for ATP can be attributed to the formation of ATP by an endogenous myokinase activity since a specific inhibitor of this enzyme, bis-adenosine-pentaphosphate (AP5A), prevented the action of ADP (but not of ATP).

3.2. Transport of Preprocecropin Depends on a Microsomal Protein That Is Sensitive to Photoaffinity Labeling with Azido-ATP

In order to independently demonstrate that the microsomes contain an ATP-binding site that is involved in preprocecropin transport, we asked whether irreversible binding of ATP renders the microsomes inactive for subsequent transport of purified preprocecropin (Figure 2B). Microsomes were photoaffinity labeled with 8-azido-ATP at different concentrations of the ATP analogue. Microsomes treated in this way were analyzed with respect to their ability to transport preprocecropin in the presence of ATP. Photoaffinity labeling of microsomes with 8-azido-ATP for 1 min at 0°C at concentrations higher than 100 μM led to inactivation of the microsomes (ppcec in Figure 2B). The effect was on membrane insertion since processing was affected. The concentration that was necessary to result in half-maximal inhibition was on the order of 500 μM. Therefore, we concluded that there is a microsomal component that is limiting for membrane insertion of preprocecropin and that this component is sensitive to photoaffinity labeling with 8-azido-ATP (Klappa *et al.*, 1991).

3.3. The Azido-ATP-Sensitive Protein Is Distinct from BiP

Studies on protein transport into yeast microsomes have suggested a role for BiP in protein transport across the yeast microsomal membrane (Vogel *et al.*, 1990; Sanders *et al.*, 1992). Since BiP has a high-affinity binding site for ATP, an ATP effect on protein transport that is related to BiP would not be unexpected (Clairmont *et al.*, 1992). Although BiP had been shown not to be a component limiting for transport of ribonucleoparticle-dependent precursor proteins (i.e., at nanomolar concentrations of precursor proteins) into mammalian microsomes (Yu *et al.*, 1989; Zimmerman and Walter, 1990; Nicchitta and Blobel, 1990), we asked whether BiP is limiting with respect to the transport of 100- to 1000-fold higher concentrations of a purified precursor protein, such as preprocecropin. Microsomes were depleted of their lumenal proteins by treatment with increasing concentrations of octyl glucoside as described by Zimmerman and Walter (1990). After removal of the detergent, the reconstituted microsomes were analyzed with respect to their BiP content as well as with respect to their ability to process and sequester purified preprocecropin. When a concentration of octyl glucoside of 22.5 mM was used during the pretreatment of microsomes, more than 90% of

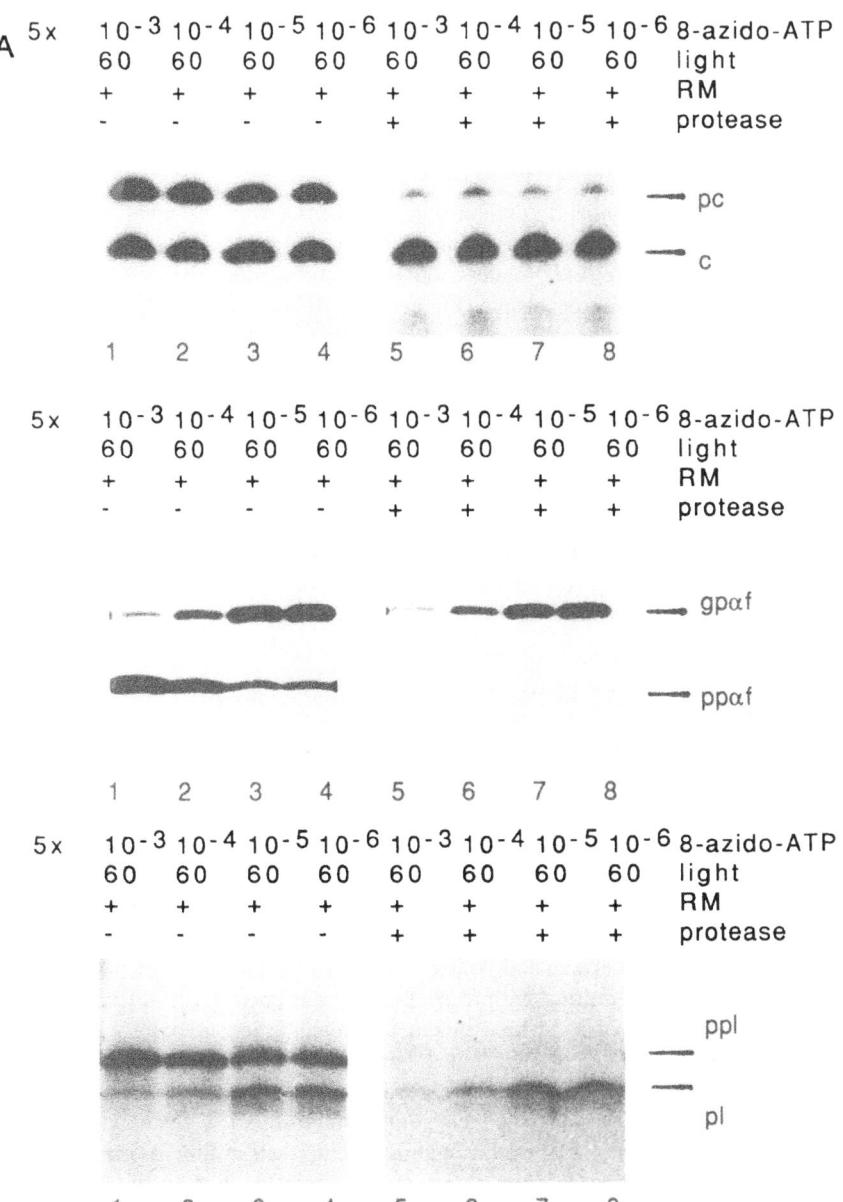

FIGURE 3. Photoaffinity modification of microsomes with 8-azido-ATP leads to inactivation of the

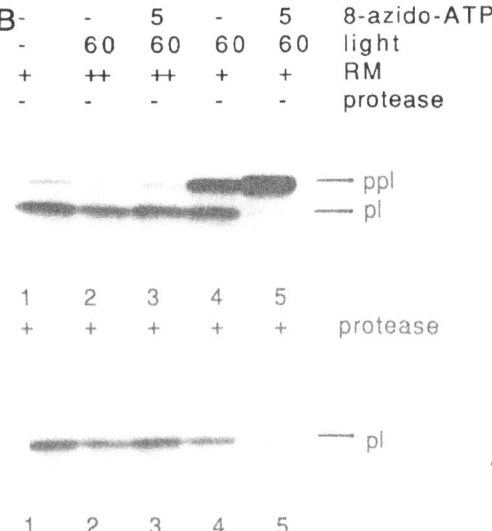

microsomes with respect to membrane insertion of preprolactin and prepro-α-factor. (A) M13 procoat protein (pc), bovine preprolactin (ppl), and yeast prepro-α-factor (ppαf), respectively, were synthesized in rabbit reticulocyte lysates in the presence of [^{35}S]methionine and of different microsomes (RM) that were pretreated with 8-azido-ATP and UV light as described (Klappa *et al.*, 1991). The concentrations of azido-ATP are given in moles per liter. c, coat protein; gpαf, glycosylated pro-α-factor; pl, prolactin. (B) Bovine preprolactin (ppl) was synthesized in rabbit reticulocyte lysates in the presence of [^{35}S]methionine and of either untreated microsomes, or microsomes that were pretreated with UV light or 8-azido-ATP plus UV light or combinations of untreated and pretreated microsomes (+ +). The concentration of azido-ATP is given in millimoles per liter. Each translation reaction was divided into two halves; one half was incubated further in the absence of protease, the other one in the presence of protease. The samples were analyzed by gel electrophoresis and fluorography. The times of UV irradiation (light) are given in seconds.

BiP was removed. However, the activity of the microsomes with respect to processing of preprocecropin and sequestration of procecropin was unaffected. In a typical experiment the concentration of preprocecropin was kept constant at about 1 μM while the concentration of BiP was varied between approximately 200 nM and 20 nM. Thus, BiP was not limiting even under conditions where the BiP content was reduced 10-fold. Since it is unlikely that photoaffinity labeling of microsomes with azido-ATP led to more than 90% derivatization of BiP, it seemed reasonable to conclude that photoaffinity labeling of BiP is not responsible for the observed inactivation of microsomes by photoaffinity labeling with azido-ATP (Klappa *et al.*, 1991).

3.4. Transport of Preprolactin Depends on a Microsomal Protein That Is Sensitive to Photoaffinity Labeling with Azido-ATP

In view of the observed inhibition of preprocecropin transport by photo-affinity labeling of microsomes with azido-ATP, we asked whether an azido-ATP-sensitive site is involved in the ribonucleoparticle-dependent transport of presecretory proteins, such as bovine preprolactin or yeast prepro-α-factor (Figure 2B) (Garcia and Walter, 1988). The same photoaffinity-labeled microsomes that were used for preprocecropin transport were analyzed with respect to their ability to transport the two ribonucleoparticle-dependent presecretory proteins. Membrane insertion of both precursor proteins was affected by photoaffinity labeling of microsomes to a similar extent (ppl in Figure 2B; ppl and ppαf in Figure 3A). Again, the concentration that was necessary to result in half-maximal inhibition was on the order of 500 μM. To rule out any nonspecific effects of the photoaffinity labeling with respect to transport of ribonucleoparticle-dependent precursor proteins, the following control experiments were performed: (1) M13 procoat protein is known to become inserted into NEM-pretreated microsomes (Watts *et al.*, 1983; Zimmermann *et al.*, 1990) as well as into protein-free liposomes (Geller and Wickner, 1985). Hence, membrane insertion of procoat protein and protection of coat protein against externally added protease served as controls for the integrity of the microsomes after photoaffinity labeling with 8-azido-ATP. Since in this respect there was no effect detectable for the various treatments (pc in Figure 3A), we conclude that the photoaffinity labeling with azido-ATP did not damage the microsomes to any significant extent. (2) The photoaffinity inactivation of microsomes by azido-ATP was not caused by any secondary effects of products of the photoactivation of azido-ATP, i.e., mixing of untreated microsomes with microsomes that had been photoaffinity labeled with 8-azido-ATP did not affect the transport activity of the untreated microsomes significantly (Figure 3B). Therefore, we conclude that transport of both preprolactin and prepro-α-factor depend on a microsomal protein that is sensitive to photoaffinity labeling with ATP analogues (Klappa *et al.*, 1991). As in the case of preprocecropin transport, membrane insertion is affected. Because of the similarities between the dose/response curves for ribonucleoparticle-dependent as well as -independent precursor proteins, we suggested that the same microsomal protein with an affinity for ATP is involved in both pathways (Klappa *et al.*, 1991).

3.5. The Azido-ATP-Sensitive Protein Is Distinct from the α Subunit of Docking Protein

The α subunit of docking protein has been shown to have an affinity for GTP as well as ATP (Connolly and Gilmore, 1989). Furthermore, the 54-kDa subunit of SRP has been suggested to have a GTP-binding site. SRP could be ruled out as a potential candidate for the observed photoaffinity inactivation with respect to preprolactin transport because the transport experiments were carried out in the reticulocyte lysate (containing intact SRP in excess) and because the azido-ATP inactivation was observed when SRP-depleted microsomes were employed (data not shown). However, the docking protein α subunit was still a potential target. Therefore, we addressed the question whether docking protein α subunit is sensitive to photoaffinity labeling with azido-ATP and, if this is the case, whether there is an additional target of photoaffinity labeling with azido-ATP (Figure 2). Trypsin-pretreated microsomes and the elastase fragment of docking protein α subunit were employed in a reconstituted assay that had been developed by Meyer and Dobberstein (1980a,b). In this approach, trypsinized microsomes, which are unable to transport preprolactin, are reconstituted in their activity by the addition of the cytosolic domain of the docking protein α subunit. Specifically, we asked whether the docking protein fragment, or the pretrypsinized microsomes, or both partners of the reconstituted system are sensitive to photoaffinity labeling with azido-ATP. The fact that the elastase fragment of docking protein as well as the trypsinized microsomes were inactivated by photoaffinity labeling with azido-ATP allows us to conclude that there is an ATP requirement in ribonucleoparticle-dependent protein transport into mammalian microsomes that is distinct from α subunit of docking protein (ppl in Figure 2C).

3.6. The Azido-ATP-Sensitive Protein Acts Prior to TRAM Protein

The observed inhibition of preprolactin transport by photoaffinity labeling of microsomes with azido-ATP allowed us to ask at what stage ribonucleoparticle-dependent transport is affected (Zimmermann et al., 1991). It was shown previously that translation of a truncated mRNA, coding for preprolactin, in the presence of microsomes leads to the binding of ribosomes which contain a nascent preprolactin chain, termed preprolactin-86mer (Connolly and Gilmore, 1986). Furthermore, it was shown that the nascent chain becomes associated with the TRAM protein under these conditions (Görlich et al., 1992) and subsequently can be translocated across the microsomal membrane, i.e., can be converted to sequestered pl-56mer, by release of the nascent chain from the ribosome with puromycin (Connolly and Gilmore, 1986). We investigated the transport of the preprolactin-86mer after photoaffinity labeling of pretrypsinized microsomes with azido-ATP and subsequent reconstitution with the cytosolic domain of the α subunit of docking protein. The results demonstrated that productive binding of the preprolactin-86mer was no longer possible. 8-Azido-ATP prevented protease-resistant binding of the preprolactin-86mer, cross-linking to what in retrospect (see above) turns out to be the TRAM protein, and chase to sequestered prolactin-56mer to a comparable extent. According to quantification by laser densitometry, there was a 60% inhibition for all three events. Therefore, we conclude that the nucleotide requirement in ribonucleoparticle-dependent protein transport into mammalian microsomes acts prior to the TRAM protein. This conclusion was supported by our observations that photoaffinity labeling with azido-ATP, carried out after binding of the preprolactin-86mer, did not give rise to protease sensitivity of the preprolactin-86mer and did not inhibit subsequent chase to sequestered prolactin-56mer.

4. MODEL FOR RIBONUCLEOPARTICLE-INDEPENDENT TRANSPORT OF PROTEINS INTO MICROSOMES

There are two alternatively acting mechanisms preserving transport competence of precursor proteins in the cytosol (Figure 1, Table I). In the mammalian system, the chain length of the precursor protein is the decisive feature with respect to which of the two mechanisms is operative. This conclusion is based on our observation that C-terminal extension of a small precursor protein, typically, leads to the phenotype of a large precursor protein (Müller and Zimmermann, 1987; Schlenstedt and Zimmermann, 1987). If one takes into account that approximately 40 amino acid residues of a nascent polypeptide chain are buried in the ribosome (Malkin and Rich, 1967; Blobel and Sabatini, 1970; Bernabeu and Lake, 1982) and that a signal peptide contains 20–30 amino acid residues (von Heijne, 1981, 1983; Perlman and Halvorson, 1983; von Heijne, 1984, 1985) and, furthermore, that SRP can bind to signal peptides only as long as they are presented by a ribosome (Ainger and Meyer, 1986; Wiedmann *et al.*, 1987a), one can imagine that precursor proteins with less than 60 to 70 amino acids cannot make use of the two ribonucleoparticles: they are released from the ribosome before SRP can bind to the signal peptide. However, the ribonucleoparticle-independent mechanism can be used by a large precursor protein, too (Schlenstedt *et al.*, 1990). A synthetic hybrid between preprocecropin and dihydrofolate reductase translocates without the involvement of SRP and ribosome. This was directly demonstrated by adding methotrexate to the translocation reaction. Methotrexate and related drugs bind to this hybrid precursor protein after it is completed and released from the ribosome and stabilize the native conformation of the dihydrofolate reductase domain. In this state, membrane insertion of the preprocecropin part is possible but completion of translocation is blocked.

There are consecutive steps of nucleotide hydrolysis involved in protein transport into the ER. In ribonucleoparticle-independent transport, there is a first ATP-dependent system involved in preserving transport competence (related to hsc70) and a second one in facilitating membrane insertion (related to translocase). At least in certain cases there may be a third one on the *trans* side (related to BiP). It is clear that precursor proteins are not transported into the ER in their folded state and that signal peptides and molecular chaperones are involved in preserving the transport-competent state (most likely a molten globular state) (Wiech *et al.*, 1990). Our current working model proposes that hsc70, hsp90, and dnaJ (Blumberg and Silver, 1991; Caplan and Douglas, 1991; Luke *et al.*, 1991) may be the molecular chaperones involved in this process. We find this to be an attractive hypothesis because a similar set of molecular chaperones, i.e., BiP (a member of the hsp70 family), grp94 (a member of the hsp90 family), and a dnaJ homologue (i.e., sec63 protein), are present in the microsomal lumen where they are assumed to be involved in protein folding. Membrane association of the precursor proteins occurs via a putative signal peptide receptor. At this stage the precursor may be in a molten globular state; it may be free or bound to molecular chaperones (Figure 4). With the help of the translocase the signal peptides are inserted into the membrane, most likely in the form of a loop structure that is made up by the signal peptide plus the N-terminus of the mature part. The ATP hydrolysis at the microsomal level seems to be directly providing the energy for membrane insertion. The question is where the energy for initial unfolding comes from. The energy for complete unfolding of a precursor protein may be as low as 10 kcal/mol. Thus, the initial hydrolysis of one ATP could be sufficient to drive such an unfolding reaction. In order for translocation to progress, the protein on the *cis* side has to unfold further. Again, the question is where the energy for unfolding comes from. It is tempting to speculate that binding of the polypeptide chain, as it emerges in the lumen, to the *trans*-acting molecular chaperone BiP at least in some cases provides the energy. Alternatively, completion of translocation may be driven by spontaneous refolding on the *trans* side of the target membrane. In either case, the energy gained by inter- and intramolecular interactions in the lumen would drive unfolding on the opposite side of the membrane.

Future experiments in this context will be directed toward (1) the exact nature of the transport-competent conformation of precursor proteins, (2) the mode of action of molecular chaperones, and (3) the nature and (4) molecular mechanism of translocase. We expect that combinations of biochemical and biophysical analyses will eventually pave the way to a detailed picture of the molecular events involved. With respect to precursor conformation and the role of molecular chaperones in this context, the protein renaturation studies (e.g., see Buchner *et al.*, 1991; Wiech *et al.*, 1992) will have to be extended to precursor proteins and a correlation with folding under synthesis and transport conditions will have to be established. With respect to translocase, the reconstituted system (Nicchitta and Blobel, 1990; Nicchitta *et al.*, 1991) in combination with the powerful yeast genetics, introduced and still most elegantly applied to this field by Schekman and his co-workers (e.g., see Deshaies and Schekman, 1987), should allow the identification of the various essential components and lead to the characterization of the exact function of the different subunits. In this

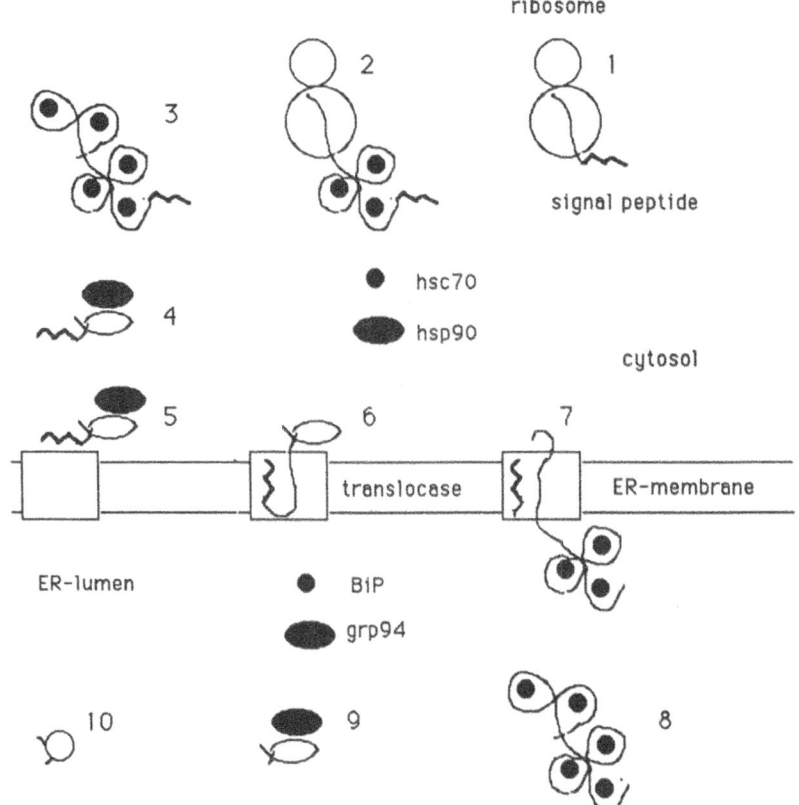

FIGURE 4. Model for ribonucleoparticle-independent transport of presecretory proteins into the endoplasmic reticulum. Refer to the text for details.

context, too, biophysical analyses of the type carried out by Simon and Blobel (1991, this volume) will eventually have to be combined with the biochemical approach.

ACKNOWLEDGMENTS. The authors' work on this subject was supported by the Deutsche Forschungsgemeinschaft, by the Fonds der Chemischen Industrie, and by the Human Frontier Science Program Organization.

5.   REFERENCES

Adelman, M. R., Sabatini, D. D., and Blobel, G., 1973, Ribosome–membrane interaction. Non-destructive disassembly of rat liver rough microsomes into ribosomes and membranous components, *J. Cell Biol.* **56:**206–229.

Ainger, K. J., and Meyer, D. I., 1986, Translocation of nascent secretory proteins across membranes can occur late in translation, *EMBO J.* **5**:951–955.

Amaya, Y., Nakano, A., Ito, K., and Mori, M., 1990, Isolation of a yeast gene, SRH1, that encodes a homologue of the 54k subunit of mammalian signal recognition particle, *J. Biochem.* **107**:457–463.

Austen, B. M., Hermon-Taylor, J., Kaderbhai, M. A., and Ridd, D. H., 1984, Design and synthesis of a consensus signal sequence that inhibits protein translocation into rough microsomal vesicles, *Biochem. J.* **224**:317–325.

Baker, R. K., and Lively, M. O., 1987, Purification and characterization of hen oviduct microsomal signal peptidase, *Biochemistry* **26**:8561–8567.

Bernabeu, C., and Lake, J. A., 1982, Nascent polypeptide chains emerge from the exit domain of the large ribosomal subunit: Immune mapping of the nascent chain, *Proc. Natl. Acad. Sci. USA* **79**:3111–3115.

Blobel, G., and Sabatini, D. D., 1970, Controlled proteolysis of nascent polypeptides in rat liver cell fractions, *J. Cell Biol.* **45**:130–145.

Blumberg, H., and Silver, P. A., 1991, A homologue of the bacterial heat shock gene DnaJ that alters protein sorting in yeast, *Nature* **349**:627–629.

Böhni, P. C., Deshaies, R. J., and Schekman, R. W., 1988, Sec11 is required for signal peptide processing and yeast cell growth, *J. Cell Biol.* **106**:1035–1042.

Borgese, N., Mok, W., Kreibich, G., and Sabatini, D. D., 1974, Ribosome–membrane interaction: *In vitro* binding of ribosomes to microsomal membranes, *J. Mol. Biol.* **88**:559–580.

Buchner, J., Schmidt, M., Fuchs, M., Jaenicke, R., Rudolph, R., Schmid, F. X., and Kiefhaber, T., 1991, GroE facilitates refolding of citrate synthase by suppressing aggregation, *Biochemistry* **30**:1586–1591.

Caplan, A. J., and Douglas, M. G., 1991, Characterization of YDJ1: A yeast homologue of the bacterial dnaJ protein, *J. Cell Biol.* **114**:609–621.

Chirico, W. J., Waters, G. M., and Blobel, G., 1988, 70K heat shock related proteins stimulate protein translocation into microsomes, *Nature* **332**:805–810.

Clairmont, C. A., De Maio, A., and Hirschberg, C. B., 1992, Translocation of ATP into the lumen of rough endoplasmic reticulum-derived vesicles and its binding to luminal proteins including BiP (GRP 78) and GRP 94, *J. Biol. Chem.* **267**:3983–3990.

Collins, P. G., and Gilmore, R., 1991, Ribosome binding to the endoplasmic reticulum: A 180-kD protein identified by crosslinking to membrane-bound ribosomes is not required for ribosome binding activity, *J. Cell Biol.* **114**:639–649.

Connolly, T., and Gilmore, R., 1986, Formation of a functional ribosome–membrane junction during translocation requires the participation of a GTP-binding protein, *J. Cell Biol.* **103**:2253–2261.

Connolly, T., and Gilmore, R., 1989, The signal recognition particle receptor mediates the GTP-dependent displacement of SRP from the signal sequence of the nascent polypeptide, *Cell* **57**:599–610.

Connolly, T., Rapiejko, P. J., and Gilmore, R., 1991, Requirement of GTP hydrolysis for dissociation of the signal recognition particle from its receptor, *Science* **252**:1171–1173.

Crimaudo, C., Hortsch, M., Gausepohl, H., and Meyer, D. I., 1987, Human ribophorins I and II: The primary structure and membrane topology of two highly conserved rough endoplasmic reticulum-specific glycoproteins, *EMBO J.* **6**:75–82.

Deshaies, R. J., and Schekman, R., 1987, A yeast mutant defective at an early stage in import of secretory protein precursors into the endoplasmic reticulum, *J. Cell Biol.* **105**:633–645.

Deshaies, R. J., and Schekman, R., 1989, Sec62 encodes a putative membrane protein required for protein translocation into the yeast endoplasmic reticulum, *J. Cell Biol.* **109**:2653–2664.

Deshaies, R. J., Koch, B. D., Werner-Washburne, M., Craig, E. A., and Schekman, R., 1988, A subfamily of stress proteins facilitates translocation of secretory and mitochondrial precursor polypeptides, *Nature* **332**:800–805.

Deshaies, R. J., Sanders, S. L., Feldheim, D. A., and Schekman, R., 1991, Assembly of yeast Sec proteins involved in translocation into the endoplasmic reticulum into a membrane-bound multisubunit complex, *Nature* **349:**806–808.

Garcia, P. D., and Walter, P., 1988, Full-length prepro-α-factor can be translocated across the mammalian microsomal membrane only if translation has not terminated, *J. Cell Biol.* **106:**1043–1048.

Geller, B. L., and Wickner, W., 1985, M13 procoat protein inserts into liposomes in the absence of other membrane proteins, *J. Biol. Chem.* **260:**13281–13285.

Görlich, D., Prehn, S., Hartmann, E., Herz, J., Otto, A., Kraft, R., Wiedmann, M., Knespel, S., Dobberstein, B., and Rapoport, T., 1990, The signal sequence receptor has a second subunit and is part of a translocation complex in the endoplasmic reticulum as probed by bifunctional reagents, *J. Cell Biol.* **111:**2283–2294.

Görlich, D., Hartmann, E., Prehn, S., and Rapoport, T. A., 1992, A protein of the endoplasmic reticulum involved early in polypeptide translocation, *Nature* **357:**47–52.

Green, N., Fang, H., and Walter, P., 1992, Mutants in three novel complementation groups inhibit membrane protein insertion into and soluble protein translocation across the endoplasmic reticulum membrane of *Saccharomyces cerevisiae*, *J. Cell Biol.* **116:**597–604.

Greenburg, G., Shelness, G. S., and Blobel, G., 1989, A subunit of mammalian signal peptidase is homologous to yeast sec11 protein, *J. Biol. Chem.* **264:**15762–15765.

Hann, B. C., and Walter, P., 1991, The signal recognition particle in *S. cerevisiae*, *Cell* **67:**131–144.

Hann, B. C., Poritz, M. A., and Walter, P., 1989, *Saccharomyces cerevisiae* and *Schizosaccharomyces pombe* contain a homologue to the 54-kD subunit of the signal recognition particle that in *S. cerevisiae* is essential for growth, *J. Cell Biol.* **109:**3223–3235.

Hann, B. C., Stirling, C. J., and Walter, P., 1992, Sec65 gene product is a subunit of the yeast signal recognition particle required for its integrity, *Nature* **356:**532–533.

Hansen, W., Garcia, P. D., and Walter, P., 1986, *In vitro* protein translocation across the yeast endoplasmic reticulum: ATP dependent post-translational translocation of the prepro-α-factor, *Cell* **45:**397–406.

Hartmann, E., Wiedmann, M., and Rapoport, T. A., 1989, A membrane component of the endoplasmic reticulum that may be essential for protein translocation, *EMBO J.* **8:**2225–2229.

High, S., Görlich, D., Wiedmann, M., Rapoport, T. A., and Dobberstein, B., 1991, The identification of proteins in the proximity of signal anchor sequences during their targeting to and insertion into the membrane of the ER, *J. Cell Biol.* **113:**35–44.

Hortsch, M., Avossa, D., and Meyer, D. I., 1986, Characterization of secretory protein translocation: Ribosome–membrane interaction in endoplasmic reticulum, *J. Cell Biol.* **103:**241–253.

Kellaris, K. V., Bowen, S., and Gilmore, R., 1991, ER translocation intermediates are adjacent to a nonglycosylated 34-kD integral membrane protein, *J. Cell Biol.* **114:**21–33.

Kelleher, D. J., Kreibich, G., and Gilmore, R., 1992, Oligosaccharyltransferase activity is associated with a protein complex composed of ribophorins I and II and a 48 kd protein, *Cell* **69:**55–65.

Klappa, P., Mayinger, P., Pipkorn, R., Zimmermann, M., and Zimmermann, R., 1991, A microsomal protein is involved in ATP-dependent transport of presecretory proteins into mammalian microsomes, *EMBO J.* **10:**2795–2803.

Koppelman, B., Zimmerman, D. L., Walter, P., and Brodsky, F. M., 1992, Evidence for peptide transport across microsomal membranes, *Proc. Natl. Acad. Sci. USA* **89:**3908–3912.

Krieg, U. C., Walter, P., and Johnson, A. E., 1986, Photocrosslinking of the signal sequence of nascent preprolactin to the 54-kilodalton polypeptide of the signal recognition particle, *Proc. Natl. Acad. Sci. USA* **83:**8604–8608.

Krieg, U. C., Johnson, A. E., and Walter, P., 1989, Protein translocation across the endoplasmic reticulum membrane: Identification by photocross-linking of a 39-kD integral membrane glycoprotein as part of a putative translocation tunnel, *J. Cell Biol.* **109:**2033–2043.

Kuchler, K., Sterne, R. E., and Thorner, J., 1989, *Saccharomyces cerevisiae* STE6 gene product: A novel pathway for protein export in eukaryotic cells, *EMBO J.* **8**:3973–3984.

Kurzchalia, T. V., Wiedmann, M., Girshovich, A. S., Bochkareva, E. S., Bielka, H., and Rapoport, T. A., 1986, The signal sequence of nascent preprolactin interacts with the 54K polypeptide of the signal recognition particle, *Nature* **320**:634–636.

Lauffer, L., Garcia, P. D., Harkins, R. N., Coussens, L., Ullrich, A., and Walter, P., 1985, Topology of signal recognition particle receptor in the endoplasmic reticulum membrane, *Nature* **318**:334–338.

Levy, F., Gabathuler, R., Larsson, R., and Kvist, S., 1991, ATP is required for in vitro assembly of MHC class I antigens but not for transfer of peptides across the ER membrane, *Cell* **67**:265–274.

Luke, M. M., Sutton, A., and Arndt, K. T., 1991, Characterization of SIS1, a *Saccharomyces cerevisiae* homologue of bacterial dnaJ proteins, *J. Cell Biol.* **114**:623–638.

Lütcke, H., High, S., Römisch, K., Ashford, A. J., and Dobberstein, B., 1992, The methionine-rich domain of the 54 kDa subunit of signal recognition particle is sufficient for the interaction with signal sequences, *EMBO J.* **11**:1543–1551.

McGrath, J. P., and Varshavsky, A., 1989, The yeast STE6 gene encodes a homologue of the mammalian multidrug resistance P-glycoprotein, *Nature* **340**:400–404.

Malkin, L. I., and Rich, A., 1967, Partial resistance of nascent polypeptide chains to proteolytic digestion due to ribosomal shielding, *J. Mol. Biol.* **26**:329–346.

Meyer, D. I., and Dobberstein, B., 1980a, A membrane component essential for vectorial translocation of nascent proteins across the endoplasmic reticulum: Requirements for its extraction and reassociation with the membrane, *J. Cell Biol.* **87**:498–502.

Meyer, D. I., and Dobberstein, B., 1980b, Identification and characterization of a membrane component essential for the translocation of nascent proteins across the membrane of the endoplasmic reticulum, *J. Cell Biol.* **87**:503–508.

Meyer, D. I., Krause, E., and Dobberstein, B., 1982, Secretory protein translocation across membranes—The role of the 'docking protein,' *Nature* **297**:647–650.

Migliaccio, G., Nicchitta, C. V., and Blobel, G., 1992, The signal sequence receptor, unlike the signal recognition particle receptor, is not essential for protein translocation, *J. Cell Biol.* **117**:15–25.

Müller, G., and Zimmermann, R., 1987, Import of honeybee prepromelittin into the endoplasmic reticulum: Structural basis for independence of SRP and docking protein, *EMBO J.* **6**:2099–2107.

Müller, G., and Zimmermann, R., 1988, Import of honeybee prepromelittin into the endoplasmic reticulum: Energy requirements for membrane insertion, *EMBO J.* **7**:639–648.

Müsch, A., Wiedmann, M., and Rapoport, T. A., 1992, Yeast Sec proteins interact with polypeptides traversing the endoplasmic reticulum membrane, *Cell* **69**:343–352.

Nguyen, T. H., Law, D. T. S., and Williams, D. B., 1991, Binding protein BiP is required for translocation of secretory proteins into the endoplasmic reticulum in *Saccharomyces cerevisiae*, *Proc. Natl. Acad. Sci. USA* **88**:1565–1569.

Nicchitta, C. V., and Blobel, G., 1989, Nascent secretory binding and translocation are distinct processes: Differentiation by chemical alkylation, *J. Cell Biol.* **108**:789–795.

Nicchitta, C. V., and Blobel, G., 1990, Assembly of translocation competent proteoliposomes from detergent-solubilized rough microsomes, *Cell* **60**:259–266.

Nicchitta, C. V., Migliaccio, G., and Blobel, G., 1991, Biochemical fractionation and assembly of the membrane components that mediate nascent chain targeting and translocation, *Cell* **65**:587–598.

Nunnari, J. M., Zimmerman, D. L., and Walter, P., 1991, Characterization of the rough endoplasmic reticulum ribosome-binding activity, *Nature* **352**:638–640.

Parham, P., 1991, Half of a peptide pump, *Nature* **351**:271–272.

Parham, P., 1992, Flying the first class flag, *Nature* **357**:193–194.

Perara, E., Rothman, R. E., and Lingappa, V. R., 1986, Uncoupling translocation from translation: Implications for transport of proteins across membranes, *Science* **232**:348–352.

Perlman, D., and Halvorson, H. O., 1983, A putative signal peptidase recognition site and sequence in eucaryotic and procaryotic signal peptides, *J. Mol. Biol.* **167**:391–409.

Poritz, M. A., Siegel, V., Hansen, W., and Walter, P., 1988, Small ribonucleoproteins in *Schizosaccharomyces pombe* and *Yarrowia lipolytica* homologous to signal recognition particle, *Proc. Natl. Acad. Sci. USA* **85**:4315–4319.

Prehn, S., Herz, J., Hartmann, E., Kurzchalia, T. V., Frank, R., Roemisch, K., Dobberstein, B., and Rapoport, T. A., 1990, Structure and biosynthesis of the signal sequence receptor, *Eur. J. Biochem.* **188**:439–445.

Rapiejko, P. J., and Gilmore, R., 1992, Protein translocation across the ER requires a functional GTP binding site in the α subunit of the signal recognition particle receptor, *J. Cell Biol.* **117**:493–503.

Ribes, V., Dehaux, P., and Tollervey, D., 1988, 7SL RNA from *Schizosaccharomyces pombe* is encoded by a single copy essential gene, *EMBO J.* **7**:231–237.

Robinson, A., Kaderbhai, M. A., and Austen, B. A., 1987, Identification of signal sequence binding proteins integrated into the rough endoplasmic reticulum membrane, *Biochem. J.* **242**:767–777.

Römisch, K., Webb, J., Lingelbach, K., Gausepohl, H., and Dobberstein, B., 1990, The 54-kD protein of signal recognition particle contains a methionine-rich RNA binding domain, *J. Cell Biol.* **111**:1793–1802.

Rothblatt, J. A., and Meyer, D. I., 1986, Secretion in yeast: Translocation and glycosylation of prepro-α-factor *in vitro* can occur via an ATP-dependent post-translational mechanism, *EMBO J.* **5**:1031–1036.

Rubartelli, A., Cozzolino, F., Talio, M., and Sitia, R., 1990, A novel secretory pathway for interleukin-1β, a protein lacking a signal sequence, *EMBO J.* **9**:1503–1510.

Sadler, I., Chiang, A., Kurihara, T., Rothblatt, J., Way, J., and Silver, P., 1989, A yeast gene important for protein assembly into the endoplasmic reticulum and the nucleus has homology to DnaJ, an *Escherichia coli* heat shock protein, *J. Cell Biol.* **109**:2665–2675.

Sanders, S. L., Whitfield, K. M., Vogel, J. P., Rose, M. D., and Schekman, R., 1992, Sec61p and BiP directly facilitate polypeptide translocation into the ER, *Cell* **69**:353–365.

Sanz, P., and Meyer, D. I., 1989, Secretion in yeast: Preprotein binding to a membrane receptor and ATP-dependent translocation are sequential and separable events *in vitro*, *J. Cell Biol.* **108**:2101–2106.

Savitz, A. J., and Meyer, D. I., 1990, Identification of a ribosome receptor in the rough endoplasmic reticulum, *Nature* **346**:540–544.

Schlenstedt, G., and Zimmermann, R., 1987, Import of frog prepropeptide GLa into microsomes requires ATP but does not involve docking protein or ribosomes, *EMBO J.* **6**:699–703.

Schlenstedt, G., Gudmundsson, G. H., Boman, H. G., and Zimmermann, R., 1990, A large presecretory protein translocates both cotranslationally, using signal recognition particle and ribosome, and posttranslationally, without these ribonucleoparticles, when synthesized in the presence of mammalian microsomes, *J. Biol. Chem.* **265**:13960–13968.

Shelness, G., and Blobel, G., 1990, Two subunits of the canine signal peptidase complex are homologous to yeast sec 11 protein, *J. Biol. Chem.* **265**:9512–9519.

Shelness, G. S., Kanwar, Y. S., and Blobel, G., 1988, cDNA-derived primary structure of the glycoprotein component of canine microsomal signal peptidase complex, *J. Biol. Chem.* **263**:17063–17070.

Simon, S. M., and Blobel, G., 1991, A protein-conducting channel in the endoplasmic reticulum, *Cell* **65**:371–380.

Spies, T., and DeMars, R., 1991, Restored expression of major histocompatibility class I molecules by gene transfer of a putative peptide transporter, *Nature* **351**:323–324.

Stirling, C. J., and Hewitt, E. W., 1992, The *S. cerevisiae* Sec65 gene encodes a component of yeast signal recognition particle with homology to human SRP19, *Nature* **356**:534–537.

Tajima, S., Lauffer, L., Rath, V. L., and Walter, P., 1986, The signal recognition particle receptor is a complex that contains two distinct polypeptide chains, *J. Cell Biol.* **103**:1167–1178.

Tazawa, S., Unuma, M., Tondokoro, J., Asano, Y., Ohsumi, T., Ichimura, T., and Sugano, H., 1991, Identification of a membrane protein responsible for ribosome binding in rough microsomal membranes, *J. Biochem.* **109**:89–98.

te Heesen, S., Rauhut, R., Aebersold, R., Abelson, J., Aebi, M., and Clark, M. W., 1991, An essential 45 kDa yeast transmembrane protein reacts with anti-nuclear pore antibodies: Purification of the protein, immunolocalization and cloning of the gene, *Eur. J. Cell Biol.* **56**:8–18.

te Heesen, S., Janetzky, B., Lehle, L., and Aebi, M., 1992, The yeast WBP1 is essential for oligosaccharyl transferase activity in vivo and in vitro, *EMBO J.* **11**:2071–2075.

Thrift, R. N., Andrews, D. W., Walter, P., and Johnson, A. E., 1991, A nascent membrane protein is located adjacent to ER membrane proteins throughout its integration and translation, *J. Cell Biol.* **112**:809–821.

Toyn, J., Hibbs, A. R., Sanz, P., Crowe, J., and Meyer, D. I., 1988, *In vivo* and *in vitro* analysis of ptl1, a yeast ts mutant with a membrane associated defect in protein translocation, *EMBO J.* **7**:4347–4353.

Vogel, J. P., Misra, L. M., and Rose, M. D., 1990, Loss of BiP/GRP78 function blocks translocation of secretory proteins in yeast, *J. Cell Biol.* **110**:1885–1895.

von Heijne, G., 1981, On the hydrophobic nature of signal sequences, *Eur. J. Biochem.* **116**:419–422.

von Heijne, G., 1983, Patterns of amino acids near signal sequence cleavage sites, *Eur. J. Biochem.* **113**:17–21.

von Heijne, G., 1984, Analysis of the distribution of charged residues in the N-terminal region of signal sequences: Implications for protein export in prokaryotic and eukaryotic cells, *EMBO J.* **3**:2315–2318.

von Heijne, G., 1985, Signal sequences: The limits of variation, *J. Mol. Biol.* **184**:99–105.

Walter, P., and Blobel, G., 1981a, Translocation of proteins across the endoplasmic reticulum. II. Signal recognition protein, SRP, mediates the selective binding to microsomal membranes of *in-vitro*-assembled polysomes synthesizing secretory protein, *J. Cell Biol.* **91**:551–556.

Walter, P., and Blobel, G., 1981b, Translocation of proteins across the endoplasmic reticulum. III. Signal recognition protein, SRP, causes signal sequence-dependent and site-specific arrest of chain elongation that is released by microsomal membranes, *J. Cell Biol.* **91**:557–561.

Walter, P., and Blobel, G., 1982, Signal recognition particle contains a 7S RNA essential for protein translocation across the endoplasmic reticulum, *Nature* **299**:691–698.

Walter, P., Ibrahimi, I., and Blobel, G., 1981, Translocation of proteins across the endoplasmic reticulum. I. Signal recognition protein, SRP, binds to *in-vitro*-assembled polysomes synthesizing secretory protein, *J. Cell Biol.* **91**:545–550.

Waters, M. G., and Blobel, G., 1986, Secretory protein translocation in a yeast cell-free system can occur posttranslationally and requires ATP hydrolysis, *J. Cell Biol.* **102**:1543–1550.

Waters, M. G., Chirico, W. J., and Blobel, G., 1986, Protein translocation across the yeast microsomal membrane is stimulated by a soluble factor, *J. Cell Biol.* **103**:2629–2636.

Watts, C., Wickner, W., and Zimmermann, R., 1983, M13 procoat and a pre-immunoglobulin share processing specificity but use different membrane receptor systems, *Proc. Natl. Acad. Sci. USA* **80**:2809–2813.

Wiech, H., Sagstetter, M., Müller, G., and Zimmermann, R., 1987, The ATP requiring step in assembly of M13 procoat protein into microsomes is related to preservation of transport competence of the precursor protein, *EMBO J.* **6**:1011–1016.

Wiech, H., Stuart, R., and Zimmermann, R., 1990, Role of cytosolic factors in the transport of proteins across membranes, *Semin. Cell Biol.* **1**:55–63.

Wiech, H., Klappa, P., and Zimmermann, R., 1991, Protein export in prokaryotes and eukaryotes, *FEBS Lett.* **285**:182–188.

Wiech, H., Buchner, J., Zimmermann, R., and Jakob, U., 1992, Hsp90 chaperones protein folding in vitro, *Nature* **358**:169–170.

Wiech, H., Buchner, J., Zimmermann, M., Zimmermann, R., and Jacob, U. 1993, Hsc70, immunoglobulin heavy chain binding protein, and Hsp90 differ in their ability to stimulate transport of precursor proteins into mammalian microsomes, *J. Biol. Chem.* **268**:7414–7421.

Wiedmann, M., Kurzchalia, T. V., Bielka, H., and Rapoport, T. A., 1987a, Direct probing of the interaction between the signal sequence of nascent preprolactin and the signal recognition particle by specific crosslinking, *J. Cell Biol.* **104**:201–208.

Wiedmann, M., Kurzchalia, T. V., Hartmann, E., and Rapoport, T. A., 1987b, A signal sequence receptor in the endoplasmic reticulum membrane, *Nature* **328**:830–833.

Wolin, S. L., and Walter, P., 1988, Ribosome pausing and stacking during translation of a eukaryotic mRNA, *EMBO J.* **7**:3559–3569.

Wolin, S. L., and Walter, P., 1989, Signal recognition particle mediates a transient elongation arrest of preprolactin in reticulocyte lysate, *J. Cell Biol.* **109**:2617–2622.

YaDeau, J. T., and Blobel, G., 1989, Solubilization and characterization of yeast signal peptidase, *J. Biol. Chem.* **264**:2928–2934.

YaDeau, J. T., Klein, C., and Blobel, G., 1991, Yeast signal peptidase contains a glycoprotein and the Sec11 gene product, *Proc. Natl. Acad. Sci. USA* **88**:517–521.

Yu, Y., Zhang, Y., Sabatini, D. D., and Kreibich, G., 1989, Reconstitution of translocation-competent vesicles from detergent-solubilized dog pancreas microsomes, *Proc. Natl. Acad. Sci. USA* **86**:9931–9935.

Zimmerman, D. L., and Walter, P., 1990, Reconstitution of protein translocation activity from partially solubilized microsomal vesicles, *J. Biol. Chem.* **265**:4048–4053.

Zimmermann, R., and Meyer, D. I., 1986, 1986: A year of new insights into how proteins cross membranes, *Trends Biochem. Sci.* **11**:512–515.

Zimmermann, R., and Mollay, C., 1986, Import of honeybee prepromelittin into the endoplasmic reticulum. Requirements for membrane insertion, processing and sequestration, *J. Biol. Chem.* **261**:12889–12895.

Zimmermann, R., Sagstetter, M., Lewis, M. J., and Pelham, H. R. B., 1988, Seventy kilodalton heat shock proteins and an additional component from reticulocyte lysate stimulate import of M13 procoat protein into microsomes, *EMBO J.* **7**:2875–2880.

Zimmermann, R., Sagstetter, M., and Schlenstedt, G., 1990, Ribonucleoparticle-independent import of proteins into mammalian microsomes involves a membrane protein which is sensitive to chemical alkylation, *Biochimie* **72**:95–101.

Zimmermann, R., Zimmermann, M., Mayinger, P., and Klappa, P., 1991, Photoaffinity labeling of dog pancreas microsomes with 8-azido-ATP inhibits association of nascent preprolactin with the signal sequence receptor complex, *FEBS Lett.* **286**:95–99.

Zopf, D., Bernstein, H. D., Johnson, A. E., and Walter, P., 1990, The methionine-rich domain of the 54 kd protein subunit of the signal recognition particle contains an RNA binding site and can be crosslinked to a signal sequence, *EMBO J.* **9**:4511–4517.

Chapter 3

Folding, Assembly, and Posttranslational Modification of Proteins within the Lumen of the Endoplasmic Reticulum

Pamela J. E. Rowling and Robert B. Freedman

1. INTRODUCTION

The lumen of the endoplasmic reticulum (ER) is a protein-folding compartment. Of course, protein folding occurs in other subcellular compartments but this is incidental to their other major functions. In the case of the lumen of the ER, protein folding comes close to being its *raison d'être*. Proteins entering this compartment do so cotranslationally (see Klappa *et al.* and Simon and Blobel, this volume) and in a physical state that is certainly not folded. Most models show them as extended polypeptide chains but there is little real evidence to

---

Abbreviations used in this chapter: BiP, heavy-chain binding protein; BPTI, bovine pancreatic trypsin inhibitor; DTT, dithiothreitol; ER, endoplasmic reticulum; GPI, glycosylphosphatidylinositol; grp, glucose-regulated proteins; GSBP, glycosylation site binding protein; HA, hemagglutinin; hsp, heat shock proteins; MHC, major histocompatibility complex; OT, oligosacchary1 transferase; PDI, protein disulfide isomerase; PPI, peptidyl-prolyl *cis/trans* isomerase; sPPI, secretory PPI.

Pamela J. E. Rowling and Robert B. Freedman   Biological Laboratory, University of Kent, Canterbury, Kent CT2 7NJ, United Kingdom.

Subcellular Biochemistry, Volume 21: Endoplasmic Reticulum, edited by N. Borgese and J. R. Harris. Plenum Press, New York, 1993.

describe their physical properties at the point of translocation; it has been suggested (Bychkova *et al.*, 1988) that the "translocation-competent" state is close to the "molten globule" or "collapsed intermediate" state described by students of protein refolding *in vitro*. On the other hand, proteins leaving the ER lumen, and proceeding to subsequent compartments of the secretory pathway, are usually fully folded, as judged by their biological activity and recognition by conformation-specific antibodies. Indeed, correct folding and assembly is the major criterion used by the "quality control" system that selectively permits exit of secretory and cell-surface proteins from the ER lumen (Hurtley and Helenius, 1989). Thus, proteins destined for secretion enter the lumen unfolded and leave folded and active.

In this book, the process of entry to the lumenal compartment is described by Klappa *et al.* and by Simon and Blobel, whereas exit and its relationship to the dynamic processes of sorting in, and passage through, the various subsequent endomembrane compartments are described by Lippincott-Schwartz and by Bonatti and Torrisi. These chapters fill the gap and discuss the lumen, as a compartment in its own right, and the processes that occur within it.

Such a discussion, in the early 1990s, differs significantly in two ways from the approach that would have been taken 5–10 years earlier. In the first place, earlier treatments pictured the ER as a series or network of hollow tubes through which newly synthesized proteins passed en route to the cell surface and secretion; the lumen of the ER was simply the first transit lounge on the protein's journey. Now we must regard the lumen as a significant subcellular compartment in its own right, a compartment with its own unique set of resident proteins (Koch, 1987) and its own characteristic composition of ions, cofactors, and other small molecules, and with its own communication mechanisms with both the cytoplasm and the nucleus. In the second place, our overall picture of protein folding in the cell has been transformed. Whereas formerly protein folding and assembly following biosynthesis was pictured as an unmediated process that could be comprehensively modeled by studies on protein unfolding and refolding *in vitro*, we now recognize a host of protein-folding factors or "chaperones" that facilitate the process in the cell (Freedman, 1992; Gething and Sambrook, 1992).

These two minor revolutions have resulted in new insights on the mechanisms of protein folding and assembly in the ER. In this context, protein folding and assembly occupy key positions in a series of processes—elongation, translocation, modification, folding and assembly, sorting—that are overlapping, rather than being simply successive. And within this set of processes, folding and assembly are no longer phenomena that have a chemistry but no biology; there is cellular machinery associated and involved.

2. FOLDING AND ASSEMBLY

2.1. Studies on the Refolding of Denatured Proteins

That the amino acid sequence of a protein determines its final native structure, and that no other factors or proteins are necessary to attain the final structure, was first suggested by Anson (1945), and amplified by Anfinsen (1973). Studies on protein folding following denaturation of mature proteins found that many could (under appropriate conditions) indeed refold *in vitro* in high yield. The demonstration that primary structure specifies tertiary structure has inspired many attempts to solve what is sometimes called "the protein folding problem," namely the prediction of tertiary structure from sequence. It has also been the source of an extensive program of physical studies on protein refolding aimed at characterizing the process in kinetic and structural terms (see Creighton, 1992).

Such kinetic studies, employing fluorescence, circular dichroism, NMR, and immunological techniques, have recently begun to create a coherent picture of the rapid processes involved in the generation of the native tertiary structure. They show that when transferred to refolding conditions, an unfolded protein collapses on a millisecond time scale to a compact form that contains significant elements of the secondary structure of the native protein. In this "collapsed intermediate state" or "molten globule," many of the hydrophobic residues of the protein are buried, but there remain some hydrophobic surface patches, a significant amount of water is bound within the interior of the protein, amino acid side chains are mobile and do not have fixed conformations, and there are no specific side chain/side chain interactions. Most notably, this collapsed intermediate is not greatly different in energy from the unfolded state and equilibrates rapidly with it. The key slow step is then the transition from this intermediate to a native, or nativelike, state, which is accompanied by loss of internal water and the assumption by buried side chains of definite conformations and interactions, generating the native tertiary structure (for a review see Pittsyn, 1992).

The emergence of this picture of protein folding, and the controversy over details such as the sequence in which collapse and secondary structure formation occur, has tended to obscure earlier insights indicating that except in simple cases, the observed kinetics of protein folding are dominated by other, slower processes. Most proteins are comprised of more than one folded domain and these refold as quasi-independent units, so that the slowest phase of refolding is not their refolding as such, but the docking of the refolded domains on each other to form the native interdomain contacts. Similarly, the kinetics of refolding of oligomeric proteins are dominated under most conditions by the process of assembly to form native quaternary contacts.

Even within a single domain, two significant structural isomerizations introduce complications into the folding process. The more recently recognized of the two is nevertheless the more general. Nearly all proteins contain the imino acid proline, and in folded proteins a significant fraction of imide bonds to the imino group of prolyl residues are in the *cis* configuration, in contrast to the overwhelming majority of peptide bonds, which are in the *trans* configuration (Stewart *et al.*, 1990). In unfolded proteins, all imide bonds to prolyl residues equilibrate between the *cis* and *trans* configurations and refolding to the native state requires isomerization to the correct conformer either before folding begins or during the folding process; this isomerization is intrinsically a slow process at physiological pH. The significance of this was first appreciated by Brandts *et al.* (1975) and it is now recognized as the dominant kinetic feature of the refolding of most small proteins (Schmid, 1992).

The other significant chemical isomerization in protein refolding is disulfide isomerization. The importance of this process was recognized earlier because many previous studies concentrated on the refolding of reduced, unfolded secretory proteins. This work is particularly relevant to a consideration of protein folding in the ER since disulfide bonds are characteristic of secretory proteins and exocellular domains of plasma membrane proteins. In their classic studies on the refolding of reduced bovine pancreatic ribonuclease, Anfinsen and colleagues (Epstein *et al.*, 1963) showed that the protein rapidly formed a mixed population of incorrectly disulfide-bonded species and that regeneration of activity was dependent on disulfide shuffling or isomerization. Hantgan *et al.* (1974) extended this by showing that the slowest, spectroscopically detectable phase of refolding was associated with the appearance of enzyme activity.

One of the best-characterized folding pathways *in vitro* is that of bovine pancreatic trypsin inhibitor (BPTI), a small 59-residue protein with one domain that contains three disulfide bonds. A feature that is unusual and makes BPTI a very amenable protein to work with is its solubility in the reduced and denatured state. Creighton (1978) determined the folding pathway of BPTI by trapping the intermediates, using alkylating agents or acidification to block free thiols during refolding and reoxidation. These experiments showed that BPTI sampled only a limited number of the possible disulfide-bonded intermediates, but that some of the intermediates contained nonnative disulfide bonds. The rate-limiting step in the formation of the native molecule is isomerization to the favorable intermediate for the formation of the third native disulfide bond. Using different techniques to quench disulfide interchange and resolve the quenched intermediates, Weissman and Kim (1991) showed that the predominant intermediates at the one- and two-disulfide states were those with native disulfides. This quantitative refinement does not alter the qualitative conclusion that the major folding pathway involves rearrangement of disulfide bonds. Indeed, Goldenberg (1992) argues that it is the formation of nativelike structure in the folding intermediates that ". . . causes steric inhibition of direct sequential formation of the three disulfides found in the native protein, thus accounting for the role of intramolecular rearrangements in the folding mechanism."

In summary, studies on protein refolding *in vitro* have identified disulfide isomerization, prolyl-peptidyl *cis/trans* isomerization, and the docking of domains and subunits as the dominant slow processes. But these studies have also revealed another phenomenon, which until recently was rarely emphasized and was the subject of some embarrassment. Although refolding under some conditions can regenerate the native state, the yields obtained are often disappointingly low; in many cases, productive refolding competes unfavorably with nonspecific aggregation. These considerations are crucial for an understanding of protein folding and assembly in the lumen of the ER; the processes that limit the rate and yield of productive refolding are disulfide- and prolyl-isomerization, specific formation of native interdomain and intersubunit contacts, and the competing processes of nonspecific aggregation of unfolded chains or partially folded intermediates.

2.2. Catalysis of Protein Refolding *in Vitro*

For over 20 years, the only known example of a cellular catalyst of protein refolding was protein disulfide isomerase (PDI); Anfinsen and colleagues showed that microsomal fractions from secretory tissues such as liver and pancreas catalyzed the renaturation and reoxidation of reduced denatured ribonuclease. This activity was purified and partially characterized (Goldberger *et al.*, 1963; De Lorenzo *et al.*, 1966). PDI is now known to be a major protein component of the ER lumen and there is general acceptance of the hypothesis (Epstein *et al.*, 1963; Freedman, 1984) that its physiological role is as a catalyst of disulfide formation and folding in protein biosynthesis (see Section 3.1).

For the other rate-limiting isomerization reaction, there was no evidence of biological catalysis until the 1980s when Fischer and Bang (1985) demonstrated a peptidyl-prolyl *cis/trans* isomerase (PPI) activity toward peptide substrates in mammalian kidney cytosol and subsequently the purified enzyme was shown to catalyze such isomerizations in the slow refolding of some proteins (Lang *et al.*, 1987). (For a fuller account of the catalytic properties of PDI and PPI, see Freedman, 1992.)

The recent appreciation of the more general role of protein factors (molecular chaperones) in protein folding and assembly in the cell has led to studies aiming to characterize the actions of the purified factors on protein refolding *in vitro*. The actions observed are mainly suppression of aggregation and misfolding (see Figure 1). BiP (see below) prevents aggregation of denatured proteins

Pamela J. E. Rowling and Robert B. Freedman

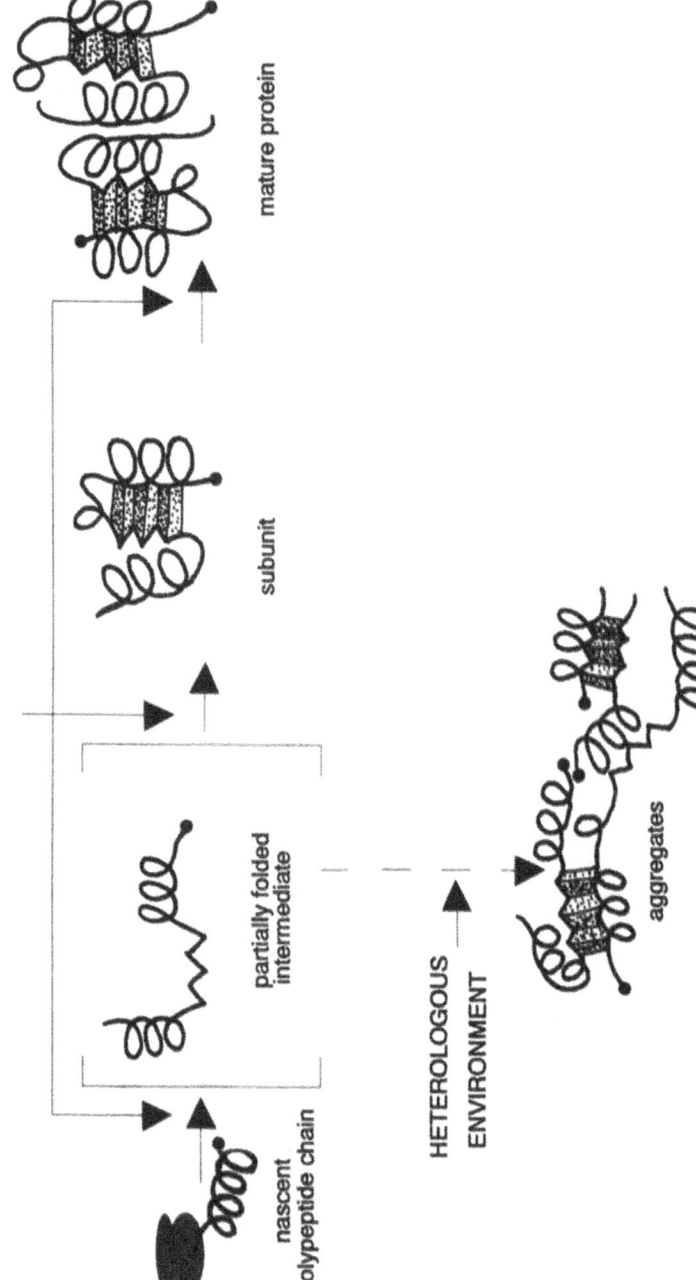

FIGURE 1. Folding in native and heterologous conditions. The folding and maturation of a dimeric protein is dependent on its environment. A milieu that does not favor the native state results in the formation of protein aggregates. *In vivo*, chaperones suppress the formation of aggregates. Adapted from Mitraki and King (1989).

during refolding *in vitro* but has no effect on the rate of folding. Its bacterial homologue, dnaK, actively refolds RNA polymerase, which has been aggregated by heat shock, in an ATP-dependent reaction (Skowyra *et al.*, 1990). Recent work by Wiech *et al.* (1992) using hsp90, the cytoplasmic homologue of the ER-resident protein endoplasmin, has shown that a fourfold excess of hsp90 completely suppresses the aggregation observed during the refolding of citrate synthase. This dimeric enzyme shows a very low yield of spontaneous refolding and a high yield of aggregation, except at very low concentrations. In refolding studies with immunoglobulin Fab fragments, the presence of hsp90 has a slight effect in increasing the yield of refolded protein.

Recently we have used a preparation of total ER lumenal proteins from a plasmacytoma cell line to examine the effect of ER proteins on the refolding of BPTI, which was previously shown to be catalyzed by PDI alone (Creighton *et al.*, 1980). The lumenal protein preparation catalyzed BPTI refolding, but a quantity of PDI equivalent to that found in the preparation catalyzed refolding to the same degree (Zapun *et al.*, 1992). This indicated that other proteins present in the lumenal preparation did not act synergistically with PDI or otherwise contribute to the catalysis in this case.

2.3. Studies on Cotranslational Protein Folding

Refolding of denatured reduced mature proteins is not a realistic model of secretory protein folding in the ER. It takes no account of the cotranslational translocation of the nascent protein, nor of the cleavage of the signal peptide (see Klappa *et al.*, this volume), nor of the other chemical modifications that the nascent protein undergoes (see Section 4). In order to integrate protein folding with these other processes, it is necessary to follow protein folding cotranslationally, which limits the analytical methods that can be employed. In contrast to the powerful physical methods that can be used in refolding studies, studies on cotranslational folding are restricted to methods that can operate on minimal quantities of proteins labeled by incorporation of radioactive amino acids at biosynthesis. Such methods exploit simple differences in properties between the native and unfolded states such as (1) electrophoretic mobility, (2) protease susceptibility, (3) solubility or aggregation, (4) formation of specific oligomers, (5) recognition by conformation-specific antibodies, or (6) affinity for native ligands.

These methods are relatively crude in the information they provide and in general only distinguish between "unfolded" and "native" states. Nevertheless, they have the advantage that they can be applied in cell-free and whole-cell systems developed to analyze other aspects of secretion such as the classic cell-free system for analysis of segregation of nascent secretory proteins employing an added mRNA, a translation system based on rabbit reticulocyte lysate, and microsomal membrane vesicles from dog pancreas (see Klappa *et al.*, this volume). The first analysis of nascent protein folding in such a system was that of Scheele and Jacoby (1982) who demonstrated that the reducing conditions conventionally used in such cell-free systems did not permit disulfide bond formation and native folding in translation products, but that the products had properties characteristic of the native state when a disulfide oxidant was added to titrate the system to the appropriate redox potential. This was subsequently confirmed by Kaderbhai and Austen (1985) who used the difference in mobility of reduced and oxidized forms of prolactin when analyzed by nonreducing SDS-PAGE to monitor cotranslational disulfide formation.

In cases where the biological activity of the translation product can be detected with high sensitivity, it is possible to assay directly for folding to the native state. Production of enzymatically active translation products in such systems has been observed for the dimeric enzyme β-hexosaminidase B (Sonderfeld-Fresko and Proia, 1988) and for the multidomain protease tissue plasminogen activator (Bulleid *et al.*, 1992). In both cases, folding to generate native activity was dependent on the presence of oxidized conditions in the translation system to permit native disulfide bond formation.

The requirements for native folding in such cell-free translation systems have been analyzed by the classic biochemical approach of resolution and reconstitution. It is possible to deplete dog pancreas microsomes of lumenal content proteins by washing at pH 9 or with low concentrations of the detergent saponin; the depleted microsomes nevertheless retain the ability to translocate and segregate nascent secretory proteins (Paver *et al.*, 1989; Bulleid and Freedman, 1990). Using such microsomes, Bulleid and Freedman (1988) examined cotranslational disulfide formation in a model protein, a truncated form of the wheat storage protein γ-gliadin, which contains four disulfide bonds. As disulfides form, the structure of the protein becomes more compact, so that when resolved on a nonreducing SDS-PAGE gel, the disulfide-bonded protein has a greater mobility than the reduced polypeptide. Microsomes that had been depleted of lumenal content were defective in cotranslational disulfide formation, but this could be restored by the incorporation of purified PDI (see Section 3.1) into the lumenal volume of the depleted vesicles (Figure 2). This is a direct indication of a requirement for PDI of the process of native disulfide formation.

The effective translocation of nascent proteins into lumenally depleted microsomal vesicles and indeed into vesicles reconstituted from solubilized microsomal membrane proteins (Nicchitta and Blobel, 1989; Yu *et al.*, 1989) clearly suggests that the functions of lumenal proteins are entirely posttranslational. Most work on the potential functions in protein folding of lumenal proteins other than PDI has used intact cells. However, Kassenbrock *et al.* (1988) used an *in vitro* translation system to examine the interaction of translocated translation products with BiP. Newly synthesized proteins could only be found associated with BiP if

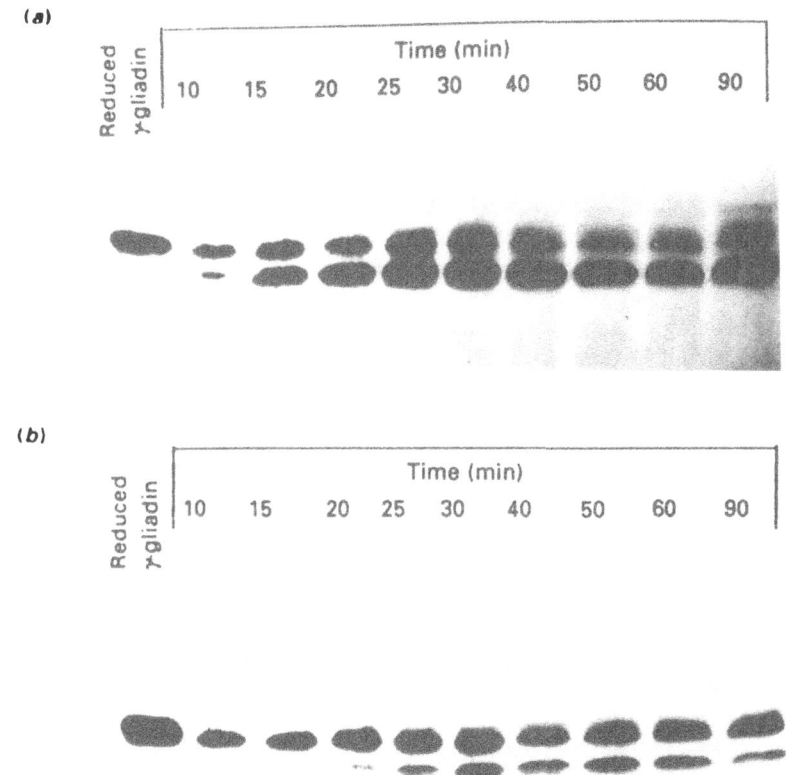

FIGURE 2. Defective disulfide formation in PDI-depleted microsomes. SDS-PAGE analysis under nonreducing conditions was used to examine the formation of the more mobile oxidized disulfide-bonded protein from the reduced molecule in the presence of (a) control dog pancreas microsomes and (b) pH 9 washed microsomes. From Bulleid and Freedman (1988).

the system had been depleted of ATP prior to the immunoprecipitation with anti-BiP antibodies. Using prolactin as the model protein, reduced or incorrectly disulfide-bonded full-length forms of the protein could be found associated with BiP, but not nascent chains or the native disulfide-bonded protein. In this system, unglycosylated forms of invertase were found associated with BiP, but not the mature glycosylated form. Other studies on cotranslational *N*-glycosylation are discussed below (Section 4.2).

Working with microsomal vesicles allows a focus on early processes in the secretory pathway, but the formation of the membrane fragments is potentially damaging, and the complex interactions between the various secretory compartments are lost. The employment of an *in vitro* translation system that uses a more gentle preparation of the secretory pathway might have advantages under some circumstances. Several techniques have recently been developed and applied to cultured cells to produce "semipermeabilized" cells; the methods include hypotonic shock (Beckers et al., 1986), treatment with streptolysin O (Miller and Moore, 1991), or wet cleavage (Brands and Feltkamp, 1988). The cells lose cytoplasmic content, but retain organelles, and the architecture of ER, Golgi, nucleus, and mitochondria appears to be maintained within a cytoskeletal matrix. Cells prepared by these methods, when supplemented with cytosolic components, are able to translate and translocate proteins into the ER. The newly synthesized proteins exit the ER and move to the Golgi by the standard process of vesicle budding, which has been shown to be dependent on Ca^{2+} (Beckers and Balch, 1990), ATP (Beckers et al., 1986), and GTP (Miller and Moore, 1991). The vesicle transfer steps appear to have the same temperature-dependence as in whole cells (Wingrove and Freedman, unpublished), suggesting that this system has considerable promise for more detailed analysis.

2.4. Secretory Protein Folding in Intact Cells

The lumen of the ER is the primary site of folding of proteins that enter the secretory pathway, whatever their final destination. However, proteins show a wide variation in the time taken to exit this compartment, indicating some selectivity in the process of transfer to distal compartments. It is now clear that this variation depends to a great extent on differences in the rate of folding and assembly to generate the native tertiary and quaternary structure (Rose and Doms, 1988; Hurtley and Helenius, 1989). This has mainly derived from work on immunoglobulins and the membrane proteins of enveloped viruses, especially VSV G protein and the influenza virus hemagglutinin (HA).

Work on immunoglobulin biosynthesis in myeloma cells produced two highly influential early results. Bergman and Kuehl (1979a,b) showed that the disulfide bond in the N-terminal domain of immunoglobulin light chains forms cotranslationally, essentially as soon as the whole domain has been translocated into the ER lumen: they also showed that assembly begins cotranslationally, in that nascent heavy chains form intermolecular disulfides either to form heavy-chain dimers (in the case of MOPC21 cells) or heavy–light dimers (in the case of MPC11 cells). Second, in myeloma cell lines synthesizing only heavy chains, these chains were retained in the ER lumen, rather than being secreted and were found in association with another protein, which was termed heavy-chain binding protein or BiP (Haas and Wabl, 1983). It was subsequently shown that BiP also interacts with immunoglobulin heavy chains in cells that make both heavy and light chains and secrete intact immunoglobulins (Bole et al., 1986). In this case it was found that BiP is associated with intermediates in the assembly process (individual chains, dimers, etc.) but not with the fully assembled product. Furthermore, if only heavy chains are synthesized, and complexes of heavy chains and BiP build up in the ER, initiation of the synthesis of light chains leads to dissociation of the BiP/heavy-chain complexes and the formation and secretion of intact immunoglobulins (Hendershot, 1990). Conversely, inhibition of N-glycosylation prolongs the interaction between heavy chains and BiP, which decreases assembly of heavy and light chains and lowers immunoglobulin secretion (Bole et al., 1986). These data indicate the complexity of interactions between folding, modification, and assembly in the biosynthesis of native oligomeric proteins.

This is borne out by studies on the assembly of the trimeric influenza virus surface protein HA. Two groups characterized this process using a wide range of techniques to follow folding and assembly within the secretory pathway (Gething et al., 1986; Copeland et al., 1986, 1988). They showed that folding and subsequent trimerization were requirements for exit from the ER and that many mutations that blocked surface expression of HA did so through interfering with folding and hence blocking exit from the ER (Figure 3).

The observation that complex folding and oligomerization events occur in the ER and are necessary for exit from this compartment is quite general (Hurtley and Helenius, 1989). The mechanisms involved in such a system also appear to be common. Many secretory proteins fail to fold when subjected to mutation or inhibition of glycosylation or when expressed at high temperatures; such proteins are not secreted but are found in the ER associated with BiP. In the case of HA, up to 10% of molecules misfold, are retained in the ER and are subsequently degraded, even under normal conditions. This percentage increases with temperature. This is comparable to similar effects seen on protein refolding in the test tube. The role of glycosylation here is expressed through its general effect on protein solubility during folding (see below). For example, in the case of VSV G protein, mutants lacking glycosylation sites do not fold and proceed to the cell surface but glycosylation at either of the two native glycosylation sites is sufficient to ensure normal maturation and transport from the ER. In mutants lacking both sites, the introduction of glycosylation sites elsewhere in the protein can restore secretion competence (Machamer et al., 1985; Machamer and Rose, 1988a,b).

The details of folding, disulfide formation, and assembly within the ER are now open to detailed analysis in intact cell systems. In a complex multisubunit protein, the mouse muscle nicotinic receptor, which contains five subunits, association of the α subunit with the γ subunit was dependent on prior intramolecular disulfide bond formation within the α subunit (Blount and Merlie, 1990). In the case of HA, careful examination of disulfide formation has been made possible by the manipulation of the effective redox potential within the ER by incubating the cells in the presence of the dithiol reductant, dithiothreitol (DTT) (Braakman et al., 1992a,b). In pulse–chase studies when DTT was present at various stages,

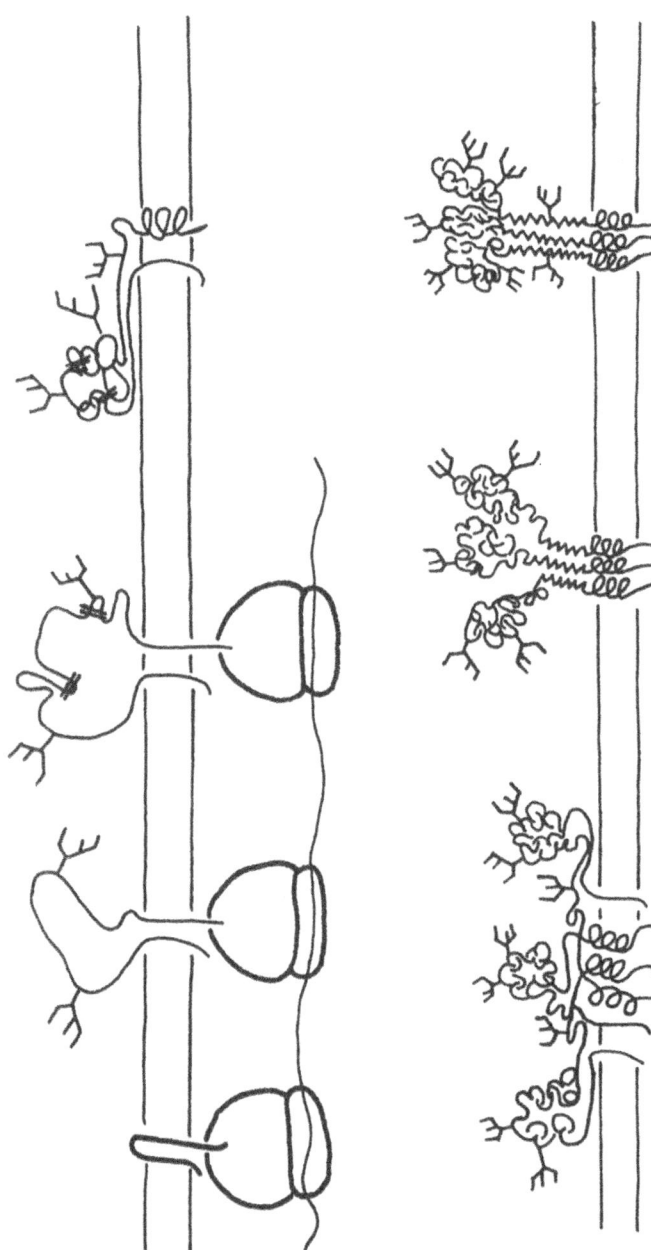

FIGURE 3. A schematic pathway of folding of hemagglutinin. Cotranslational glycosylation and disulfide formation occur in the globular domain which can fold as an independent domain. Oligomerization may occur by association of the hydrophobic transmembrane domains (Gething *et al.*, 1986).

DTT prevented disulfide formation in newly synthesized HA and reduced HA in the ER which had previously been disulfide-bonded. This reduced HA was incorrectly folded, failed to trimerize, and was retained for long periods in the ER. When the DTT was washed out, the protein was rapidly oxidized, correctly folded, trimerized, and transported to the Golgi complex, indicating that disulfide bond formation can occur posttranslationally as well as cotranslationally.

There have been few studies on the process of prolyl-peptidyl *cis/trans* isomerization during protein biosynthesis in intact cells. However, when fibroblasts were treated with cyclosporin A, an inhibitor of PPI, there was a small delay in the folding of newly synthesized procollagen to the triple-helical, protease-resistant conformation (Steinmann *et al.*, 1990); as a result, the newly synthesized protein was over hydroxylated (see Section 4.4).

3. LUMENAL PROTEINS OF THE ENDOPLASMIC RETICULUM

Analysis by two-dimensional electrophoresis of proteins derived from the ER lumen reveals that there is a particular pattern of proteins separated, which is similar for extracts from different organisms and tissues (see Figure 4; Kaderbhai

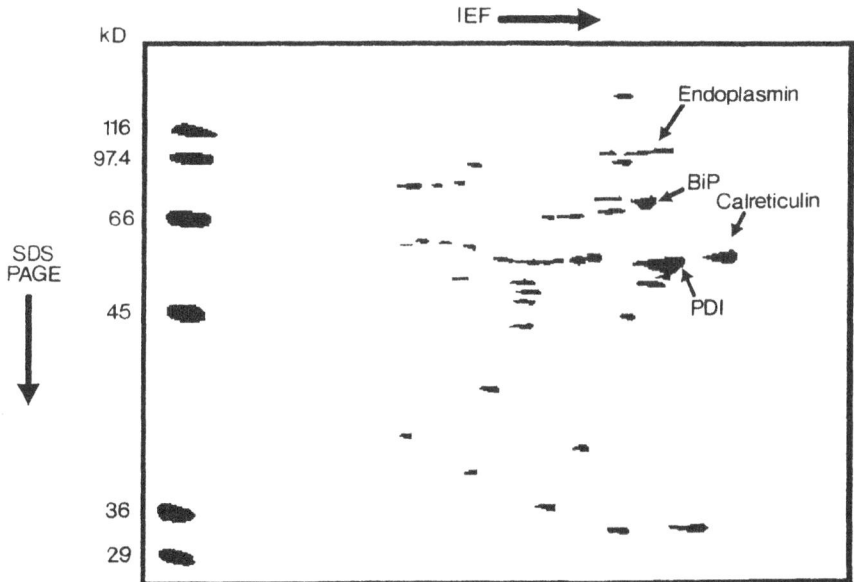

FIGURE 4. Two-dimensional electrophoresis of dog pancreas microsomes. A distinct pattern of proteins derived from the ER can be resolved using this technique. The major ER-resident proteins have been labeled.

and Austen, 1984). Some of these proteins have been found to have a role in aiding protein folding of the newly translocated polypeptides in the ER (see Sections 2.2–2.4; for review see Freedman, 1992; Gething and Sambrook, 1992).

Early investigations found that if cells were stressed, e.g. by heat treatment, glucose starvation, calcium ionophores, or inhibitors of glycosylation, a subset of proteins within the cell was induced. These proteins were termed the heat shock proteins (hsp) (reviewed by Welch, 1990) or glucose-regulated proteins (grp) (Lee, 1987) depending on the stress the cells received. The heat shock proteins are divided into groups on the basis of their molecular weight and sequence homology, e.g., hsp70 and hsp90 families. Distinct members of these families are often associated with different subcellular organelles. These proteins are highly conserved between higher and lower eukaryotes. E. coli also has analogues with considerable sequence similarity and functional homology to those of eukaryotes. Pelham (1986) suggested that stress response proteins were induced when misfolded and aggregated proteins accumulated, and the hsp's either prevented their aggregation or "solubilized" these aggregates within the cell. The ER contains several such proteins, including a member of the ubiquitous hsp70 family (BiP; see Section 3.2) and a member of the hsp90 family, endoplasmin (Section 3.4).

Proteins exit from the ER by bulk flow as vesicles bud off the ER and fuse with the Golgi stack and then leave the Golgi to their final destination (Rothman, 1987). This would mean that proteins that aid protein folding in the ER would be eventually secreted, but fractionation of microsomes revealed that these ER proteins are not found in any of the Golgi stacks or in distal vesicles. Biochemical analysis of the resident proteins of the ER indicates that they only undergo posttranslational modifications associated with the cis-Golgi (Pelham, 1988; Dean and Pelham, 1990). The basis of this selective retention was first suggested when established resident proteins of the ER (PDI and BiP) were shown to have an identical C-terminal sequence, -KDEL. This is characteristic in vertebrates and similar sequences have been found in other sources, e.g., -HDEL in Saccharomyces cerevisiae and a mixture of the two sequences in plants (reviewed by Pelham, 1990).

Pelham's group created mutants in yeast and screened for those that secreted ER proteins and then analyzed the gene product. A mutant was isolated and termed ERD2 (for ER retention deficient), which appeared to be responsible for the retention of proteins in the ER. The ERD2 gene encoded a 25-kDa product and is predicted to contain seven transmembrane regions (Semenza et al., 1990). A similar receptor has been found in humans that has 50% homology with the ERD2 gene product. When expressed in COS cells, it localized to an intermediate compartment between the ER and Golgi (Lewis and Pelham, 1992). The process of sorting from the ER and selective retention of resident ER lumenal proteins is discussed further by Lippincott-Schwartz (this volume).

3.1. Protein Disulfide Isomerase

The most thoroughly characterized process of protein folding in the ER is that of disulfide formation catalyzed by PDI. Homogeneous PDI catalyzes *in vitro* thiol:disulfide interchange reactions, which under the appropriate conditions can lead to the interchange of disulfide until the stable native structure of the protein is achieved (Lambert and Freedman, 1983b).

PDI is characteristic of lumenal content proteins in containing the ER C-terminal retention signal sequence allowing it to be recycled back to the ER. It is an abundant protein of the ER; in dog pancreas microsomes it constitutes 14% of the total lumenal content protein and in liver it has been estimated to be 0.4% of the total cell protein (Lambert and Freedman, 1983a). The amount of PDI in a cell correlates with the amount of secretory protein synthesized by that cell, and undergoes physiological changes in parallel with changes in the rate of secretory protein synthesis (Freedman *et al.*, 1989).

The purified mammalian enzyme is a homodimer of 57-kDa subunits; it is acidic with an isoelectric point of 4.2 and has no glycosylation sites. The enzyme from yeast is glycosylated but is otherwise similar. In yeast, PDI has been found to be a single-copy gene that is essential for cell viability (LaMantia *et al.*, 1991; Farquhar *et al.*, 1991).

The first complete sequence of PDI was deduced by Edman *et al.* (1985) from rat cDNA. The sequence showed that PDI had regions of internal homology and on the basis of this could be divided into six domains: a, e, b, b', a', c. Comparison of the a and a' domains showed that they were 47% identical. Strikingly, these two domains of PDI showed similarity with thioredoxin, a small oxidoreductase found in all organisms from prokaryotes to higher eukaryotes.

Models of the a and a' domains have been based on the X-ray crystal structure of oxidized *E. coli* thioredoxin, with which they show 60% sequence identity (Freedman *et al.*, 1988). These models suggest that the active site of PDI is on an exposed loop. In thioredoxin the active-site vicinal dithiol:disulfide couple is formed by two cysteines in the sequence Trp-Cys-Gly-Pro-Cys-Lys. PDI contains two regions analogous to this in the a and a' domains, Trp-Cys-Gly-His-Cys-Lys. Chemical modification studies by Hawkins and Freedman (1991) showed that modification of one thiol group in each of the two dithiols present in a PDI subunit led to the inactivation of the molecule. The modified reactive cysteines had an unusually low pK_a of 6.7, compared with a pK_a of around 8.5 in the majority of proteins. One theory to account for this stabilization of the thiolate ion is the positive helix dipole that lies above the active site in both thioredoxin and the models of PDI. Chemical modification has also shown that PDI active sites are much stronger oxidants than those of thioredoxin (and consequently weaker reductants). The redox properties of the PDI active site dithiols make them at least 300-fold more oxidizing than structural disulfides in known proteins (Hawkins et al., 1991).

An interesting study by Vuori *et al.* (1992a) expressing human PDI and mutants in *E. coli* found that, when either of the active sites was mutated to Ser-Gly-His-Cys, activity of the molecule was 50% that of the wild-type molecule. If both active sites were mutated in the same molecule, the activity was abolished. This work suggests the active sites act independently. Similarly, if the active site of *E. coli* thioredoxin was mutated to that of PDI, the active site became more oxidizing and showed a disulfide isomerase activity between that of PDI and thioredoxin (Krause *et al.*, 1991). This suggests that both local chemical effects and the multisubunit nature of PDI are important for the catalytic properties of the molecule.

Morjana and Gilbert (1991) used peptides as inhibitors of PDI activity; increasing the length of the peptide produced greater inhibition, suggesting that a large part of the protein backbone may be interacting with PDI. Peptides that contained a cysteine were bound with a four- to eightfold greater affinity; no preference for any other residue was found.

PDI has been shown to be a multifunctional protein and therefore its function may not be limited to disulfide isomerization. The best established case is that of prolyl-4-hydroxylase, which catalyzes the hydroxylation of prolines, a unique posttranslational modification of collagen. Prolyl-4-hydroxylase is an $\alpha_2\beta_2$ tetrameric enzyme. The β subunits were found to be identical to PDI in sequence and also a purified prolyl-4-hydroxylase had PDI activity (Pihlajaniemi *et al.*, 1987; Koivu *et al.*, 1987). It was long known that the β subunits were synthesized in vast excess and there was a free pool of them within cells; this pool is now recognized to be free PDI. The mechanism of PDI in the holoenzyme is unclear, but since PDI has been shown to contain a dehydroascorbate reductase activity, it may be that its function is to produce ascorbic acid, which is necessary as a cofactor of the hydroxylation reaction (Wells *et al.*, 1990).

There is also compelling evidence that PDI forms part of the microsomal triglyceride transfer complex. This complex catalyzes the transfer of triglyceride into nascent plasma lipoproteins. The complex consists of an 88-kDa protein and a 58-kDa protein, with a combined molecular mass of 150 kDa as determined by sedimentation. The 58-kDa protein has been identified as PDI by N-terminal sequence, immunological cross-reactivity, and PDI activity of the dissociated complex (Wetterau *et al.*, 1990). It appears that the two molecules form a highly stable complex that does not readily dissociate (Wetterau *et al.*, 1991a). When the complex was dissociated by chaotropic agents and nondenaturing detergents, the activity of the triglyceride transfer complex was lost; when renaturation was attempted, PDI renatured as an active free enzyme, and the 88-kDa large subunit formed insoluble aggregates (Wetterau *et al.*, 1991b). This suggests that the role of the PDI subunit is to retain the large subunit soluble and in a folded conformation. PDI may play a similar role in prolyl-4-hydroxylase, since studies on expression in insect cells show that coexpression of PDI is necessary to prevent aggregation of expressed α subunits (Vuori *et al.*, 1992b).

3.2. BiP

BiP, otherwise known as hsp74 or grp 78 (Hendershot *et al.*, 1988; Kozutsumi *et al.*, 1989), is a member of the heat shock protein-70 family. The proteins of the hsp70 family are produced constitutively but may be induced by stresses that lead to the accumulation of malfolded protein. Members of this family are expressed in different subcellular locations. There is an overall 50% to 95% identity between each one of these family members (Lindquist and Craig, 1988). BiP is a major component of the ER comprising 7% of the protein in the lumen of the plasmacytoma cell line MOPC-315. Mutation of the BiP gene is lethal in *S. cerevisiae* (Normington *et al.*, 1989; Rose *et al.*, 1989).

BiP, like other members of the hsp70 family, contains a weak ATPase activity with a K_m for ATP of 1 μM and a k_{cat} of less than 1 min^{-1} (Kassenbrock and Kelly, 1989). Flynn *et al.* (1989) demonstrated that this activity could be stimulated fivefold by peptides. ATP hydrolysis by BiP is linked to its release of bound proteins or peptides. The use of nonhydrolyzable ATP analogues does not seem to allow protein release. The N-terminal 450 amino acids of the hsp70s are a highly conserved domain. An N-terminal 44-kDa proteolytic fragment with ATPase activity can be generated from both BiP (Kassenbrock and Kelly, 1989) and hsc70 (Chappell *et al.*, 1987). The N-terminal fragment from bovine hsc70 has been crystallized and the structure resolved to 2.2Å (Figure 5). There are two lobes to the structure formed chiefly by α helices with a deep cleft in which the ATP binds. The structure resembles that of the monomer of G-actin, which binds ATP and hydrolyzes it to generate bound ADP and free phosphate (Flaherty *et al.*, 1990, 1991).

Sequence analysis of the C-terminal region of BiP and other members of the hsp70 family showed little homology with proteins of known structure. However, comparison of its predicted secondary structure with the data base revealed that the C-terminal domain could be aligned with the peptide binding domain of the MHC class I molecule (Rippmann *et al.*, 1991; Flajnik *et al.*, 1991). Both groups produced models of the C-terminus based on these data. These models indicate that the proposed peptide binding cleft of BiP contains both polar and hydrophobic residues (Figure 5).

BiP was originally found to bind heavy chains of immunoglobulin in myeloma cell lines that did not produce light chains (Haas and Wabl, 1983). By

a

b   **c**

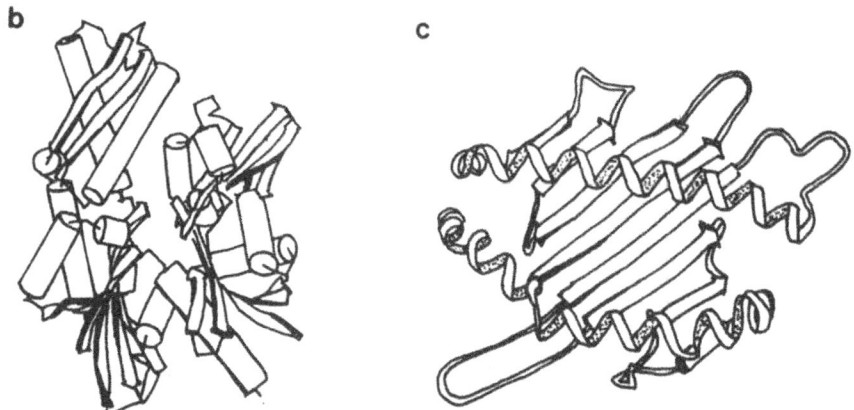

FIGURE 5. Structure of the stress-70 proteins. (a) A linear diagram indicating the highly conserved N-terminal ATPase domain and the peptide binding domain. (b) The three-dimensional structure of the hsp70 ATPase, which is composed of two subdomains connected by a deep cleft in which ATP binds. (c) A model of the peptide binding domain based on the structure of MHC antigen class I. From Gething and Sambrook (1992).

deletion of various domains of the heavy chain, it was found that BiP interacted with the CH$_1$ domain (Hendershot *et al.*, 1987). This is the domain that binds the immunoglobulin light chain. Similar studies reported by Gething and Sambrook (1992) found BiP bound to a region of the HA protein where the three molecules of the trimer regions are in contact. Hence, it appears that BiP binds to segments of protein that are not usually exposed in the native protein and this may prevent the protein from aggregating.

Flynn *et al.* (1989) examined the requirements of peptide binding to BiP using a range of peptides derived from a number of proteins known to bind to BiP. All bound to BiP but with a wide range of affinities from 10 μM to > 1 mM. In a second study (Flynn *et al.*, 1991), this group used random peptides produced by solid-phase synthesizer. Peptides of at least seven amino acids promote maximal ATP turnover relative to smaller peptides, suggesting that the binding site of BiP is capable of binding seven amino acids. Analysis of the amino acids from peptides that bound to BiP indicated that BiP had a preference for aliphatic side chains at all positions along the seven-mer peptide; aromatic residues are tolerated but polar and charged amino acids and proline appear to be excluded at every position, except for a preference for arginine at the C-terminus.

There appears to be a regulation of BiP activity that has not been fully elucidated and manifests itself in the adenylation of BiP (Hendershot et al., 1988). BiP was found to be labeled on its threonine and serine residues. The labeling of BiP decreased after treatments that lead to the accumulation of non-glycosylated forms and also in heavy-chain oversecreting cell lines, where there is an increase in BiP synthesis. It appears that only unmodified forms of BiP were able to bind proteins. BiP is able to autophosphorylate itself and this can be stimulated twofold by calcium (Carlino et al., 1992). The autophosphorylation leads to three more acidic forms of BiP as revealed by two-dimensional electrophoresis, suggesting multiple phosphorylation of the molecule. The modified forms of BiP were predominately monomers but the unmodified forms of BiP were dimers, indicating the active form of BiP is a dimer.

BiP has also been indicated to be an important part of the translocation machinery in the cell. Experiments in which BiP levels were decreased by the use of mutants in S. cerevisiae (Vogel et al., 1990; Nguyen et al., 1991) found that nontranslocated forms of normally secreted proteins accumulated on the cytoplasmic face of the ER. This block appears to occur at a late stage of protein translocation as newly synthesized proteins could only be found associated with the membrane. Recent work by Sanders et al. (1992) has clarified BiP's role in translocation; BiP was efficiently cross-linked to molecules that had been arrested in translocation. Their work suggests that BiP interacts with the translocating protein after its interaction with sec61p, a membrane protein necessary for translocation in yeast. BiP may bind the protein as it initially traverses the ER membrane, and releases it shortly after translocation. ATP increased the amount of protein associating the sec61p; the only protein in the translocation system known to have an ATPase activity is BiP and therefore it may catalyze the transfer of protein across the membrane. This contrasts with the results on translocation in vitro, which show no requirement for BiP or other lumenal proteins (see Section 2.3).

3.3. Prolyl-Peptidyl cis/trans Isomerase

Proteins are synthesized on the ribosomes with all of their proline peptide bonds in a trans configuration, but in native proteins cis-prolines occur commonly. This cis/trans isomerization is a major rate-limiting step in protein folding (Figure 6). An enzyme capable of cis/trans isomerization of proline was originally detected in cytoplasmic extracts of porcine kidney by Fischer et al. (1985) and was called prolyl-peptidyl cis/trans isomerase (PPI). This enzyme was purified to homogeneity from the same source and found to be a small protein of 17 kDa. Since then, a number of PPIs have been identified in different species:

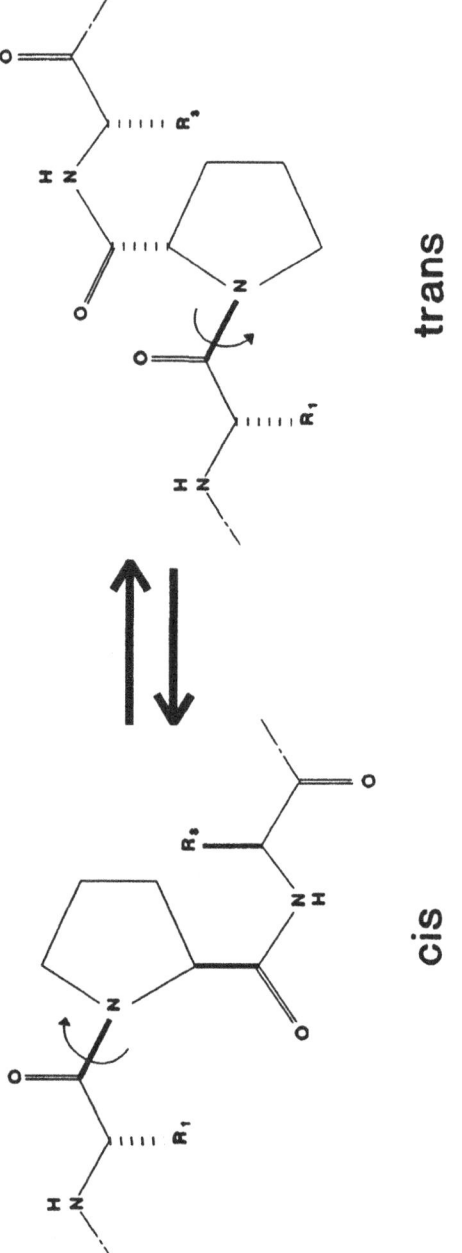

FIGURE 6. *Cis* and *trans* isomers of prolyl peptide bonds. The majority of peptide bonds between amino acids favor the *trans* isomer to avoid steric clashes. In the case of bonds to prolyl residues, a significant minority of bonds are in the *cis* conformation.

E. coli (Hayano *et al.*, 1991), lower eukaryotes, yeast (Haendler *et al.*, 1989; McLaughlin *et al.*, 1992), and Candida albicans (Kosher *et al.*, 1990). PPI is widely distributed in tissues (Koletsky *et al.*, 1986; Hasel *et al.*, 1991).

Studies on PPI have shown that it refolds a wide range of proteins but with different efficiencies. The best catalysis of protein refolding by PPI is a 100-fold increase in the rate constant for ribonuclease T_1; a similar degree of catalysis is seen with immunoglobulin light chains (Freedman, 1992). When the first PPI was sequenced, it was discovered to be identical to an already known protein, cyclophilin. Cyclophilin is the major binding protein for the immunosuppressant cyclosporin A and there is a vast body of work concerning cyclophilin's pharmacological function (reviewed by Schreiber, 1991).

Recent work has been directed at trying to establish whether a member of the PPI family is present in the secretory pathway. By screening of human and mouse (Price *et al.*, 1991; Hasel *et al.*, 1991) and chicken (Caroni *et al.*, 1991) cDNA libraries, a PPI clone has been detected that contains a putative signal sequence of 25 amino acids. The sequence of the human secretory PPI (sPPI) shows 64% homology with its cytoplasmic counterparts and encodes a protein of 20 kDa (Hasel *et al.*, 1991).

Hasel *et al.* (1991) found that their mouse clone, when expressed in *E. coli*, had PPI activity that could be inhibited by cyclosporin A. Antibodies raised against sPPI were found to react only with heavy and light microsomes corresponding to those fractions containing ER, with very little associating with those from the Golgi and plasma membrane; no staining was detected in any of the other subcellular rat liver fractions. More compelling evidence as to the sorting of sPPI has come from immunohistochemical work on L6 lymphoma cultured cells (Arber *et al.*, 1992). This work showed that sPPI had a similar but not identical stain distribution to that of ER markers; the staining suggested colocalization with the calcium storage protein calsequestrin (see Meldolesi and Villa, this volume).

Bose and Freedman (1992), employing a biochemical rather than a genetic approach, have isolated a microsomal PPI activity using subcellular fractionation. Sucrose density gradient analysis of these microsomes revealed that PPI copurified with markers of the smooth and rough ER. The activity was not released by high-salt washes and was protected from proteases unless the microsomes were disrupted by sonication or detergents. This lumenal microsomal PPI activity could be inhibited by cyclosporin A. The protein has now been purified and the N-terminal sequence confirms its identity to the sPPI demonstrated by cDNA sequencing (Bose and Freedman, unpublished results).

The DNA sequences of the secretory pathway-associated PPIs do not display the ER retention sequence, KDEL. However, the C-terminal extensions of all sPPIs in chick, man, mouse, and rat are identical, VEKPFAIAKE. When this sequence was removed from PPI and a control ten amino acids put in its place, the molecule was found in the Golgi and at no time localized with the normal distribution of the secretory pathway PPI. The C-terminal extension sequence of PPI was attached to a soluble secretory protein, glia-derived nexin, an inhibitor of serine proteases; it distributed in a manner similar to that of PPI, indicating that the C-terminal sequence was enough to direct targeting of the molecule (Arber *et al.*, 1992).

Another protein that has PPI activity and a homologous sequence is the product of the *nina-A* gene of *Drosophila*. The gene product has a cleavable signal sequence and is thought to be an integral membrane protein located in the ER and also in small particles elsewhere in the cell. Nina-A is essential for a processing step in the biosynthesis of rhodopsin; if the *nina-A* gene is nonfunctional, then rhodopsin accumulates in the ER (Colley *et al.*, 1991). The role of PPI *in vivo* is still unclear. It may play a role in the catalysis of protein folding by proline *cis/trans* isomerization (see evidence in Section 2.4), but there is also much evidence linking it to Ca^{2+}-mediated signal transduction pathways (Schreiber, 1991).

3.4. Endoplasmin

Endoplasmin is a member of the glucose-regulated proteins (grp 94) and of the heat shock 90 family. Koch *et al.* (1986) originally purified it from MOPC-315 cells as the most abundant glycoprotein present in the ER. Analysis of the lumenal contents of a plasmacytoma cell line revealed that its endoplasmin content was 7%. Like other resident proteins of the mammalian ER, it displays the KDEL retention motif (Maki *et al.*, 1990).

Endoplasmin can be induced by various stresses, e.g., calcium ionophores, inhibitors of glycosylation, and glucose starvation. Like other proteins in the ER that have an acidic pI, endoplasmin has an affinity for calcium; it contains four high-affinity calcium binding sites, K_d 0.4 mM, and eight low-affinity binding sites, K_d 6 mM (Koch *et al.*, 1986).

No other function of endoplasmin has been demonstrated, but work on its homologue, the cytoplasmic hsp90, has shown that an important function is to maintain the native structure of proteins. For example, hsp90 binds estrogen receptors in the cytosol which are released from the hsp90 and then move into the nucleus when estrogen is bound to the receptor. A recent report (Wiech *et al.*, 1992) suggests that hsp90 can prevent aggregation of denatured proteins.

3.5. Calcium-Binding Proteins of the Endoplasmic Reticulum

In all cells, calcium is used to regulate a number of cell functions in the cytoplasm (see Meldolesi and Villa, this volume). In nonmuscle cells, calcium is stored in the ER. Calcium is taken up from the cytoplasm by a Ca^{2+}/Mg^{2+}-ATPase pump (Berridge, 1987) and can be released by an inositol-1,4,5-

trisphosphate-sensitive calcium channel (Streb *et al.*, 1983). The calcium within the ER is sequestered by proteins that reduce the free concentration of calcium within the lumen.

Two groups sought to identify the major calcium binding proteins of the ER; these proteins were termed "reticuloplasms" by Koch's group (Macer and Koch, 1988) and calcium binding proteins (CaBP) (Van *et al.*, 1989). Using different screening methods, both groups detected calreticulin and endoplasmin; in addition, Koch's group identified PDI and BiP as being able to bind calcium. Van *et al.* (1989) also identified two other proteins with apparent molecular masses of 59 and 80 kDa. All of these proteins contained high-affinity sites for calcium with K_d between 1 and 5 μM. Endoplasmin and calreticulin also contain about 12 low-affinity sites for calcium. All of the calcium binding proteins have acidic pI's; it is well known that acidic amino acids are usually responsible for binding calcium.

The calcium binding properties of calreticulin are the best characterized and have recently been reviewed by Michalak *et al.* (1992) (see also Meldolesi and Villa, this volume). Calreticulin has the ability to bind the greatest amount of calcium (25 mol Ca^{2+}/mol calreticulin) relative to the other proteins. The high- and low-affinity calcium binding sites are located in different parts of the molecule: the high-affinity binding sites are found in the P-domain region of the molecule whereas the low-affinity binding sites are found in the acidic C-terminal region (Baksh and Michalak, 1991).

The nucleotide sequence of rabbit calreticulin cDNA has been determined (Fliegel *et al.*, 1989a). It can be divided into domains some of which may have a role in the regulation and release of calcium from the molecule since its sequence has putative recognition signals for phosphorylation and protein kinase C active site (Hanks *et al.*, 1988). Calreticulin has the KDEL ER retention sequence and the C-terminal regions also display some homology to those of BiP, PDI, and endoplasmin (Fliegel *et al.*, 1989b).

4. EARLY CO- AND POSTTRANSLATIONAL MODIFICATIONS

Secreted proteins are subject to a surprisingly diverse set of possible posttranslational modifications as they proceed through the secretory pathway. These modifications take place in different compartments and are characteristic of those compartments; hence, posttranslational modifications can be used as indirect indicators of protein traffic through the secretory pathway (see Lippincott-Schwartz, this volume; Bonatti and Torrisi, this volume). A small subset of modifications occur on nascent polypeptides and newly synthesized proteins within the lumen of the rough ER; these modifications are unusual in that they function on protein substrates that are not yet fully folded to their native state.

These modifications therefore may directly or indirectly affect the folding process.

4.1. Signal Peptidase

Proteins destined for the secretory pathway contain an N-terminal signal sequence that targets the newly translocated protein to the ER but is not a necessary part of the functional protein. Signal sequences are usually between 15 and 30 amino acids long and overall have similar features although their sequences are comparatively degenerate. They have a positively charged N-terminus, a central hydrophobic region, and a C-terminal region that contains polar amino acids which are frequently negatively charged (Gierasch, 1989; von Heijne, 1985; Watson, 1984).

Cleavage of the signal peptide occurs cotranslationally after translocation of the N-terminus of the nascent protein into the lumen of the ER. The hydrolysis of the signal peptide is catalyzed by an integral membrane enzyme known as a signal or leader peptidase. The integral nature of the activity has made its purification in an active state difficult. A complex of six glycosylated proteins that retains signal peptidase activity has been purified from dog pancreas microsomes. The activity has not been resolved further, so it is not clear whether it can be ascribed to one specific protein within the complex (Evans *et al.*, 1986). The signal peptidase of yeast has been cloned and the sequence determined; this encodes a protein of 18,800 Da (Deshaies *et al.*, 1989).

4.2. *N*-Glycosylation

Three different types of glycosylation occur in the ER (Abeijon and Hirschberg, 1992): core *N*-glycosylation, in which an oligosaccharide chain is transferred to the amide group of asparagine residues; *O*-glycosylation, in which *N*-acetylglucosamine residues are linked to the proteins via a threonine or serine side chain; and the addition of glucosamine and mannose residues to a glycosylphosphatidylinositol (GPI) anchor of some membrane glycoproteins (see Tartakoff, this volume).

N-glycosylation is highly characteristic of proteins that have passed through the secretory pathway, and its mechanism has been explored for many years (Hanover and Lennarz, 1981; Kornfeld and Kornfeld, 1985). The early events of *N*-linked glycosylation that occur in the ER can be divided into three distinct phases: (1) the synthesis of the dolichol-phosphate oligosaccharide moiety, (2) the transfer of the oligosaccharide from the dolichol phosphate to the nascent polypeptide chain, and (3) the trimming of mannose and glucose residues from the oligosaccharide chain after its transfer to the newly synthesized glycoprotein.

4.2.1. Synthesis of the Dolichol-Phosphate Oligosaccharide

Dolichol is a lipid molecule that spans the ER membrane (see Ericsson and Dallner, this volume). The oligosaccharide moiety is attached via a pyrophosphate bridge to the dolichol and is built up sugar by sugar while attached to the ER membrane by the dolichol phosphate. The sugar precursors are the UDP derivatives (or GDP in the case of mannose); these precursors are synthesized in the cytoplasm and are transferred to the dolichol phosphate by specific membrane-bound glycosyl transferases.

The oligosaccharide transferred to the proteins in the ER is identical for all proteins, and is composed of $(N\text{-acetylglucosamine})_2$, $(\text{mannose})_9$, $(\text{glucose})_3$. Individuality of glycosylation occurs by the trimming and addition of sugars at a later stage, mainly in the Golgi. Recent work has clarified the topography of glycosylation reactions in the ER as shown in Figure 7 (Abeijon and Hirschberg, 1992). The dolichol-PP oligosaccharide bearing $(N\text{-acetylglucosamine})_2$, $(\text{mannose})_5$ is synthesized on the cytoplasmic side of the ER membrane and the

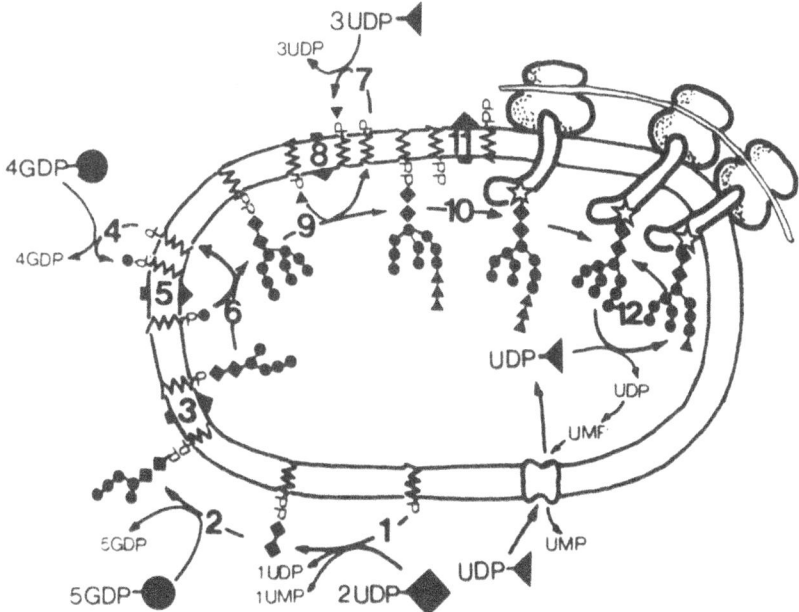

FIGURE 7. Topography of N-glycosylation in the rough ER. The order of formation of the dolichol-phosphate oligosaccharide residues, and their subsequent transfer to a protein and trimming of the moiety are indicated by the stages 1 to 12. ∿ dolichol phosphate, ● mannose, ▼ glucose, ◆ N-acetylglucosamine, ☆ glycosylation site. From Abeijon and Hirschberg (1992).

glycosyl transferases specific for these stages have their active sites on the cytoplasmic face of the membrane (Hirschberg and Snider, 1987). This intermediate oligosaccharide dolichol-phosphate is then translocated across the lipid bilayer. The final four mannoses are donated by dolichol-phosphate mannose, which is itself translocated from the cytoplasmic to the lumenal face of the ER membrane (Tanner and Lehle, 1987). UDP-glucose can cross the ER membrane and is thought to be added to the chain by lumenally facing glycosyl transferases, although the exact details of this stage have not been elucidated.

4.2.2. Oligosaccharyl Transfer

The assembled oligosaccharide chain is transferred by oligosaccharyl transferase (OT) to an asparagine residue of a nascent protein in the tripeptide sequence Asn-X-Ser/Thr; transfer is to lumenally located proteins, peptides, or regions of nascent polypeptides. The purification of OT has been a major challenge because of the integral membrane character of the enzyme, and there have been a number of false starts. The enzyme was initially partially purified from hen oviduct (Das and Heath, 1980) but with no evidence of its homogeneity. An elegant affinity labeling method was then developed using photoactivatable derivatives of the Asn-X-Thr tripeptide; this approach labeled a soluble, lumenally located 58-kDa protein, which was termed "glycosylation site binding protein" (GSBP) and subsequently shown to be PDI (Geetha-Habib et al., 1988). It was proposed that OT comprised an integral membrane catalytic component and the soluble GSBP which presented the nascent protein for modification. Further work established that GSBP/PDI could not be a component of the OT, since depletion studies show that it is not required for N-glycosylation of peptides and nascent proteins (Bulleid and Freedman, 1990; Noiva et al., 1991).

OT has now been solubilized from dog pancreas microsomes with retention of activity and purified to yield a complex of three proteins (Kelleher et al., 1992); two of the proteins present are ribophorins I and II, membrane-spanning glycoproteins located exclusively in the rough ER and originally identified because of their close association with membrane-bound ribosomes. A striking feature supporting the identification of these proteins with OT is that the membrane-spanning segment of ribophorin I is homologous to a sequence motif proposed as the recognition site for dolichol (see p. 71).

The close association between ribophorins and bound ribosomes emphasizes that acquisition of core oligosaccharide is a very early modification that occurs on nascent chains. Not all Asn-X-Ser/Thr sequences in secreted proteins are sites at which glycosylation occurs and there is evidence that modification at these sites is blocked by partial folding which makes the sites inaccessible. The competition between folding and N-glycosylation is emphasized by in vitro translation work showing that translation under reducing conditions (thereby preventing disulfide formation and folding) leads to glycosylation at sites that are normally unmodified (Bulleid *et al.*, 1992).

Although folding competes with glycosylation, paradoxically a major effect of glycosylation is to assist folding. Studies on unfolding and refolding of glycosylated and nonglycosylated forms of the same protein (e.g., Schulke and Schmid, 1988) have shown that glycosylation does not affect the stability of the protein, but enhances the solubility of unfolded and partially folded forms, so that the yield of refolded protein is improved.

4.2.3. Trimming of Glucose and Mannose Residues

After transfer of the oligosaccharide to the polypeptide, the glycan chain is trimmed prior to exit from the ER. The two membrane-bound α-glucosidases remove all three glucose residues. Glucosidase I removes the first glucose residues and glucosidase II removes the final two (Kilker *et al.*, 1980).

Two mannosidases are required for oligosaccharide processing, an endomannosidase and the specific (Man_8-Man_9) α-mannosidase (Kornfeld, 1982; Jelinek-Kelly and Herscovics, 1988). The endomannosidase cleaves the bond between the α1,2-mannose on a 3-mannose branch. These initial processing steps of the oligosaccharide side chain generate the core on which complex carbohydrate side chains are built by subsequent enzymes located in the Golgi stacks (see Lippincott-Schwartz, this volume). Not all proteins or all glycosylation sites are trimmed by either of the mannosidases and why some are left as "high-mannose" structures is not known. There is some correlation with protein type and position of the glycosylation sites in the protein; complex oligosaccharide side chains are more commonly found near the N-terminus of proteins, and high-mannose side chains are rarely found on the N-terminal side of a site with a complex side chain (Pollack and Atkinson, 1983). This suggests a subtle relationship between folding, modification, and movement through the pathway, but the elusiveness of a clear explanation emphasizes our limited knowledge.

4.3. Carboxylation

γ-Carboxylation is a specific modification of glutamyl residues in blood-clotting proteins (e.g., prothrombin) and other Ca^{2+}-binding proteins (Vermeer, 1990). Like prolyl-hydroxylation and unlike *N*-glycosylation, it is a specific rather than a universal modification. The dibasic acid side chain formed has an affinity for Ca^{2+}, and in the blood-clotting proteins, the binding of this ion is essential in that it brings about a concentration of clotting factors on phospholipid surfaces (Tai *et al.*, 1980).

No reproducible local specificity feature around Glu residues is identifiable in sites that undergo γ-carboxylation, which clearly distinguishes them from nonmodified residues. It is striking that modified residues in the blood-clotting proteins are clustered near the N-terminus of the protein, but the natural substrates show such a wide variation in local sequence that it seemed unlikely that local primary or secondary structure in the mature protein could be the determinant of specificity. Now complete sequencing of several blood-clotting proteins has indicated that they are synthesized as precursors; immediately following the conventional N-terminal secretory signal sequence (pre-sequence) is a pro-sequence of similar length, which is also not found in the mature protein. Site-directed mutagenesis studies (Jorgensen *et al.*, 1987a) show that it is this pro-sequence that confers the specificity for γ-carboxylation on the adjacent region (which forms the N-terminal region in the mature protein). For this reason, γ-carboxylation occurs cotranslationally *in vivo* before the cleavage of the pro-sequence. *In vitro,* some small peptides are alternative low-affinity substrates.

Carboxylation is vitamin K-dependent and is catalyzed by vitamin K-dependent carboxylase (Esmon *et al.*, 1975). This enzyme, an integral protein of the ER membrane, requires NADH as well as vitamin K for activity. The integral nature of γ-carboxylase and its insolubility frustrated its purification and characterization until recently. However, the enzyme has now been purified from detergent-solubilized microsomes by affinity chromatography on a synthetic peptide ligand corresponding to the pro-sequence which contains the specificity site (Hubbard *et al.*, 1989). The purified enzyme is a polypeptide of 77 kDa and has both vitamin K-dependent carboxylase and vitamin K epoxidase activities.

The carboxylation reaction is associated with the interconversion of the hydroquinone form of vitamin K (KH_2) and its epoxide (KO) (Figure 8; Suttie, 1980). The recycling of the epoxide requires two reduction steps (to the quinone and then to the hydroquinone). The conventional reductant for enzymatic studies is DTT but this can be replaced by thioredoxin, which has been proposed to be the reductant *in vivo* (van Haarlem *et al.*, 1987). A limited amount of data also suggest that PDI may be involved as a component of the vitamin K-associated redox system (Soute *et al.*, 1992).

The high value of blood-clotting factors as protein pharmaceuticals has led to considerable work on their expression as recombinant products. The uniqueness of γ-carboxylation as a modification is a problem here, as bacteria and yeast cannot carry out the modification and only a few potential host animal cells have significant levels of the γ-carboxylase. In such a case, authentic modification may be a function of the level at which the protein is expressed. When factor IX or prothrombin is highly amplified and expressed from CHO cells, the extent of γ-carboxylation is significantly decreased over that seen at lower levels of expression, suggesting that the carboxylase is limiting and may be saturated when its substrate is overexpressed (Jorgensen *et al.*, 1987b).

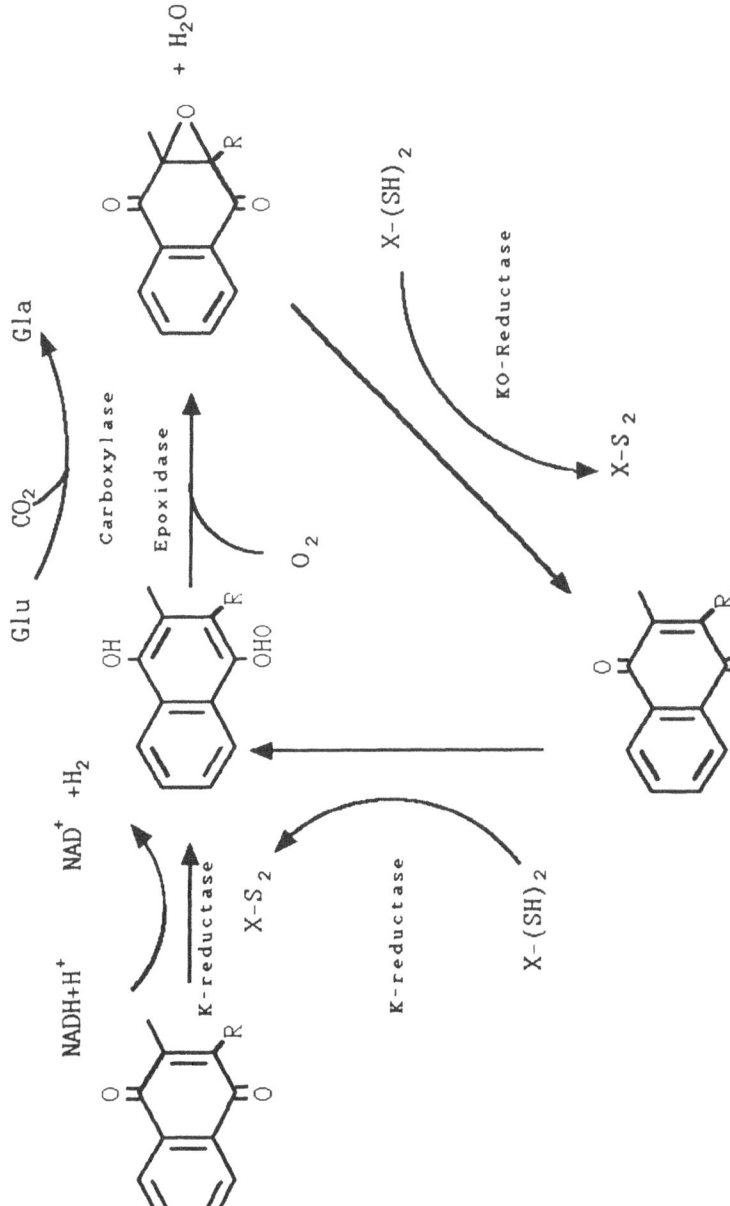

FIGURE 8. The vitamin K cycle, indicating the enzymes that catalyze each step of the pathway. Glu and Gla represent original and modified glutamate residues in the protein substrate. X-(SH)$_2$ and X-S$_2$ represent the reduced and oxidized thiols, respectively; the possible *in vivo* dithiols are discussed in the text. From Vermeer (1990).

4.4. Hydroxylation

Hydroxylation is a modification that is characteristic of molecules that are members of the collagen family or proteins that have a triple-helical region similar to that of the collagen family; e.g., acetylcholine esterase, which has a collagenlike tail. The triple-helical region of these proteins is formed by a repeating tripeptide sequence Gly-X-Y, where X and Y may be any amino acid but are frequently proline and hydroxyproline, respectively. Lysines in the collagen family are also modified to hydroxylysine (reviewed by Kivirikko and Myllyla, 1980).

The process of hydroxylation occurs on nascent chains and unfolded protocollagen polypeptides within the ER lumen (Kivirikko and Myllyla, 1980; Kivirikko *et al.*, 1989; Figure 9). The modification to generate 4-hydroxyprolyl residues, in particular, is important for collagen structure; without this modification the collagen triple helix has a melting temperature below 37°C, so that the folded collagen trimer is not stable at physiological temperature (Fessler and Fessler, 1974). The function of 3-hydroxyprolyl residues is not known. The hydroxylysyl residues have two important functions; these residues serve as attachment sites for carbohydrate units (Kivirikko and Myllyla, 1979) and they

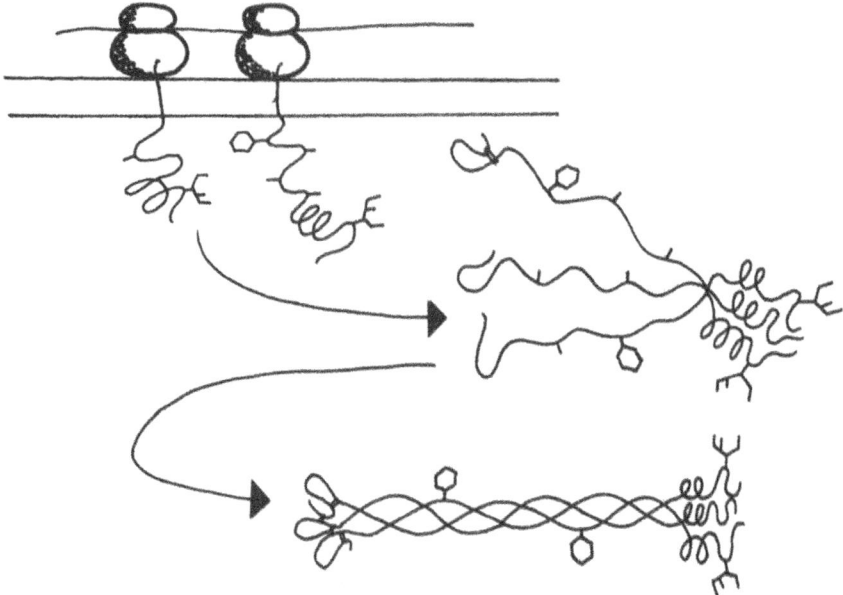

FIGURE 9. Collagen biosynthesis within the rough ER. This scheme indicates that hydroxylation and glycosylation of collagen is initiated cotranslationally and continues posttranslationally until triple helix formation is complete.

are also essential for the formation of intermolecular collagen cross-links (Eyre, 1987).

The enzymes that catalyze protein hydroxylation are prolyl-4-hydroxylase, prolyl-3-hydroxylase, and lysyl hydroxylase; they are all members of the group of enzymes known as 2-oxoglutarate dioxygenases. These enzymes require Fe^{2+}, 2-oxoglutarate, O_2, and ascorbate. The reaction mechanism is such that 2-oxoglutarate is decarboxylated, one oxygen atom from O_2 is incorporated into the succinate while the other is incorporated into the substrate to form the hydroxyl group. The activity of the enzymes can be inhibited *in vivo* by the absence of ascorbate or the presence of iron chelators, leading to the formation of underhydroxylated chains, which form interchain disulfides in the normal way but fail to fold to the secretion-competent triple-helical molecule. Conversely, collagens in the triple-helical conformation are not substrates for hydroxylation; formation of the triple helix completely prevents hydroxylation, indicating that hydroxylation of the polypeptides *in vivo* must occur before triple helix formation. Mutations or other factors that delay folding extend the period during which the unfolded chains are accessible to hydroxylation and increase the extent of hydroxylation that occurs (Kivirikko *et al.*, 1991).

All three hydroxylases are soluble and located in the lumen of the ER. The prolyl-4-hydroxylase is a tetrameric enzyme, $\alpha_2\beta_2$, of which the β subunit has been identified as PDI (see Section 3.1). The purified α subunit has a molecular mass of 64 kDa, an acidic pI of 5.7, and is glycosylated (Kivirikko and Myllyla, 1987). Electron microscopy reveals that the α and β subunits are rod-shaped while the tetramer is thought to consist of two "V" shapes that are interlocked (Olsen *et al.*, 1973). cDNAs encoding human and chicken α subunits of prolyl-4-hydroxylase have been cloned and sequenced (Helaakoski *et al.*, 1989; Bassuk *et al.*, 1989); they do not contain the C-terminal -KDEL ER retention sequence, indicating that one function of the β subunits in the tetrameric enzyme is to retain it within the ER compartment. Work on the expression and assembly of prolyl-4-hydroxylase tetramers was described above (Section 3.1).

Lysyl hydroxylase isolated from chick embryo is a homodimer, 2×85 kDa (Turpeeniemi-Hansen *et al.*, 1980). cDNA encoding the chick enzyme has been cloned and sequenced; surprisingly, there is little overall homology between its primary structure and that of either subunit of prolyl-4-hydroxylase, despite the marked similarities between the enzymes in kinetic properties (Myllyla *et al.*, 1991).

NOTE ADDED IN PROOF. A cDNA clone encoding the third component of this complex, a 48-kDa protein, has recently been isolated and sequenced from a canine kidney cell library and shown to be homologous to a yeast ER membrane protein, WBP1 (Silberstein *et al.*, 1992).

5. REFERENCES

Abeijon, C., and Hirschberg, C. B., 1992, Topography of glycosylation reactions in the endoplasmic reticulum, *Trends Biochem. Sci.* **17:**32–36.

Anfinsen, C. B., 1973, Principles that govern the folding of protein chains, *Science* **181:**223–230.

Anson, M. L., 1945, Protein denaturation and the properties of protein groups, *Adv. Protein Chem.* **2:**361–386.

Arber, S., Krause, K. -H., and Caroni, P., 1992, S-Cyclophilin is retained intracellularly via a unique COOH-terminal sequence and co-localises with the calcium storage protein calreticulin, *J. Cell Biol.* **116:**113–125.

Baksh, S., and Michalak, M., 1991, Expression of calreticulin in *Escherichia coli* and identification of its Ca^{2+} binding domains, *J. Biol. Chem.* **266:**21458–21465.

Bassuk, J. A., Kao, W. W. -Y., Herzer, P., Kedersha, N. L., Seyer, J., DeMartino, J. A., Daugherty, B. L., Mark, G. E., III, and Berg, R. A., 1989, Prolyl-4-hydroxylase: Molecular cloning and the primary structure of the α-subunit from chick embryo, *Proc. Natl. Acad. Sci. USA* **86:**7382–7386.

Beckers, C. J. M., and Balch, W. E., 1990, Calcium and GTP: Essential components in vesicular trafficking between the endoplasmic reticulum and Golgi apparatus, *J. Cell Biol.* **108:**1245–1249.

Beckers, C. J. M., Keller, D. S., and Balch, W. E., 1986, Semi-intact cells permeable to macromolecules: Use in reconstitution of protein transport from the endoplasmic reticulum to the Golgi complex, *Cell* **50:**523–534.

Bergman, L. W., and Kuehl, W. M., 1979a, Formation of intermolecular disulphide bonds on nascent immunoglobulin polypeptides, *J. Biol. Chem.* **254:**5690–5694.

Bergman, L. W., and Kuehl, W. M., 1979b, Formation of an interchain disulfide bond on nascent immunoglobulin light chains, *J. Biol. Chem.* **254:**8869–8876.

Berridge, M. J., 1987, Inositol triphosphate and diacylglycerol—Two interacting secondary messengers, *Annu. Rev. Biochem.* **56:**159–193.

Blount, P., and Merlie, J. P., 1990, Mutational analysis of mouse muscle nicotinic acetylcholine receptor subunit assembly, *J. Cell Biol.* **111:**2613–2622.

Bole, D. G., Hendershot, L. M., and Kearney, J. F., 1986, Post-translational association of immunoglobulin heavy-chain binding protein with nascent heavy chain in non-secreting and secreting hybridomas, *J. Cell Biol.* **102:**1558–1566.

Bose, S., Freedman, R. B., 1992, Characterization of peptidyl-prolyl cis-trans isomerase (PPI) activity associated with the endoplasmic reticulum, *Biochem. Soc. Trans.* **20:**256S.

Braakman, I., Helenius, J., and Helenius, A., 1992a, Manipulating disulfide bond formation and protein folding in the endoplasmic reticulum, *EMBO J.* **11:**1717–1722.

Braakman, I., Helenius, J., and Helenius, A., 1992b, Role of ATP and disulfide bonds during protein folding in the endoplasmic reticulum, *Nature* **356:**260–262.

Brands, R., and Feltkamp, C. A., 1988, Wet-cleaving of cells: A method to introduce macromolecules into the cytoplasm, *Exp. Cell Res.* **176:**309–318.

Brandts, J. F., Halvorsen, H. R., and Brennan, M., 1975, Consideration of the possibility that the slow step in protein denaturation reactions is due to cis-trans isomerization of proline residues, *Biochemistry* **14:**4953–4963.

Bulleid, N. J., and Freedman, R. B., 1988, Defective co-translational formation of disulphide bonds in protein disulphide isomerase-deficient microsomes, *Nature* **335:**649–651.

Bulleid, N. J., and Freedman, R. B., 1990, Co-translational glycosylation of proteins in systems depleted of protein disulphide isomerase, *EMBO J.* **9:**3527–3532.

Bulleid, N. J., Bassel-Duby, R. S., Freedman, R. B., Sambrook, J. F., and Gething, M. -J. H., 1992, Cell-free synthesis of enzymically active tissue-type plasminogen activator, *Biochem. J.* **286:**275–280.

Bychkova, V. E., Pain, R. H., and Pittsyn, O., 1988, The molten globule state is involved in the translocation of proteins across membranes, *FEBS Lett.* **238**:231–234.

Carlino, A., Toledo, H., Skaleris, D., DeLisio, R., Weissbach, H., and Brot, N., 1992, Interactions of liver GRP-78 and *Escherichia coli* recombinant GRP-78 with ATP: Multiple species and disaggregation, *Proc. Natl. Acad. Sci. USA* **89**:2081–2085.

Caroni, P., Rothenfluh, A., McGlynn, E., and Schneider, C., 1991, S-Cyclophilin: A new member of the cyclophilin family associated with the secretory pathway, *J. Biol. Chem.* **266**:10739–10742.

Chappell, T. G., Konforti, B. B., Schmid, S. L., and Rothman, J. E., 1987, The ATP-ase core of a clathrin uncoating protein, *J. Biol. Chem.* **262**:746–751.

Colley, N. J., Baker, N. J., Stamnes, M. A., and Zuker, C. A., 1991, The cyclophilin homolog NinaA is required in the secretory pathway, *Cell* **67**:255–263.

Copeland, C. S., Doms, R. W., Bolzau, E. M., Webster, R. G., and Helenius, A., 1986, Assembly of influenza hemagglutinin trimers and its role in intracellular transport, *J. Cell Biol.* **103**:1179–1191.

Copeland, C. S., Zimmer, K. -P., Wagner, K. R., Healey, G. A., Mellman, I., and Helenius, A., 1988, Folding and trimerisation and transport are sequential events in the biogenesis of influenza hemagglutinin, *Cell* **53**:197–209.

Creighton, T. E., 1978, Experimental studies of protein folding and unfolding, *Prog. Biophys. Mol. Biol.* **33**:231–297.

Creighton, T. E., 1992, *Protein Folding*, Freeman, New York.

Creighton, T. E., Hillson, D. E., and Freedman, R. B., 1980, Catalysis by protein disulphide isomerase of the unfolding and refolding of proteins with disulphide bonds, *J. Mol. Biol.* **142**:43–62.

Das, R. C., and Heath, E. C., 1980, Dolichyldiphosphoryl oligosaccharide–protein oligosaccharyltransferase: Solubilisation, purification and properties, *Proc. Natl. Acad. Sci. USA* **77**:3811–3815.

Dean, N., and Pelham, H. R. B., 1990, Recycling of proteins from the Golgi compartment to the endoplasmic reticulum in yeast, *J. Cell Biol.* **111**:369–377.

De Lorenzo, F., Goldberger, R. F., Steers, E., Jr., Givol, D., and Anfinsen, C. B., 1966, Purification and properties of an enzyme from beef liver which catalyses sulfhydryl–disulfide interchange in proteins, *J. Biol. Chem.* **241**:1562–1567.

Deshaies, R. J., Kepes, F., and Bohni, P. C., 1989, Genetic dissection of the early stages of protein secretion in yeast, *Trends Genet.* **5**:87–93.

Edman, J. C., Ellis, L., Blacher, R. W., Roth, R. A., and Rutter, W. J., 1985, Sequence of protein disulphide isomerase and implications of its relation to thioredoxin, *Nature* **317**:267–270.

Epstein, C. J., Goldberger, R. F., and Anfinsen, C. B., 1963, The genetic control of tertiary protein structure: Studies with model systems, *Cold Spring Harbor Symp. Quant. Biol.* **28**:439–449.

Esmon, C. T., Sadowski, J. A., and Suttie, J. W., 1975, A new carboxylation reaction: The vitamin K-dependent incorporation of $H^{14}CO_3$ into prothrombin, *J. Biol. Chem.* **250**:4744–4748.

Evans, E. A., Gilmore, R., and Blobel, G., 1986, Purification of microsomal signal peptidase as a complex, *Proc. Natl. Acad. Sci. USA* **83**:581–585.

Eyre, D., 1987, Collagen cross-linking amino acids, *Methods Enzymol.* **144**:115–139.

Farquhar, R., Honey, N., Murant, S. J., Bossier, P., Schultz, L., Montgomery, D., Ellis, R. W., Freedman, R. B., and Tuite, M. F., 1991, Protein disulphide isomerase is essential for viability in *Saccharomyces cerevisiae*, *Gene* **108**:81–89.

Fessler, L. I., and Fessler, J. H., 1974, Protein assembly of procollagen and the effects of hydroxylation, *J. Biol. Chem.* **249**:7637–7646.

Fischer, G., and Bang, H., 1983, The refolding of urea denatured ribonuclease-A is catalysed by peptidyl-prolyl isomerase, *Biochim. Biophys. Acta* **828**:39–42.

Fischer, G., Bang, H., Berger, E., and Schellenberger, A., 1985, Conformational specificity of chymotrypsin toward proline-containing substrates, *Biochim. Biophys. Acta* **791**:87–97.

Flaherty, K. M., DeLuca-Flaherty, C., and McKay, D. B., 1990, Three dimensional structure of the ATPase fragment of a 70 000Da heat shock protein, *Nature* **346**:623–628.

Flaherty, K. M., McKay, D. B., Kabsch, W., and Holmes, K. C., 1991, Similarity of the three dimensional structures of actin and the ATPase fragment of the 70kDa heat shock protein, *Proc. Natl. Acad. Sci. USA* **88**:5041–5045.

Flajnik, M. F., Canel, C., Kramer, J., and Kasabara, M., 1991, Evolution of the major histocompatibility complex—molecular cloning of the major histocompatibility complex class I from the amphibian Xenopus, *Proc. Natl. Acad. Sci. USA* **88**:537–541.

Fliegel, L., Burns, K., MacLennan, D. H., Rathmeier, R. A. F., and Michalak, M., 1989a, Molecular cloning of the high affinity calcium binding protein (calreticulin) of skeletal muscle sarcoplasmic reticulin, *J. Biol. Chem.* **264**:21522–21528.

Fliegel, L., Burns, K., Wlasichuk, K., and Michalak, M., 1989b, Peripheral membrane proteins of the sarcoplasmic and endoplasmic reticulum—Comparison of the carboxy-terminal amino acid sequences, *Biochem. Cell Biol.* **67**:696–702.

Flynn, G. C., Chapell, T. G., and Rothman, J. E., 1989, Peptide binding and release by proteins implicated as catalysts of protein assembly, *Science* **245**:385–390.

Flynn, G. C., Pohl, J., Flocco, M. T., and Rothman, J. E., 1991, Peptide binding specificity of the molecular chaperone BiP, *Nature* **353**:726–730.

Freedman, R. B., 1984, Native disulphide bond formation in protein synthesis: Evidence for the role of protein disulphide isomerase, *Trends Biochem. Sci.* **9**:438–441.

Freedman, R. B., 1992, Protein folding in the cell, in *Protein Folding* (T. E. Creighton, ed.), pp. 455–539, Freeman, New York.

Freedman, R. B., Hawkins, H. C., Murant, S. J., and Reid, L., 1988, Protein disulphide isomerase: A homologue of thioredoxin implicated in the biosynthesis of secretory proteins, *Biochem. Soc. Trans.* **16**:96–99.

Freedman, R. B., Bulleid, N. J., Hawkins, H. C., and Paver, J. L., 1989, Role of protein disulphide isomerase in the expression of native proteins, *Biochem. Soc. Symp.* **55**:167–192.

Geetha-Habib, M., Noiva, R., Kaplan, H. A., and Lennarz, W. J., 1988, Glycosylation site binding protein, a component of oligosaccharyl transferase, is highly similar to three other 57kD luminal proteins of the endoplasmic reticulum, *Cell* **54**:1053–1060.

Gething, M. -J. H., and Sambrook, J., 1992, Protein folding in the cell, *Nature* **355**:33–45.

Gething, M. -J. H., McCammon, K., and Sambrook, J., 1986, Expression of wild-type and mutant forms of the influenza hemagglutinin: The role of intracellular folding, *Cell* **46**:939–950.

Gierasch, L. M., 1989, Signal sequences, *Biochemistry* **28**:924–930.

Goldberger, R. F., Epstein, C. J., and Anfinsen, C. B., 1963, Acceleration of reactivation of reduced bovine pancreatic ribonuclease by a microsomal system from rat liver, *J. Biol. Chem.* **238**:628–635.

Goldenberg, D. P., 1992, Native and non-native intermediates in the BPTI folding pathway, *Trends Biochem. Sci.* **17**:257–261.

Haas, I., and Wabl, M., 1983, Immunoglobulin heavy chain binding protein, *Nature* **306**:387–389.

Haendler, B., Kelder, R., Hiestand, P. C., Kocher, H. P., Wegmann, G., and Moiva, N. R., 1989, Yeast cyclophilin: Isolation and characterisation of the protein, cDNA and gene, *Gene* **83**:39–46.

Hanks, S. K., Quinn, A. M., and Hunter, T., 1988, The protein kinase family—Conserved features and deduced phylogeny of the catalytic domains, *Science* **241**:42–52.

Hanover, J. A., and Lennarz, W. J., 1981, Transmembrane assembly of membrane and secretory glycoproteins, *Arch. Biochem. Biophys.* **211**:1-19.

Hantgan, R. R., Hammes, G. G., and Scheraga, H. A., 1974, Pathways of folding of reduced pancreatic ribonuclease, *Biochemistry* **13**:3421-3431.

Hasel, K. W., Glass, J. R., Godbout, M., and Sutcliffe, J. G., 1991, An endoplasmic reticulum specific cyclophilin, *Mol. Cell Biol.* **11**:3484-3491.

Hawkins, H. C., and Freedman, R. B., 1991, The reactivities and ionisation properties of the active site dithiol groups of mammalian protein disulphide isomerase, *Biochem. J.* **275**:335-339.

Hawkins, H. C., de Nardi, M. and Freedman, R. B. 1991, Redon properties and cross-linking of the dithid/disulphide active sites of mammalian protein disulphide-isomerase *Biochem, J.* **275**:341-348.

Hayano, T., Takahashi, N., Kato, S., Maki, N., and Suzuki, M., 1991, Two distinct forms of peptidyl-prolyl cis-trans isomerase are expressed separately in the periplasmic and cytoplasmic compartments in *E. coli* cells, *Biochemistry* **30**:3041-3048.

Helaakoski, T., Vuori, K., Myllyla, R., Kivirikko, K. I., and Pihlajaniemi, T., 1989, Molecular cloning of the α-subunit of human prolyl 4-hydroxylase: The complete cDNA-derived amino acid sequence and evidence for alternative splicing of transcripts, *Proc. Natl. Acad. Sci. USA* **86**:4392-4396.

Hendershot, L. M., 1990, Immunoglobulin heavy chain and binding protein complexes are dissociated *in vivo* by light chain addition, *J. Cell Biol.* **111**:829-837.

Hendershot, L. M., Bole, D. G., Kohler, G., and Kearney, J. F., 1987, Assembly and secretion of heavy chains that do not associate with immunoglobulin heavy-chain binding protein, *J. Cell Biol.* **104**:761-767.

Hendershot, L. M., Ting, J., and Lee, A. S., 1988, Identity of immunoglobulin heavy chain binding protein with the 78 000 dalton glucose-regulated protein and the role of post-translational modifications in its binding function, *Mol. Cell Biol.* **8**:4250-4256.

Hirschberg, C. B., and Snider, M. D., 1987, Topography of glycosylation in the rough endoplasmic reticulum and Golgi apparatus, *Annu. Rev. Biochem.* **56**:63-88.

Hubbard, B. R., Ulrich, M. W. M., Jacobs, M., Vermeer, C., Walsh, C., Furie, B., and Furie, B. C., 1989, Vitamin K-dependent carboxylase: Affinity purification from bovine liver using a synthetic propeptide containing the γ-carboxylation recognition site, *Proc. Natl. Acad. Sci. USA* **86**:6893-6897.

Hurtley, S. M., and Helenius, A., 1989, Protein oligomerisation in the endoplasmic reticulum, *Annu. Rev. Cell Biol.* **5**:277-307.

Jelinek-Kelly, S., and Herscovics, A., 1988, Glycoprotein biosynthesis in *Saccharomyces cerevisiae*; purification of the α-mannosidase which removes one specific mannose residue from Man_9-Man_8, *J. Biol. Chem.* **263**:14757-14763.

Jorgensen, M. J., Cantor, A. B., Furie, B. C., Brown, C. L., Shoemaker, C. B., and Furie, B., 1987a, Recognition site directing vitamin-K dependent γ-carboxylation residues on the propeptide of factor IX, *Cell* **48**:185-191.

Jorgensen, M. J., Cantor, A. B., Furie, B. C., and Furie, B., 1987b, Expression of completely γ-carboxylated recombinant human prothrombin, *J. Biol. Chem.* **262**:6729-6734.

Kaderbhai, M. A., and Austen, B. M., 1984, Dog pancreatic microsomal-membrane polypeptides analysed by two-dimensional electrophoresis, *Biochem. J.* **217**:145-157.

Kaderbhai, M. A., and Austen, B. M., 1985, Studies on the formation of intrachain disulphide bonds in newly synthesised bovine prolactin—role of protein disulphide isomerase, *Eur. J. Biochem.* **153**:167-170.

Kassenbrock, C. K., and Kelly, R. B., 1989, Interaction of heavy chain binding protein (BiP/GRP78) with adenine nucleotides, *EMBO J.* **8**:1461-1467.

Kassenbrock, C. K., Garcia, P. D., Walter, P., and Kelly, R. B., 1988, Heavy-chain binding protein recognises aberrant polypeptides translocated *in vitro, Nature* **333:**90–93.

Kelleher, D. J., Kreibich, G., and Gilmore, R., 1992, Oligosaccharyltransferase activity is associated with a complex composed of ribophorins I and II and a 48kD protein, *Cell* **69:**55–65.

Kilker, R. D., Saunier, B., Tkacz, J. S., and Herscovics, A., 1980, Partial purification from *Saccharomyces cerevisiae* of a soluble glucosidase which removes the terminal galactose from the oligosaccharide Glc$_3$Man$_9$GlcNac$_2$, *J. Biol. Chem.* **256:**5299–5303.

Kivirikko, K. I., and Myllyla, R., 1979, Collagen glycosyltransferases, *Int. Rev. Connect. Tissue Res.* **8:**23–37.

Kivirikko, K. I., and Myllyla, R., 1980, Hydroxylation of prolyl and lysyl residues, in: *The Enzymology of Post-Translational Modifications of Proteins* (R. B. Freedman and H. C. Hawkins, eds.), pp. 53–104, Academic Press, New York.

Kivirikko, K. I., and Myllyla, R., 1987, Recent developments in post-translational modification of collagen: Intracellular processing, *Methods Enzymol.* **144:**96–114.

Kivirikko, K. I., Myllyla, R., and Pihlajaniemi, T., 1991, Hydroxylation of proline and lysine residues in collagens and other plant and animal proteins, in *Post-Translational Modifications of Proteins* (J. J. Harding and M. J. C. Crabbe, eds.), pp. 1–51, CRC Press, Boca Raton, Fla.

Koch, G. L. E., 1987, Reticuloplasmins: A novel group of proteins in the endoplasmic reticulum, *J. Cell Sci.* **87:**491–492.

Koch, G. L. E., Smith, M., Macer, D., Webster, P., and Mortara, R., 1986, Endoplasmic reticulum contains a common abundant calcium-binding glycoprotein, endoplasmin, *J. Cell Sci.* **86:**217–232.

Koivu, J., Myllyla, R., Helakoski, T., Pihlajaniemi, T., Tasanen, K., and Kivirikko, K. I., 1987, A single polypeptide acts as both the β-subunit of prolyl-4-hydroxylase and as protein disulphide isomerase, *J. Biol. Chem.* **262:**6447–6449.

Koletsky, A. J., Harding, M. W., and Handschumaker, R. E., 1986, Cyclophilin: distribution and variant properties in normal and neoplastic tissues, *J. Immunol.* **137:**1054–1059.

Kornfeld, S., 1982, Oligosaccharide processing during glycoprotein biosynthesis, in: *The Glycoconjugates*, Vol. III (M. Horowitz, ed.), pp. 3–23, Academic Press, New York.

Kornfeld, S., and Kornfeld, R., 1985, Assembly of asparagine linked oligosaccharides, *Annu. Rev. Biochem.* **54:**631–664.

Koser, P. L., Livi, G. P., Levy, M. A., Rosenberg, M., and Bergsma, D. J., 1990, A *Candida albicans* homolog of a human cyclophilin gene encodes peptidyl-prolyl cis-trans isomerase, *Gene* **96:**189–195.

Kozutsumi, Y., Normington, K., Press, E., Slaughter, C., Sambrook, J., and Gething, M. -J., 1989, Identification of immunoglobulin heavy chain binding protein as glucose regulated protein-78 on the basis of amino acid sequence, immunological cross-reactivity and functional activity, *J. Cell Sci. Suppl.* **11:**115–137.

Krause, G., Lundstrom, J., Barea, J. L., De la Cuesta, C. P., and Holmgren, A., 1991, Mimicking the active site of protein disulphide isomerase by substitution of proline-34 in *Escherichia coli* thioredoxin, *J. Biol. Chem.* **266:**9494–9500.

LaMantia, M. L., Miura, T., Tagikawa, H., Kaplan, H. A., Lennarz, W. J., and Mizunaga, T., 1991, Glycosylation site binding protein and protein disulphide isomerase are identical and essential for cell viability in yeast, *Proc. Natl. Acad. Sci. USA* **88:**4453–4457.

Lambert, N., and Freedman, R. B., 1983a, Structural properties of homogeneous protein disulphide isomerase from bovine liver purified by a rapid high yielding procedure, *Biochem. J.* **213:**225–234.

Lambert, N., and Freedman, R. B., 1983b, Kinetics and specificity of homogeneous protein disulphide isomerase in protein disulphide isomerization and in thiol-protein-disulphide oxidoreduction, *Biochem. J.* **213:**235–243.

Lang, K., Schmid, F. X., and Fischer, G., 1987, Catalysis of protein folding by prolyl isomerase, *Nature* **329**:268–270.

Lee, A. S., 1987, Co-ordinated regulation of a set of genes by glucose and calcium ionophores in mammalian cells, *Trends Biochem. Sci.* **12**:20–23.

Lewis, M. J., and Pelham, H. R. B., 1992, Ligand induced redistribution of a human KDEL receptor from the Golgi complex to the endoplasmic reticulum, *Cell* **68**:353–364.

Lindquist, S., and Craig, E. A., 1988, The heat shock proteins, *Annu. Rev. Genet.* **22**:631–677.

Macer, D. R. J., and Koch, G. L. E., 1988, Identification of a set of calcium-binding proteins in reticuloplasm, the luminal content of the endoplasmic reticulum, *J. Cell Sci.* **91**:61–70.

Machamer, C. E., and Rose, J. K., 1988a, Influence of new glycosylation sites on expression of the vesicular stomatitus-virus G-protein at the plasma membrane, *J. Biol. Chem.* **263**:5948–5954.

Machamer, C. E., and Rose, J. K., 1988b, Vesicular stomatitus-virus G-protein with altered glycosylation sites display temperature-sensitive intracellular transport and are subject to aberrant intermolecular disulphide bonding, *J. Biol. Chem.* **263**:5955–5960.

Machamer, C. E., Florkiewicz, R. Z., and Rose, J. K., 1985, A single N-linked oligosaccharide at either of the two normal sites is sufficient for transport of vesicular stomatitus virus-G protein to the cell surface, *Mol. Cell Biol.* **5**:3074–3083.

McLaughlin, M. M., Bosard, M. J., Koser, P. L., Cafferkey, R., Morris, R. A., Miles, L. N., Strickler, J., Bergsma, D. J., Levy, M. A., and Livi, G. P., 1992, The yeast cyclophilin multigene family: Purification, cloning and characterisation of a new isoform, *Gene* **111**:85–92.

Maki, R. G., Old, L. J., and Srivastava, P. K., 1990, Human homolog of murine tumour rejection protein, GP96:5', regulatory and coding regions and relationship to stress induced proteins, *Proc. Natl. Acad. Sci. USA* **87**:5658–5662.

Michalak, M., Milner, R. E., Burns, K., and Opas, M., 1992, Calreticulin, *Biochem. J.* **285**:681–692.

Miller, S. G., and Moore, H. -P. H., 1991, Reconstitution of constitutive secretion using semi-intact cells: Regulation by GTP but not calcium, *J. Cell Biol.* **112**:39–54.

Mitraki, A., and King, J., 1989, Protein folding intermediates and inclusion body formation, *Biotechnology* **7**:690–697.

Morjana, N., and Gilbert, H. F., 1991, Effect of protein and peptide inhibitors on the activity of protein disulphide isomerase, *Biochemistry* **30**:4985–4990.

Myllyla, R., Pihlajaniemi, T., Pajunen, L., Turpeeniemi-Hujanen, T., and Kivirikko, K. I., 1991, Molecular-cloning of chick lysyl hydroxylase: Little homology in primary structure to the two types of subunit of prolyl-4-hydroxylase, *J. Biol. Chem.* **266**:2805–2810.

Nguyen, T. H., Law, D. T. S., and Williams, D. B., 1991, Binding protein BiP is required for translocation of secretory proteins into the endoplasmic reticulum in *Saccharomyces cerevisiae*, *Proc. Natl. Acad. Sci. USA* **88**:1565–1569.

Nicchitta, C. V., and Blobel, G., 1989, Nascent secretory chain binding and translocation are distinct processes: Differentiation by chemical alkylation, *J. Cell Biol.* **108**:789–795.

Noiva, R., Kaplan, H. A., and Lennarz, W. J., 1991, Glycosylation site-binding protein is not required for N-linked glycoprotein synthesis, *Proc. Natl. Acad. Sci. USA* **88**:1986–1990.

Normington, K., Kohno, K., Kozutsumi, Y., Gething, M. -J., and Sambrook, J., 1989, *S. cerevisiae* encodes an essential protein homologous in sequence and function to mammalian BiP, *Cell* **57**:1223–1236.

Olsen, B. R., Berg, R. A., Kivirikko, K. I., and Prokop, D. J., 1973, Structure of protocollagen proline hydroxylase from chick embryos, *Eur. J. Biochem.* **35**:135–147.

Paver, J. L., Hawkins, H. C., and Freedman, R. B., 1989, Preparation and characterization of dog pancreas microsomal membranes specifically depleted of protein-disulphide isomerase, *Biochem. J.* **257**:657–663.

Pelham, H. R. B., 1986, Speculations on the functions of the major heat shock and glucose regulated proteins, *Cell* **46**:959–961.

Pelham, H. R. B., 1988, Evidence that lumenal ER proteins are sorted from secreted proteins in a post-ER compartment, *EMBO J.* **7**:913–918.

Pelham, H. R. B., 1990, The retention signal for soluble proteins of the endoplasmic reticulum, *Trends Biochem. Sci.* **15**:483–486.

Pihlajaniemi, T., Helakoski, T., Tasanen, K., Myllyla, R., Huhtala, M. -L., Koivu, J., and Kivirik-ko, K. I., 1987, Molecular cloning of the β-subunit of human prolyl-4-hydroxylase. This product and protein disulphide isomerase are products of the same gene, *EMBO J.* **6**:643–649.

Pittsyn, O., 1992, The molten globule state, in: *Protein Folding* (T. E. Creighton, ed.), pp. 243–300, Freeman, New York.

Pollack, L., and Atkinson, P. H., 1983, Correlation of glycosylation forms with position in amino acid sequence, *J. Cell Biol.* **97**:293–300.

Price, E. R., Zydowsky, L. D., Jin, M., Baker, C. H., McKeon, F. D., and Walsh, C. T., 1991, Human cyclophilin B: A second gene encodes a peptidyl-prolyl isomerase with a signal sequence, *Proc. Natl. Acad. Sci. USA* **88**:1903–1907.

Rippmann, F., Taylor, W. R., Rothbard, J. B., and Green, N. M., 1991, A hypothetical model for the peptide-binding domain of HSP-70 based on the peptide binding domain of HLA, *EMBO J.* **10**:1053–1059.

Rose, J. K., and Doms, R. W., 1988, Regulation of protein export from the endoplasmic reticulum, *Annu. Rev. Cell Biol.* **4**:257–288.

Rose, M. D., Misra, L. M., and Vogel, J. P., 1989, *KAR2*, a karyogamy gene, is the yeast homolog of the mammalian BiP/GRP78 gene, *Cell* **57**:1211–1221.

Rothman, J. E., 1987, Protein sorting by selective retention in the endoplasmic reticulum and Golgi stack, *Cell* **50**:521–522.

Sanders, S. L., Whitfield, K. M., Vogel, J. P., Rose, M. D., and Schekman, R. W., 1992, Sec61p and BiP directly facilitate polypeptide translocation into the endoplasmic reticulum, *Cell* **69**:353–365.

Scheele, G., and Jacoby, R., 1982, Conformational changes associated with proteolytic processing of pre-secretory proteins allows glutathione catalysed formation of disulfide bonds, *J. Biol. Chem.* **257**:12277–12282.

Schmid, F. X., 1992, Kinetics of unfolding and refolding of single-domain proteins, in: *Protein Folding* (T. E. Creighton, ed.), pp. 97–242, Freeman, New York.

Schreiber, S. L., 1991, Chemistry and biology of the immunosuppressants and their immunosuppressive ligands, *Science* **251**:283–287.

Schulke, N., and Schmid, F. X., 1988, Effect of glycosylation on the mechanism of renaturation of invertase from yeast, *J. Biol. Chem.* **263**:8832–8837.

Semenza, J. C., Hardwick, K. G., Dean, N., and Pelham, H. R. B., 1990, *ERD2*, a yeast gene required for the receptor-mediated retrieval of luminal endoplasmic reticulum proteins from the secretory pathway, *Cell* **61**:1349–1357.

Silberstein, S. Kelleher, D. J., and Silmore, R., 1992, The 48-kDa subunit of the mammalian oligosaccharyl transferase complex is homologous to the essential yeast protein WBP1. *J. Biol. Chem.* **267**:23658–23663.

Skowyra, D., Georgopoulos, C., and Zyclicz, M., 1990, The *E. coli* dnaK gene product, the HSP-70 homolog can reactivate heat denatured RNA polymerase in an ATP hydrolysis dependent manner, *Cell* **62**:939–944.

Sonderfeld-Fresko, S., and Proia, R. L., 1988, Synthesis of a catalytically active lysosomal-enzyme beta-hexosaminidase-B, in a cell-free system, *J. Biol. Chem.* **263**:13463–13469.

Soute, B. A. M., Groenen-van-Dooren, M. M. C. L., Holmgren, A., Lundstrom, J., and Vermeer, C., 1992, Stimulation of dithiol-dependent reductases in the vitamin-K cycle by the thioredoxin system, *Biochem. J.* **281**:255–259.

Steinmann, B., Bruckner, P., and Superti-Furga, A., 1990, Cyclosporin-A slows collagen triple helix formation *in vivo:* Indirect evidence for a physiological role of peptidyl-prolyl cis-trans isomerase, *J. Biol. Chem.* **265**:1299–1303.

Stewart, D. E., Sarkar, A., and Wampler, J. E., 1990, Occurrence and role of cis peptide bonds in protein structures, *J. Mol. Biol.* **214**:253–260.

Streb, H., Irvine, H. F., Berridge, M. J., and Schulz, I., 1983, Release of calcium from a non-mitochondrial intracellular store in pancreatic acinar cells by inositol-1,4,5-triphosphate, *Nature* **306**:67–68.

Suttie, J. W., 1980, Carboxylation of glutamyl residues, in: *The Enzymology of Post-Translational Modifications of Proteins* (R. B. Freedman and H. C. Hawkins, eds.), pp. 213–258, Academic Press, New York.

Tai, M. M., Furie, B. C., and Furie, B., 1980, Conformation-specific antibodies directed against the bovine prothrombin complex, *J. Biol. Chem.* **255**:2790–2795.

Tanner, W., and Lehle, L., 1987, Protein glycosylation in yeast, *Biochim. Biophys. Acta* **906**:81–99.

Turpeeniemi-Hansen, T. M., Puistola, U., and Kivirikko, K., 1980, Isolation of lysyl hydroxylase, an enzyme of collagen synthesis, from chick embryos as a homogeneous protein, *Biochem. J.* **189**:247–253.

Van, P. N., Peter, F., and Soling, H. -D., 1989, Four intracisternal calcium-binding glycoproteins from rat liver microsomes with high affinity for calcium, *J. Biol. Chem.* **264**:17494–17501.

van Haarlem, L. J. M., Soute, B. A. M., and Vermeer, C., 1987, Vitamin-K dependent carboxylase: Possible role for thioredoxin in the reduction of vitamin-K metabolites in liver, *FEBS Lett.* **222**:353–357.

Vermeer, C., 1990, γ-Carboxyglutamate-containing proteins and the vitamin K-dependent carboxylase, *J. Biochem.* **266**:625–636.

Vogel, J. P., Misra, L. M., and Rose, M. D., 1990, Loss of Bip/GRP78 function blocks translocation of secretory proteins in yeast, *J. Cell Biol.* **110**:1885–1895.

von Heijne, G., 1985, Signal sequences: The limits of variation, *J. Mol. Biol.* **184**:99–105.

Vuori, K., Myllyla, R., Pihlajaniemi, T., and Kivirikko, K. I., 1992a, Expression and site-directed mutagenesis of human protein disulphide isomerase in *Escherichia coli*, *J. Biol. Chem.* **267**:7211–7214.

Vuori, K., Pihlajaniemi, T., Martilla, M., and Kivirikko, K., 1992b, Characterisation of the human prolyl-4-hydroxylase tetramer and its multifunctional protein disulphide-isomerase subunit synthesized in a baculovirus expression system, *Proc. Natl. Acad. Sci. USA* **89**:7467–7470.

Watson, M. E. E., 1984, Compilation of published signal sequences, *Nucleic Acid Res.* **12**:5145–5164.

Weissman, J. S., and Kim, P. S., 1991, Reexamination of the folding pathway of BPTI: Predominance of native intermediates, *Science* **253**:1386–1393.

Welch, W. J., 1990, The mammalian stress response: Cell physiology and biochemistry of stress proteins, in: *Stress Proteins in Biology and Medicine* (R. I. Morimoto, A. Tissieres, and G. Georgopoulos, eds.), 223–278, Cold Spring Harbor Press, Cold Spring Harbor, N.Y.

Wells, W. W., Xu, D. P., Yang, Y., and Rocque, P. A., 1990, Mammalian thioltransferase (glutaredoxin) and protein disulphide isomerase have dehydroascorbate reductase activity, *J. Biol. Chem.* **265**:15361–15364.

Wetterau, J. R., Combs, K. A., Spinner, S. N., and Joiner, B. J., 1990, Protein disulphide isomerase is a component of the microsomal triglyceride transfer protein complex, *J. Biol. Chem.* **265**:9800–9807.

Wetterau, J. R., Aggerback, L. P., Laplaud, P. M., and McLean, L. R., 1991a, Structural properties of the microsomal triglyceride-transfer protein complex, *Biochemistry* **30**:4406–4412.

Wetterau, J. R., Combs, K. A., McLean, L. R., Spinner, S. N., and Aggerback, L. P., 1991b, Protein disulphide isomerase appears necessary to maintain catalytically active structure of the microsomal triglyceride-transfer protein, *Biochemistry* **30**:9728–9735.

Wiech, H., Buchner, J., Zimmerman, R., and Jakob, U., 1992, HSP-90 chaperones protein folding *in vitro*, *Nature* **358**:169–170.

Yu, Y., Zhang, Y., Sabatini, D. D., and Kreibich, G., 1989, Reconstitution of translocation-competent membrane vesicles from detergent solubilised dog pancreas rough microsomes, *Proc. Natl. Acad. Sci. USA* **86**:9931–9935.

Zapun, A., Creighton, T. E., Rowling, P. J. E., and Freedman, R. B., 1992, Folding *in vitro* of bovine pancreatic trypsin inhibitor in the presence of proteins of the endoplasmic reticulum, *Proteins* **14**:10–15.

Chapter 4

Biological Functions and Biosynthesis of Glycolipid-Anchored Membrane Proteins

Alan M. Tartakoff

1. INTRODUCTION: FUNCTIONS

The glycosylinositol phospholipid (GPI) anchors of proteins such as Thy-1 of neurons and lymphocytes and the variant surface glycoprotein of African trypanosomes are posttranslationally linked to the C-terminus of the corresponding proteins (Figure 1) (Cross, 1990; Ferguson *et al.*, 1988; Low and Saltiel, 1988; Tartakoff *et al.*, 1990). The *in vivo* sensitivity of many anchor units to microbial phosphatidylinositol (PI)-specific phospholipases implies that they do not traverse the lipid bilayer. Such proteins cannot be subject to many cytoplasmic events, such as direct linkage to the cytoskeleton and phosphorylation, which impact on transmembrane proteins (Figure 2). It was therefore anticipated both that the lateral diffusion of GPI-anchored proteins would be high and that these proteins might be released by physiologically important phospholipases. The evidence in support of such conjectures with regard to function is, however,

---

Abbreviations used in this chapter: CHO, Chinese hamster ovary; GDP, guanosine diphosphate; GPI, glycosylinositol phospholipid; PI, phosphatidylinositol; PL, phospholipase; PMSF, phenyl-methyl-sulfonyl fluoride; RER, rough endoplasmic reticulum; UDP, uridine diphosphate.

Alan M. Tartakoff   Institute of Pathology, Case Western Reserve University, Cleveland, Ohio 44106.
Subcellular Biochemistry, Volume 21: Endoplasmic Reticulum, edited by N. Borgese and J. R. Harris. Plenum Press, New York, 1993.

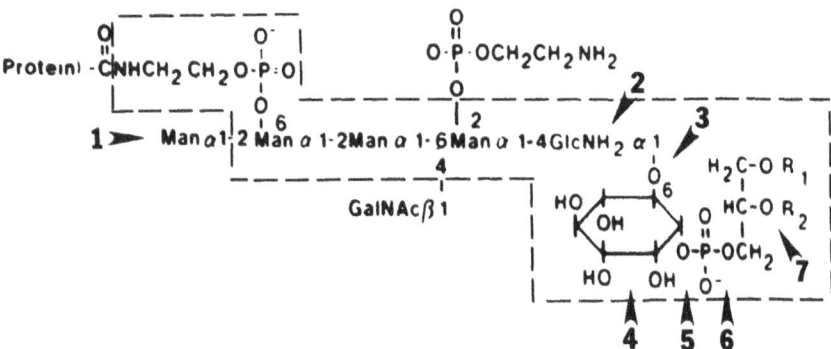

FIGURE 1. Structures of GPI anchors. The anchor structure has been determined for rat brain Thy-1 (Homans *et al.*, 1988), a trypanosome variant surface glycoprotein (Ferguson *et al.*, 1988), and human erythrocyte acetylcholinesterase (Roberts *et al.*, 1988a,b). In each case, an identical core glycan is found, consisting of ethanolamine phosphate, three mannose residues, and nonacetylated glucosamine (arrowhead 2) linked to inositol; however, there are significant differences that distinguish the three structures. Constituents that are found in some but not all anchors are outside the enclosed region. These are additional ethanolamine phosphate, mannose (arrowhead 1), galactose, and *N*-acetylgalactosamine, as well as an acyl chain linked to inositol C-2 or -3 (arrowhead 4). In animal cells, but not trypanosomes, the glycerol unit often bears an alkyl substituent on the first carbon (R_1). The figure also indicates the site of anchor cleavage by nitrous acid (arrowhead 3), PI-PLD (arrowhead 5), PI-PLC (arrowhead 6), and PLA_2 (arrowhead 7). PI-PLC is not active if the inositol-linked acyl chain (arrowhead 4) is present.

difficult to generalize. For example, GPI-anchored class I histocompatibility antigens do appear to gain access to much of the cell surface (Edidin and Stroynowski, 1991), although their lateral diffusion is not uniformly higher than for transmembrane proteins (Bulow *et al.*, 1988; Phelps *et al.*, 1988). Furthermore, although there are a few experimental reports describing release of GPI-anchored proteins (Chan *et al.*, 1988; He *et al.*, 1987; Huizinga *et al.*, 1988) and although soluble forms of several GPI-anchored proteins are found in extracellular fluids (Almqvist and Carlsson, 1988; Epstein *et al.*, 1986; Overath *et al.*, 1983; Quentmeier *et al.*, 1987), release appears to be the exception rather than the rule.

Both Thy-1 and other GPI-anchored proteins are underrepresented in coated pits and are endocytosed only very slowly (Bamezai *et al.*, 1989; Lemansky *et al.*, 1990; Lisanti *et al.*, 1990). The apparent result is that these proteins are concentrated at the plasma membrane (Lemansky *et al.*, 1990) and that at least GPI-anchored cell surface Thy-1 has an unusually long half-life (Lemansky *et al.*, 1990). In some polarized epithelial cells and in neurons, GPI-anchored proteins are preferentially enriched at the apical plasma membrane, possibly because the GPI unit itself is an apical sorting signal (Dotti *et al.*, 1991; Lisanti *et al.*, 1989). A few GPI-anchored proteins are concentrated in membranes of

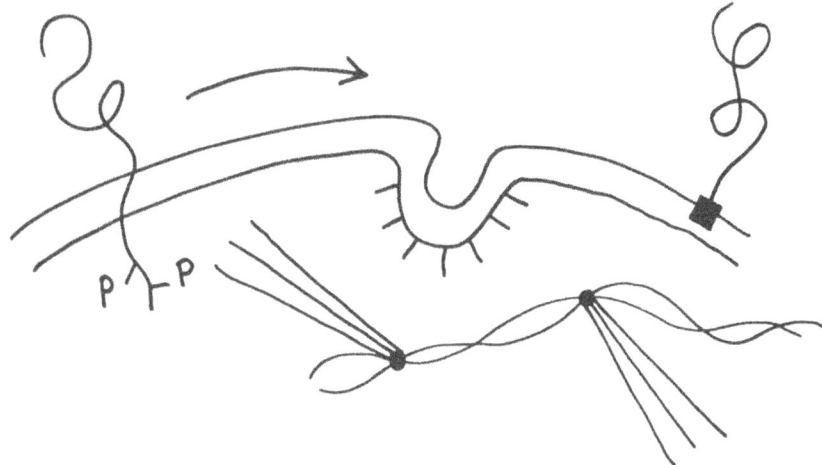

FIGURE 2. Cartoon illustrating the relation of GPI-anchored proteins (right) and a transmembrane protein (left) to the plasma membrane. Clearly, the GPI-anchored protein cannot be directly linked to the cytoskeleton or to clathrin. It also cannot be phosphorylated by cytosolic protein kinases.

secretion granules (Berger and Medof, 1987; Fukuoka *et al.*, 1991; Jost *et al.*, 1990).

Comparative data on Thy-1 expression by wild-type and mutant lymphoma cells that do not add GPI anchors call attention to the possible functional significance of GPI anchors in the context of intracellular transport (Hyman, 1988; Tartakoff *et al.*, 1990). For example, mutant cells that do not add GPI anchors do not express the corresponding proteins at the cell surface (Conzelmann *et al.*, 1986, 1988a; DeGasperi *et al.*, 1990; Fatemi and Tartakoff, 1986, 1988). Since these observations imply that the polypeptide precursors of GPI-anchored proteins (which include a putative transmembrane segment that is normally destined to be replaced by the GPI unit; see below) do not reach to the cell surface, we suggest that removal of the cytoplasmic and transmembrane domain is normally a prerequisite for allowing GPI-anchored proteins to avoid retention along the secretory path and/or degradation (Lemansky *et al.*, 1990).

The addition of the GPI moiety has been studied in yeast, African trypanosomes, and higher eukaryotes (DeGasperi *et al.*, 1990; Lemansky *et al.*, 1991; Heller *et al.*, 1990; Masterson and Ferguson, 1991; Orlean, 1990; Singh and Tartakoff, 1991; Singh *et al.*, 1991). In all cases, it is added in the RER within a couple of minutes of chain termination (Conzelmann *et al.*, 1987, 1988a; Takami *et al.*, 1988a,b). This is actually an event of anchor exchange, since the primary translation product includes a relatively orthodox putative membrane-spanning segment near its C-terminus that signals anchor addition and is removed before or concomitant with anchor addition (Bailey *et al.*, 1989; Caras, 1991; Cross, 1990; Ferguson and Williams, 1988; Heller *et al.*, 1990; Tartakoff *et al.*, 1990). The fate of this C-terminal peptide and the enzyme(s) responsible for its removal have not been identified. A transamidation may be involved (Mayor *et al.*, 1991), thus explaining why it is important to remove a C-terminal peptide, rather than simply appending the GPI unit to the preexisting C-terminus.

Both in trypanosomes and in murine T lymphoma cells, putative pre-assembled anchor units have been identified (Field *et al.*, 1991; Krakow *et al.*, 1986; Lemansky *et al.*, 1991; Mayor *et al.*, 1990a,b; Sugiyama *et al.*, 1991). These GPI units closely resemble the anchor moieties found on the corresponding proteins expressed at the cell surface. Cell-free systems have been established that transfer these GPI units to the corresponding acceptor proteins (Fasel *et al.*, 1989; Kodukula *et al.*, 1992; Mayor *et al.*, 1991).

The biosynthetic path leading to these units has been partly defined by making use of broken cell preparations from trypanosomes (Doering *et al.*, 1989, 1990; Hirose *et al.*, 1991; Masterson *et al.*, 1989, 1990; Menon *et al.*, 1990a,b) and animal cells (Hirose *et al.*, 1991; Stevens and Raetz, 1991; Sugiyama *et al.*, 1991). Supplementation of these preparations with GDP-mannose and UDP-*N*-acetylglucosamine produces a sequence of intermediates (see Figure 3). In a final step in trypanosomes, individual glycerol-linked acyl chains of the PI unit are removed and replaced (Masterson *et al.*, 1990). In animal cells and in procyclic trypanosomes, apart from the first three indicated species, the inositol ring appears to be acylated (Field *et al.*, 1991). In bloodstream forms of trypanosomes, both acylated and nonacylated GPI units exist. The biosynthetic origin of the PI, which for animal cells often has a 1-alkyl, 2-acyl structure (Roberts *et al.*, 1988b), is unknown. The terminal (linker) ethanolamine phosphate residue is thought to be derived from phosphatidylethanolamine (A. Menon, unpublished). Galactose, *N*-acetylgalactosamine, additional mannose and ethanolamine phosphate may all be added after anchor transfer (Bangs *et al.*, 1988).

Several agents have been identified that inhibit synthesis of the mature precursor unit *in vivo* (Figure 3). Mannosamine inhibits addition of the third mannose residue (Lisanti *et al.*, 1991), PMSF inhibits addition of ethanolamine phosphate (Masterson and Ferguson, 1991), and certain myristic acid analogues appear to block myristate addition in trypanosomes (Doering *et al.*, 1991).

2. GPI-NEGATIVE MUTANTS

The family of Thy-1-negative T lymphoma mutants is highly relevant to understanding GPI anchor synthesis (Hyman, 1988; Tartakoff *et al.*, 1990). It is ironic that they were first characterized before any GPI structures were known to exist. These recessive mutants have been assigned to complementation groups

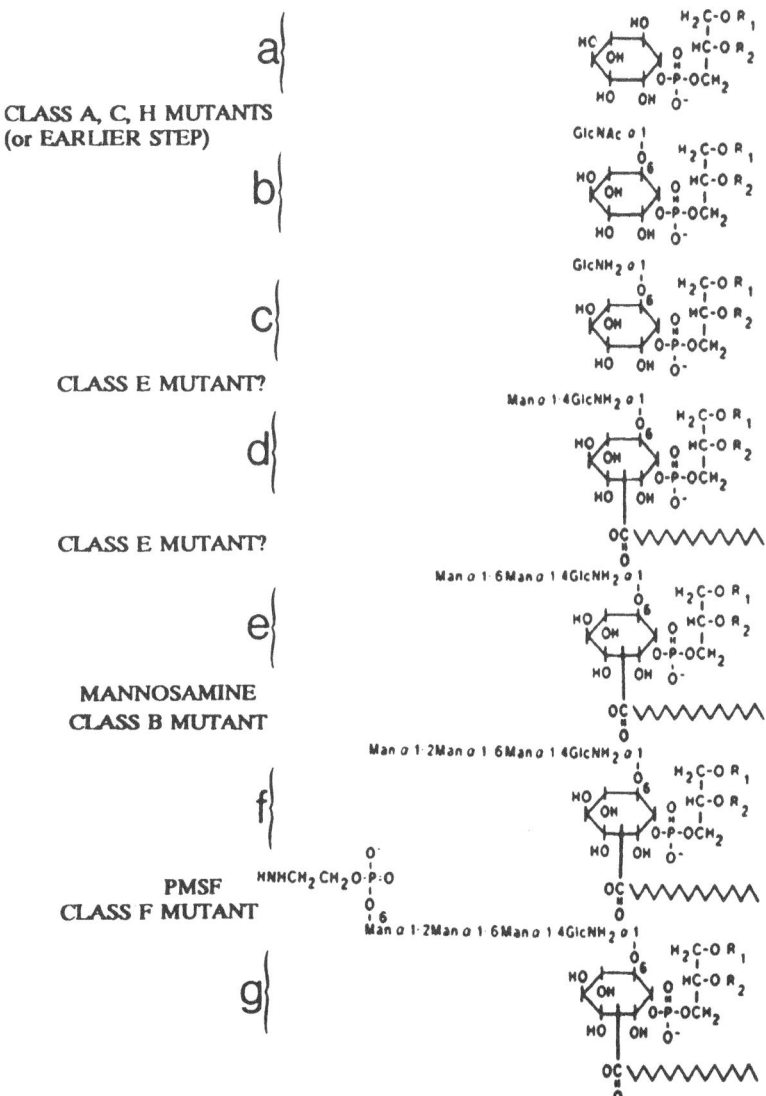

FIGURE 3. Model of synthesis of the preassembled anchor precursor in animal cells. Judging from *in vivo* and *in vitro* studies of both animal cells and trypanosomes, the sequence of intermediates appears to grow (top-to-bottom) one residue at a time. R_1 is thought to be an alkyl chain. Species d–g may include additional ethanolamine phosphate linked to mannose. Site(s) of interruption of the biosynthetic path by inhibitors and mutations are indicated. After transfer of the mature unit to protein, additional lateral substituents may be added and the inositol may be deacylated (see Figure 4).

A–F and H, after introduction of appropriate drug-resistance markers and pair-wise fusion. In all cases except for class D (where no Thy-1 is made), synthesis of an apparently normal Thy-1 peptide occurs, but (as mentioned above) neither Thy-1 nor any other GPI-anchored protein is expressed at the cell surface. This lack of surface expression is matched by the inability of each of the mutants to add a GPI anchor to the Thy-1 that it synthesizes (Conzelmann *et al.*, 1986, 1988b; Fatemi and Tartakoff, 1986, 1988). Class B and E mutants secrete much of their Thy-1 while the others degrade the Thy-1 that they synthesize, probably in lysosomes (Singh *et al.*, 1991). Only for class B is the major intracellular pool of Thy-1 hydrophilic, suggesting removal of the C-terminus.

The lack of anchor addition by these mutants appears to reflect lesions that interrupt the synthesis of GPI units which are anchor precursors (DeGasperi *et al.*, 1990; Lemansky *et al.*, 1991; Singh *et al.*, 1991; Stevens and Raetz, 1991; Sugiyama *et al.*, 1991). Thus, *in vivo* [^{3}H]mannose labeling of wild-type T lymphomas detects mannolipids with characteristic sensitivity to microbial PI-specific phospholipases and nitrous acid (which splits the amino sugar–inositol linkage; see Figure 1). The most polar of these units—the putative anchor precursor—resists exomannosidase, indicating that its nonreducing terminus is blocked, presumably with ethanolamine phosphate. By contrast, parallel *in vivo* labeling of the class A–C and E, F, and H mutant lymphomas shows that none of them produces such complete putative anchor precursor mannolipids.

For mutants A, C, and H, no mannolipids are produced that are sensitive to microbial phospholipases and nitrous acid. In these cases, taking into consideration data from *in vitro* labeling of membrane preparations of lymphoma cells (Stevens and Raetz, 1991; Sugiyama *et al.*, 1991), we propose that there is a lesion that interrupts the biosynthetic path so early that even the *N*-acetylglucosamine residue is not added. Thus, the lesion may affect synthesis or translocation of the proper variety of PI or the addition of *N*-acetylglucosamine to PI.

In vivo and *in vitro* [^{3}H]mannose labeling of the B, E, and F mutants shows that GPI units are produced that are different from the mature species in the wild type (Lemansky *et al.*, 1991; Puoti *et al.*, 1991; Sugiyama *et al.*, 1991). In each case, these units are sensitive to microbial PI-specific phospholipase D and to nitrous acid. Their glycans are smaller than those of the wild type. Furthermore, in the class F mutant (which has a lesion in ether lipid biosynthesis; Stevens and Raetz, 1990), some of these phospholipase-sensitive species are uniquely sensitive to exomannosidase. Thus, for class B, E, and F mutants, GPI biosynthesis has progressed far enough to cause some mannose addition; however, elongation is incomplete, presumably because of the lack of key enzyme activities (sugar transferases?). The class B mutant appears blocked in addition of the third mannose residue while the class F mutant is blocked in ethanolamine phosphate addition (Figure 3). Given the ability of azacytidine (Tisdale *et al.*, 1991), aminoglycosides (Gupta *et al.*, 1988), and sodium butyrate (Tisdale *et al.*, 1991) to partially correct class C, F, and H mutants, respectively, the corresponding structural genes may be present but not transcribed into active products. The class A and class F mutants have recently been complemented by expression of appropriate cDNAs (Miyata *et al.*, 1993; Inoue *et al.*, 1993).

3. ROLE OF DOLICHOL-PHOSPHORYL-MANNOSE

Dolichol-phosphoryl-mannose contributes mannose residues to the anchor. [For a detailed discussion on mevalonate pathway lipids, see Ericsson and Dallner (this volume).] Thus, mutant yeast, class E mutant lymphoma cells, and B4-2-1 mutant CHO cells, none of which make dolichol-phosphoryl-mannose, do not produce GPI anchors (DeGasperi *et al.*, 1990; Fatemi and Tartakoff, 1986; Orlean, 1990; Singh and Tartakoff, 1991). In one of the lymphoma mutants, this defect can be corrected by transfection with a plasmid encoding yeast dolichol-phosphoryl-mannose synthase (DeGasperi *et al.*, 1990). Moreover, when the CHO cell mutant B4-2-1 is treated with tunicamycin, which greatly boosts levels of dolichol-phosphoryl-mannose, the defect in anchor synthesis is corrected (Singh and Tartakoff, 1991). This involvement of dolichol-phosphoryl-mannose is important because, judging from studies of N-glycan synthesis, dolichol-phosphoryl-mannose is responsible for mannose addition at the cisternal face of the RER membrane (Hirschberg and Snider, 1987). Thus, at least some of the mannose residues of the GPI unit must be added at the cisternal face of the RER. Consistent with this genetic information on animal cells, biochemical studies of broken cell preparations of trypanosomes argue that dolichol-phosphoryl-mannose contributes the three core mannose residues to the anchor structure (Menon *et al.*, 1990a).

The observations on the class E lymphoma mutant are of particular interest since some *in vivo* studies of this mutant show that it synthesizes a GPI-related mannolipid even though dolichol-phosphoryl-mannose is not made (Lemansky *et al.*, 1991). The origin of the mannose in this lipid must therefore be GDP-mannose, suggesting that both GDP-mannose and dolichol-phosphoryl-mannose contribute to GPI anchor synthesis. By analogy to N-glycan biosynthesis (Hirschberg and Snider, 1987), the assembly of the GPI moiety may therefore begin on the endodomain and, after a mannolipid flip-flop event, continue on the cisternal face of the RER. Such flip-flop may be facilitated by the inositol-linked acyl chains. GDP-mannose is also required for cell-free synthesis of the repeating disaccharide of *Leishmania donovani* lipophosphoglycan (Carver and Turco, 1991).

4. INOSITOL ACYLATION

As mentioned above, wild-type cells have precursor GPI units that include acylinositol. One can therefore readily rationalize the observation that GPI an-

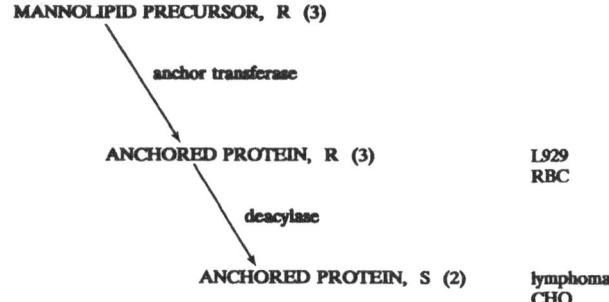

FIGURE 4. Cell fusion experiments show that PI-PLC sensitivity (the hallmark of the absence of acylinositol) behaves as a dominant trait (Singh *et al.*, 1991). When a cell that synthesizes PI-PLC-sensitive anchored proteins is fused with a cell that synthesizes PI-PLC-resistant anchored proteins, stable hybrids express PI-PLC-sensitive proteins. The simplest explanation of this observation is that the anchor precursor resists (R) PI-PLC because of the presence of an inositol-linked, third (3) acyl chain and that certain cells express a deacylase that can remove this chain (e.g., murine lymphomas, CHO cells) while others (L929 fibroblasts, RBC precursors) do not. Deacylase activity converts a 3-footed, PI-PLC-resistant anchor to 2-footed, PI-PLC-sensitive (S) form. This result is not consistent with a model in which the second class of cell produces an anchor inositol acyl transferase that is absent from the first cell type.

chors on many cells (e.g., L929 fibroblasts, human red cells) also have acylinositol (Roberts *et al.*, 1988a,b; Singh *et al.*, 1991), which causes them to resist PI-PLC hydrolysis (Walter *et al.*, 1990). Since fusion of such cells with cells whose GPI-anchored proteins lack acylinositol (murine T lymphoma cells, CHO cells) yields stable hybrids whose GPI-anchored proteins lack acylinositol (Singh *et al.*, 1991), we suggest that cells such as L929 lack a GPI deacylase that is present in the other fusion partner (Figure 4).

5. IMPLICATIONS

The broader significance of these observations concerns the expression of an abundant class of surface proteins of diverse function. Many are enzymes, some have been described as receptors [the FcIII receptor for immunoglobulin (Hollander *et al.*, 1988; Kimberly *et al.*, 1990; Ueda *et al.*, 1989); for folate (Rothberg *et al.*, 1990)], and several can—surprisingly—transduce signals to cells when they are liganded by appropriate antibodies (Fischer *et al.*, 1990; Kroczek *et al.*, 1986; MacDonald *et al.*, 1985; Pont *et al.*, 1985; Robinson *et al.*, 1989; Su *et al.*, 1991). Given the existence of mutant cells that do not express GPI-anchored proteins, synthesis and expression of such proteins is certainly not essential for survival of individual cells in culture. In the context of the organism, ectopic expression of at least one GPI-anchored protein leads to proliferation of the cells that bear it (Chen *et al.*, 1987). Moreover, lesions similar to those of the Thy-1-negative lymphoma mutants may explain the lack of GPI-anchored proteins on red and white cells in the human disease, paroxysmal nocturnal hemoglobinuria (Hirose *et al.*, 1991; Hollander *et al.*, 1988; Medof *et al.*, 1987; Ueda *et al.*, 1989; Zalman *et al.*, 1987). Finally, lesions that affect GPI biosynthesis can eliminate surface expression of multiple GPI-anchored proteins and therefore may be important for allowing malignant cells to escape immune surveillance.

6. REFERENCES

Almqvist, P., and Carlsson, S. R., 1988, Characterization of a hydrophilic form of Thy-1 purified from human cerebrospinal fluid, *J. Biol. Chem.* **263**:12709–12715.

Bailey, C. A., Gerber, L., Howard, A. D., and Udenfriend, S., 1989, Processing at the carboxyl terminus of nascent placental alkaline phosphatase in a cell-free system: Evidence for specific cleavage of a signal peptide, *Proc. Natl. Acad. Sci. USA* **86**:22-26.

Bamezai, A., Goldmacher, V., Reiser, H., and Rock, K. L., 1989, Internalization of phosphatidylinositol-anchored lymphocyte proteins, *J. Immunol.* **143**:3107–3116.

Bangs, J. D., Doering, T. L., Englund, P. T., and Hart, G. W., 1988, Biosynthesis of a variant surface glycoprotein of *Trypanosoma brucei*, *J. Biol. Chem.* **263**:17697–17705.

Berger, M., and Medof, M. E., 1987, Increased expression of complement decay accelerating factor on human neutrophils during activation, *J. Clin. Invest.* **79**:214–220.

Bulow, R., Overath, P., and Davoust, J., 1988, Rapid lateral diffusion of the variant surface glycoprotein in the coat of *Trypanosoma brucei*, *Biochemistry* **27**:2384–2388.

Caras, I., 1991, An internally positioned signal can direct attachment of a glycophospholipid membrane anchor, *J. Cell Biol.* **113**:77–85.

Carver, M. A., and Turco, S. J., 1991, Cell-free biosynthesis of lipophosphoglycan from *Leishmania donovani*, *J. Biol. Chem.* **266**:10974–10981.

Chan, B. L., Lisanti, M. P., Rodriguez-Boulan, E., and Saltiel, A. R., 1988, Insulin-stimulated release of lipoprotein lipase by metabolism of its phosphatidylinositol anchor, *Science* **241**:1670–1672.

Chen, S., Botteri, F., van der Putten, H., Landel, C. P., and Evans, G. A., 1987, A lymphoproliferative abnormality associated with inappropriate expression of the Thy-1 antigen in transgenic mice, *Cell* **51**:7–19.

Conzelmann, A., Spiazzi, A., Hyman, R., and Bron, C., 1986, Anchoring of membrane proteins via phosphatidylinositol is deficient in two classes of Thy-1 negative mutant lymphoma cells, *EMBO J.* **5**:3291–3296.

Conzelmann, A., Spiazzi, A., and Bron, C., 1987, Glycolipid anchors are attached to Thy-1 glycoprotein rapidly after translation, *Biochem. J.* **246**:605–610.

Conzelmann, A., Riezman, H., Desponds, C., and Bron, D., 1988a, A major 125-kd membrane glycoprotein of *Saccharomyces cerevisiae* is attached to the lipid bilayer through an inositol-containing phospholipid, *EMBO J.* **7**:2233–2240.

Conzelmann, A., Spiazzi, A., Bron, C., and Hyman, R., 1988b, No glycolipid anchors are added to Thy-1 glycoprotein in Thy-1-negative mutant thymoma cells of four different complementation classes, *Mol. Cell. Biol.* **8**:674–678.

Cross, G. A. M., 1990, Glycolipid anchoring of plasma membrane proteins, *Annu. Rev. Cell Biol.* **6**:1–39.

DeGasperi, R., Thomas, L. J., Sugiyama, E., Chang, H. M., Beck, P. J., Orlean, P., Albright, C., Waneck, G., Sambrook, J. F., Warren, C. D., and Yeh, E. T. H., 1990, Correlation of a defect in mammalian GPI anchor biosynthesis by a transfected veast gene, *Science* **250**:998–991.

Doering, T. L., Masterson, W. J., Englund, P. T., and Hart, G. W., 1989, Biosynthesis of the glycosyl phosphatidylinositol membrane anchor of the trypanosome variant surface glycoprotein, *J. Biol. Chem.* **264**:11168–11173.

Doering, T. L., Masterson, W. J., Hart, G. W., and Englund, P. T., 1990, Biosynthesis of glycosyl phosphatidylinositol membrane anchors, *J. Biol. Chem.* **265**:611–614.

Doering, T. L., Raper, J., Buxbaum, L. U., Adams, S. P., Gordon, J. I., Hart, G. W., and Englund, P. T., 1991, An analog of myristic acid with selective toxicity for African trypanosomes, *Science* **252**:1851–1854.

Dotti, C. G., Parton, R. G., and Simons, K., 1991, Polarized sorting of glypiated proteins in hippocampal neurons, *Nature* **349**:158–161.

Edidin, M., and Stroynowski, I., 1991, Differences between the lateral organization of conventional and inositol phospholipid-anchored membrane proteins. A further definition of micrometer scale membrane domains, *J. Cell Biol.* **112**:1143–1150.

Epstein, E., Kiechle, F. L., Artiss, J. D., and Zak, B., 1986, The clinical use of alkaline phosphatase enzymes, *Clin. Lab. Med.* **6**:491–505.

Fasel, N., Rousseaux, M., Schaerer, E., Medof, M. E., Tykocinski, M. L., and Bron, C., 1989, *In vitro* attachment of glycosyl-inositolphospholipid anchor structures to mouse Thy-1 antigen and human decay-accelerating factor, *Proc. Natl. Acad. Sci. USA* **86**:6858–6862.

Fatemi, S. H., and Tartakoff, A. M., 1986, Hydrophilic anchor-deficient Thy-1 is secreted by a class E mutant T lymphoma, *Cell* **46**:653–657.

Fatemi, S. H., and Tartakoff, A. M., 1988, The phenotype of five classes of T lymphoma mutants, *J. Biol. Chem.* **263**:1288–1294.

Ferguson, M. A. J., and Williams, A., 1988, Cell-surface anchoring of proteins via glycosyl-phosphatidylinositol structures, *Annu. Rev. Biochem.* **57**:285–320.

Ferguson, M. A. J., Homans, S. W., Dwek, R. A., and Rademacher, T. W., 1988, Glycosyl-phosphatidylinositol moiety that anchors *Trypanosoma brucei* variant surface glycoprotein to the membrane, *Science* **239**:753–759.

Field, M. C., Menon, A. K., and Cross, G. A. M., 1991, Developmental variation of glycosylphosphatidylinositol membrane anchors in *Trypanosoma brucei*, *J. Biol. Chem.* **266**:1–9.

Fischer, G. F., Majdic, O., Gadd, S., and Knapp, W., 1990, Signal transduction in lymphocytic and myeloid cells via CD24, a new member of phosphoinositol-anchored membrane molecules, *J. Immunol.* **144**:638–641.

Fukuoka, S. -I., Freedman, S. D., and Scheele, G. A., 1991, A single gene encodes membrane-bound and free forms of GP-2, the major glycoprotein in pancreatic secretory (zymogen) granule membranes, *Proc. Natl. Acad. Sci. USA* **88**:2898–2902.

Gupta, D., Tartakoff, A., and Tisdale, E., 1988, Metabolic correction of defects in the lipid anchoring of Thy-1 in lymphoma mutants, *Science*, **242**:1446–1448.

He, H. -T., Finne, J., and Goridis, C., 1987, Biosynthesis, membrane association, and release of N-CAM-120, a phosphatidylinositol-linked form of the neural cell adhesion molecule, *J. Cell Biol.* **105**:2489–2500.

Heller, M., Micanovic, R., and Udenfriend, S., 1990, Molecular biology and biosynthesis of placental alkaline phosphatase: Requirements for processing to the mature form anchored to membranes by a phosphatidylinositol-glycan, in: *Molecular and Cell Biology of Membrane Proteins, Glycolipid Anchors of Cell-Surface Proteins* (A. J. Turner, ed.), pp. 111–128, Ellis Horwood, England.

Hirose, S., Ravi, L., Hazra, S. V., and Medof, M. E., 1991, Assembly and deacetylation of

N-acetylglucosaminyl-plasmanylinositol in normal and affected paroxysmal nocturnal hemoglobinuria cells, *Proc. Natl. Acad. Sci. USA* **88**:3762–3766.

Hirschberg, C., and Snider, M., 1987, Topography of glycosylation in the rough endoplasmic reticulum and Golgi apparatus, *Annu. Rev. Biochem.* **56**:63–87.

Hollander, N., Selvaraj, P., and Springer, T. A., 1988, Biosynthesis and function of LFA-3 in human mutant cells deficient in phosphatidylinositol-anchored proteins, *J. Immunol.* **141**:4283–4290.

Homans, S. W., Ferguson, M. A. J., Dwek, R. A., Rademacher, T. W., Anand, R., and Williams, A. F., 1988, Complete structure of the glycosyl phosphatidylinositol membrane anchor of rat brain Thy-1 glycoprotein, *Nature*, **333**:269–272.

Huizinga, T. W. J., van der Schoot, C. E., Jost, C., Klaassen, R., Kleijer, M., von dem Borne, A. E. G., Roos, D., and Tetteroo, P. A. T., 1988, The PI-linked receptor FcRIII is released on stimulation of neutrophils, *Nature* **333**:667–669.

Hyman, R., 1988, Somatic genetic analysis of the expression of cell surface molecules, *Trends Genet.* **4**:5–8.

Inoue, N. Kinoshita, T., Osii, T., and Takeda, J., 1993, Cloning of a human gene, PIG-F, a component of glycosylphosphatidylinositol anchor biosynthesis, *J. Biol. Chem.* **268**:6882–6885.

Jost, C. R. Huizinga, T. W. J., de Goede, R., Fransen, J. A. M., Tetteroo, P. A. T., Daha, M. R., and Ginsel, L. A., 1990, Intracellular localization and de novo synthesis of FcRIII in human neutrophil granulocytes, *Blood* **75**:114–151.

Kimberly, R. P., Ahlstrom, J. W., Click, M. E., and Edberg, J. C., 1990, The glycosyl phosphatidylinositol-linked FcRIII mediates transmembrane signalling events distinct from FcRII, *J. Exp. Med.* **171**:1239–1255.

Kodukula, K., Cines, D., Amthauer, R., Gerber, L., and Udenfriend, S., 1992, Biosynthesis of phosphatidylinositol-glycon-anchored membrane proteins in cell-free systems, *Proc. Natl. Acad. Sci. USA* **89**:1350–1353.

Krakow, J., Herald, J., Bangs, J., Hart, G., and Englund, P., 1986, Identification of a glycolipid precursor of the *Trypanosoma brucei* variant surface glycoprotein, *J. Biol. Chem.* **261**:12147–12153.

Kroczek, R. A., Gunter, K. C., Germain, R. N., and Shevach, E. M., 1986, Thy-1 functions as a signal transduction molecule in T lymphocytes and transfected B lymphocytes, *Nature* **322**:181–184.

Lemansky, P., Fatemi, S. H., Gorican, B., Meyale, S., Rossero, R., and Tartakoff, A., 1990, Dynamics and longevity of the glycolipid-anchored membrane protein, Thy-1, *J. Cell Biol.* **110**:125–1532.

Lemansky, P., Gupta, D. K., Meyale, S., Tucker, G., and Tartakoff, A. M., 1991, Atypical mannolipids characterize Thy-1 negative lymphoma mutants, *Mol. Cell Biol.* **11**:3879–3885.

Lisanti, M. P., Caras, I. W., Davitz, M. A., and Rodriguez-Boulan, E., 1989, A glycophospholipid membrane anchor acts as an apical targeting signal in polarized epithelial cells, *J. Cell Biol.* **109**:2145–2156.

Lisanti, M. P., Caras, I. W., Gilbert, T., Hanzel, D., and Rodriguez-Boulan, E., 1990, Vectorial apical delivery and slow endocytosis of a glycolipid-anchored fusion protein in transfected MDCK cells, *Proc. Natl. Acad. Sci. USA* **87**:7419–7423.

Lisanti, M. P., Field, M. C., Caras, I. W., Menon, A. K., and Rodriguez-Boulan, E., 1991, Mannosamine, a novel inhibitor of glycosyl-phosphatidylinositol incorporation into proteins, *EMBO J.* **10**:1969–1977.

Low, M. G., and Saltiel, A. R., 1988, Structural and functional roles of glycosylphosphatidylinositol in membranes, *Science* **239**:268–275.

MacDonald, H. R., Bron, C., Rousseaux, M., Horvath, C., and Cerottini, J. -C., 1985, Production and characterization of monoclonal anti-Thy-1 antibodies that stimulate lymphokine production by cytolytic T cell clones, *Eur. J. Immunol.* **15**:495–501.

Masterson, W. J., and Ferguson, M. A. J., 1991, Phenylmethanesulphonyl fluoride inhibits GPI anchor biosynthesis in the African trypanosome, *EMBO J.* **10**:2041–2045.

Masterson, W. J., Doering, T. L., Hart, G. W., and Englund, P. T., 1989, A novel pathway for glycan assembly: Biosynthesis of the glycosyl-phosphatidylinositol anchor of the trypanosome variant surface glycoprotein, *Cell* **56**:793–800.

Masterson, W. J., Raper, J., Doering, T. L., Hart, G. W., and Englund, P. T., 1990, Fatty acid remodeling: A novel reaction sequence in the biosynthesis of trypanosome glycosyl phosphatidylinositol membrane anchors, *Cell* **62**:73–80.

Mayor, S., Menon, A. K., Cross, G. A. M., Ferguson, M. A. J., Dwek, R. A., and Rademacher, T. W., 1990a, Glycolipid precursors for the membrane anchor of *Trypanosoma brucei* variant surface glycoproteins. I. Glycan structure of the phosphatidylinositol-specific phospholipase C sensitive and resistant glycolipids, *J. Biol. Chem.* **265**:6164–6173.

Mayor, S., Menon, A. K., and Cross, G. A. M., 1990b, Glycolipid precursors for the membrane anchor of *Trypanosoma brucei* variant surface glycoproteins. II. Lipid structures of phosphatidylinositol-specific phospholipase C sensitive and resistant glycolipids, *J. Biol. Chem.* **265**:6174–6181.

Mayor, S., Menon, A. K., and Cross, G. A. M., 1991, Transfer of glycosyl-phosphatidylinositol membrane anchors to polypeptide acceptors in a cell-free system, *J. Cell Biol.* **114**:61–71.

Medof, M. E., Gottlieb, A., Kinoshita, T., Hall, S., Silber, R., Nussenzweig, V., and Rosse, W. F., 1987, Relationship between decay accelerating factor deficiency, diminished acetylcholinesterase activity, and defective terminal complement pathway restriction in paroxysmal nocturnal hemoglobinuria erythrocytes, *J. Clin. Invest.* **80**:165–174.

Menon, A. K., Schwarz, R. T., Mayor, S., and Cross, G. A. M., 1990a, Cell-free synthesis of glycosyl-phosphatidylinositol precursors for the glycolipid membrane anchor of *Trypanosoma brucei* variant surface glycoproteins, *J. Biol. Chem.* **265**:9033–9042.

Menon, A. K., Mayor, S., and Schwarz, R. T., 1990b, Biosynthesis of glycosyl-phosphatidylinositol lipids in *Trypanosoma brucei:* Involvement of mannosyl-phosphoryldolichol as the mannose donor, *EMBO J.* **9**:4249–4258.

Miyata, T., Takeda, J., Iida, Y., Yamada, N., Inoue, N., Takahashi, M., Maeda, K., Kitani, T., and Kinoshita, T., 1993, The cloning of PIG-A, a component in the early step of GPI-anchor biosynthesis, *Science* **259**:1313–1320.

Orlean, P., 1990, Dolichol phosphate mannose synthase is required *in vivo* for glycosyl phosphatidylinositol membrane anchoring, O mannosylation, and N glycosylation of protein in *Saccharomyces cerevisiae*, *Mol. Cell Biol.* **10**:5796–5805.

Overath, P., Czichos, J., Stock, U., and Nonnengaesser, C., 1983, Repression of glycoprotein synthesis and release of surface coat during transformation of *Trypanosoma brucei*, *EMBO J.* **2**:1721–1728.

Phelps, B. M., Primakoff, P., Koppel, D. E., Low, M. G., and Myles, D. G., 1988, Restricted lateral diffusion of PH-20, a PI-anchored sperm membrane protein, *Science* **240**:1780–1782.

Pont, S., Regnier-Vigouroux, A., Naquet, P., Blanc, D., Pierres, A., Marchetto, S., and Pierres, M., 1985, Analysis of the Thy-1 pathway of T cell hybridoma activation using 17 rat monoclonal antibodies reactive with distinct Thy-1 epitopes, *Eur. J. Immunol.* **15**:1222–1228.

Puoti, A., Desponds, C., Fankhauser, C., and Conzelmann, A., 1991, Characterization of a glycophospholipid intermediate in the biosynthesis of glycophosphatidylinositol anchors accumulating in the Thy-1-negative lymphoma line SIA-b, *J. Biol. Chem.* **266**:21051–21059.

Quentmeier, A., Moller, P., Schwarz, V., Abel, U., and Schlag, P., 1987, Carcinoembryonic antigen, CA 19-9, and CA 125 in normal and carcinomatous human colorectal tissue, *Cancer* **60**:2261–2266.

Roberts, W. L., Myher, J. J., Kuksis, A., Low, M. G., and Rosenberry, T. L., 1988a, Lipid analysis of the glycoinositol phospholipid membrane anchor of human erythrocyte acetylcholinesterase, *J. Biol. Chem.* **263**:18766–18775.

Roberts, W. L., Santikarn, S., Reinhold, V. N., and Rosenberry, T. L., 1988b, Structural characterization of the glycoinositol phospholipid membrane anchor of human erythrocyte acetylcholinesterase by fast atom bombardment mass spectrometry, *J. Biol. Chem.* **263**:18776–18784.

Robinson, P. J., Millrain, M., Antoniou, J., Simpson, E., and Mellor, A. L., 1989, A glycophospholipid anchor is required for Qa-2-mediated T cell activation, *Nature* **342**:85–87.

Rothberg, K. G., Ying, Y., Kolhouse, J. F., Kamen, B. A., and Anderson, R. G. W., 1990, The glycophospholipid-linked folate receptor internalizes folate without entering the clathrin-coated pit endocytic pathway, *J. Cell Biol.* **110**:637–650.

Singh, N., and Tartakoff, A., 1991, Two different mutants blocked in synthesis of dolichol-phosphoryl-mannose do not add glycophospholipid anchors to membrane proteins: Quantitative correction of the phenotype of a CHO cell mutant with tunicamycin, *Mol. Cell Biol.* **11**:391–400.

Singh, N., Singleton, D., and Tartakoff, A., 1991, Anchoring and degradation of glycolipid-anchored membrane proteins by L929 and LM-TK⁻ mouse fibroblasts: Implications for anchor biosynthesis, *Mol. Cell Biol.* **11**:2362–2374.

Stevens, V., and Raetz, C. R. H., 1990, Class F Thy-1-negative murine lymphoma cells are deficient in ether lipid biosynthesis, *J. Biol. Chem.* **265**:15653–15658.

Stevens, V. L., and Raetz, C. R. H., 1991, Defective glycosyl phosphatidylinositol biosynthesis in extracts of three Thy-1 negative lymphoma cell mutants, *J. Biol. Chem.* **266**:10039–10042.

Su, B., Waneck, G. L., Flavell, R. A., and Bothwell, A. L. M., 1991, The glycosyl phosphatidylinositol anchor is critical for Ly-6A/E-mediated T cell activation, *J. Cell Biol.* **112**:377–384.

Sugiyama, E., DeGasperi, R., Urakaze, M., Chang, H. -M., Thomas, L. J., Hyman, R., Warren, C. D., and Yeh, E. T. H., 1991, Identification of defects in glycosylphosphatidylinositol anchor biosynthesis in the Thy-1 expression mutants, *J. Biol. Chem.* **266**:12119–12122.

Takami, N., Ogata, S., Oda, K., Misumi, Y., and Ikehara, Y., 1988a, Biosynthesis of placental alkaline phosphatase and its post-translational modification by glycophospholipid for membrane-anchoring, *J. Biol. Chem.* **263**:3016–3021.

Takami, N., Misumi, Y., Kuroki, M., Matsuoka, Y., and Ikehara, Y., 1988b, Evidence for carboxyl-terminal processing and glycolipid-anchoring of human carcinoembryonic antigen, *J. Biol. Chem.* **263**:12716–12720.

Tartakoff, A., Fatemi, S., Gupta, D., Kaetzel, D., Lemansky, P., Singleton, D., Singh, N., and Tisdale, E., 1990, Biosynthesis and transport of lipid-anchored membrane proteins, in: *Molecular and Cell Biology of Membrane Proteins, Glycolipid Anchors of Cell-Surface Proteins* (A. J. Turner, ed.), pp. 111–128, Ellis Horwood, England.

Tisdale, E., Schimenti, J., and Tartakoff, A., 1991, Sodium butyrate causes reexpression of three membrane proteins on glycolipid anchoring mutants, *Som. Cell Mol. Genet.* **17**:349–357.

Ueda, E., Kinoshita, T., Nojima, J., Inoue, K., and Kitani, T., 1989, Different membrane anchors of FC RIII (CD16) on K/NK-lymphocytes and neutrophils, *J. Immunol.* **143**:1274–1277.

Walter, E. I., Roberts, W. L., Rosenberry, T. L., Ratnoff, W. D., and Medof, M. E., 1990, Structural basis for variations in the sensitivity of human decay accelerating factor to phosphatidylinositol-specific phospholipase C cleavage, *J. Immunol.* **144**:1030–1036.

Zalman, L. S., Wood, L. M., Frank, M. M., and Muller-Eberhard, H. J., 1987, Deficiency of the homologous restriction factor in paroxysmal nocturnal hemoglobinuria, *J. Exp. Med.* **165**:572–577.

Chapter 5

Membrane Cycling between the ER and Golgi Apparatus and Its Role in Biosynthetic Transport

Jennifer Lippincott-Schwartz

1. INTRODUCTION

Selective localization and transport of protein and lipid within eukaryotic cells requires the proper functioning of an intercommunicating "endomembrane" system characterized by distinct membrane-bound compartments and membrane transport pathways. Two organelles that play a key role in the generation and maintenance of this endomembrane system, as well as in membrane targeting within it, are the ER and Golgi apparatus. All newly synthesized proteins enter the endomembrane system in the ER and only move to different final destinations in the cell after passing through the Golgi apparatus. This fundamental relationship between the ER and Golgi apparatus in the regulation and sorting events of

Abbreviations used in this chapter: ARF, ADP ribosylation factor; BFA, brefeldin A; BiP heavy-chain binding protein; CGN, *cis*-Golgi network; βCOP, coatomer subunit; ER, endoplasmic reticulum; IC, intermediate compartment; MHC, major histocompatibility complex; MTOC, microtubule organizing center; NEM, *N*-ethyl-maleimide; NSF, NEM-sensitive fusion protein; SNAP, soluble NSF attachment protein; TGN, *trans*-Golgi network.

Jennifer Lippincott-Schwartz   Cell Biology and Metabolism Branch, National Institute of Child Health and Human Development, National Institutes of Health, Bethesda, Maryland 20892.
Subcellular Biochemistry, Volume 21: Endoplasmic Reticulum, edited by N. Borgese and J. R. Harris. Plenum Press, New York, 1993.

biosynthetic transport was first recognized in the 1960s (Palade, 1975). Only recently, however, has insight into the underlying mechanisms of membrane traffic between the ER and Golgi apparatus been achieved, because of the development of new biochemical, pharmacologic, genetic, and morphologic approaches.

Of particular significance to our understanding of membrane transport between the ER and Golgi apparatus has been the recognition of the role of membrane cycling in this process (Pelham, 1991; Lippincott-Schwartz et al., 1990; Hsu et al., 1991). Recent studies suggest that membrane traffic between the ER and Golgi apparatus is not unidirectional but appears to be a finely regulated bidirectional highway connecting two steady-state systems (i.e., ER and Golgi apparatus) whose respective size and structure depend on membrane input and efflux rates (Klausner et al., 1992). Given this apparent phenomenon, several important questions regarding ER/Golgi trafficking must be addressed. These include: What is the nature of the pathways connecting the ER and Golgi apparatus? What is the biochemical machinery that regulates transport within this system? How is selectivity of the membrane components moving within this system achieved?

In this review, membrane transport within the intercommunicating ER/Golgi membrane system will be discussed in light of these questions. It begins with a description of ER and Golgi structure/function and then focuses on the characteristics of anterograde (forward) and retrograde (reverse) membrane transport between the ER and Golgi apparatus. It concludes with a discussion of how regulation of transport within this bidirectional membrane transport system may be achieved.

2. CHARACTERISTICS OF THE ER/GOLGI MEMBRANE SYSTEM

2.1. Structure and Function of the ER

A number of important cellular processes occur within the ER. These include: biosynthesis and assembly of proteins, compartmentalization of the nucleus, lipid biosynthesis and metabolism, regulation of ion gradients, drug detoxification, protein degradation, and certain types of bulk cytoplasmic movements (Palade and Siekevitz, 1956; Helenius et al., 1993; Wilgram and Kennedy, 1963; Terasaki and Sardet, 1991; Bonifacino and Lippincott-Schwartz, 1991; Kachar and Reese, 1988; Terasaki and Jaffe, 1991). Since the ER is the sole site of synthesis of protein and of most lipid comprising the endomembrane system, an additional, constitutive function of this organelle is in membrane export.

The morphology of the ER reflects these diverse functions. As a network of intercommunicating tubules and lamellae extending throughout the cell, the ER consists of both smooth and rough (studded with ribosomes) portions (see Figure 1). Tubular extensions of the ER reach out to the cell periphery, utilizing microtubules to maintain their peripheral distribution (Terasaki *et al.*, 1986; Dabora and Sheetz, 1988). By contrast, regions of the cell near the microtubule organizing center (MTOC) are largely devoid of ER membrane. Membrane tubules of

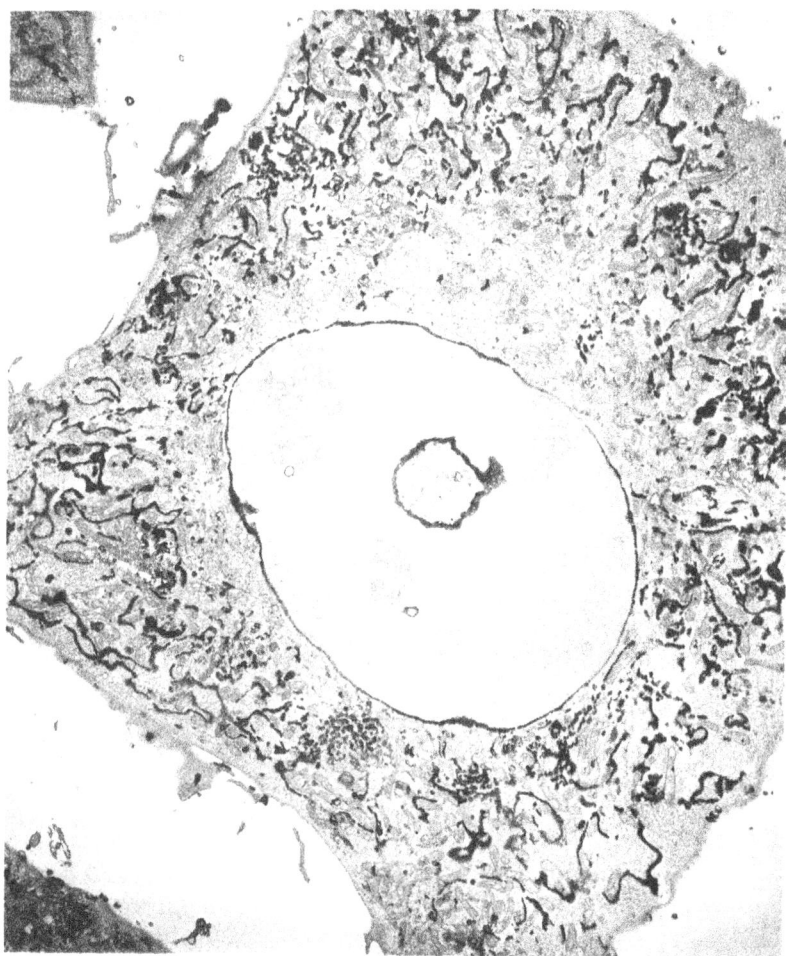

FIGURE 1. Distribution and morphology of the ER. The extensive tubular and peripheral distribution of the ER is revealed in this mouse fibroblast transfected with and immunoperoxidase stained for the ER-retained T cell antigen receptor alpha chain. Note the relative absence of ER membrane (marked by the black reaction product) in central regions of the cell adjacent to the nucleus where the Golgi resides. Photograph courtesy of Lydia Yuan (CBMB, NICHD, NIH).

the ER are extremely dynamic and selectively fuse with each other but not with other membranes (Dabora and Sheetz, 1988).

The rough ER is the site where newly synthesized proteins enter the ER from the cytoplasm. Proteins are targeted here via a complex machinery uniquely associated with the ER that recognizes hydrophobic signal sequences (see Klappa et al. and Simon and Blobel, this volume). Once within the lumen of the ER, newly synthesized proteins are enmeshed in a milieu ideal for folding and assembly into higher-order structures (see Rowling and Freedman, this volume). Protein concentrations in the lumen of the ER have been estimated to be as high as 100 mg/ml giving it a gellike consistency (Booth and Koch, 1990). Much of the ER lumenal contents (comprising the ER matrix) is made of resident ER proteins which participate in the folding and assembly of proteins (Helenius et al., 1993).

2.2. Structure and Function of the Golgi Apparatus

All membrane transported out of the ER is conveyed uniquely to the Golgi apparatus before being sorted to different final destinations in the cell. This involves routing of newly synthesized proteins from multiple peripheral sites in the ER to the centrally located Golgi complex (Saraste and Svensson, 1991; Lippincott-Schwartz et al., 1990), which consists of organized stacks of cisternae localized near the MTOC (Rambourg and Clermont, 1990).

Overall, the Golgi complex is organized into three distinct polarized domains: the cis-Golgi network (CGN), the Golgi stack, and the trans-Golgi network (TGN) (Mellman and Simons, 1992). Membrane and protein enter this complex in the CGN and then move through the Golgi stacks before entering the TGN. The CGN includes the cis-most cisternae of the Golgi stacks together with vesicles and an array of tubules extending from this region. Recent data suggest that the CGN is primarily involved in the selective recycling of protein and lipid back to the ER with only a limited role in glycosylation (Hsu et al., 1991; Pelham, 1991). The Golgi stack, comprising the cisternal and tubular structures in the middle of the Golgi complex, by contrast, functions primarily in glycosylation events, including the ordered remodeling of N-linked oligosaccharide side chains of glycoproteins. The third domain of the Golgi complex, the TGN, appearing on the trans side of the Golgi stack as a sacculotubular network, mediates the sorting and final exit of proteins (Griffiths and Simons, 1986).

A consequence of the polarized (cis-to-trans), multicisternal organization of the Golgi complex is that newly synthesized proteins must be transported from one Golgi cisternal compartment to the next. This is widely believed to occur via transport vesicles that bud from the rims of one Golgi cisterna and then fuse with the next (Rothman and Orci, 1990). Isolation of the putative Golgi transport vesicles from mammalian cells has revealed their membrane to be coated with an electron-dense material containing several unique cytosolic components ("COPs") that form a large complex called coatomer (Malhotra et al., 1989;

Serafini *et al.*, 1991). One of the COPs, βCOP, was previously identified as a 110-kDa peripheral protein associated with the Golgi apparatus (Allan and Kreis, 1986; Duden *et al.*, 1991). Coatomer (detected by antibodies to βCOP) has been shown to rapidly cycle on and off Golgi membrane in an energy- and GTP-dependent manner, and on membranes interacts with ADP ribosylation factor (ARF), a low-molecular-weight GTP binding protein (Donaldson *et al.*, 1991a,b; Serafini *et al.*, 1991). Significantly, coatomer binding is not restricted to the Golgi apparatus. ER-to-Golgi intermediate structures also contain coatomer (as judged by βCOP binding) (Duden *et al.*, 1991), suggesting a role for coatomer binding in ER-to-Golgi as well as intra-Golgi transport.

What unique functions of the Golgi apparatus dictate its separate compartmentalization within the cell? As mentioned above, the Golgi apparatus enzymatically functions as a carbohydrate factory engaged in the biosynthesis of glycolipids and of the oligosaccharide portions of glycoproteins and proteoglycans. Many of these posttranslational processing events occurring in the Golgi stack are needed for proteins to function properly. The characteristic cisternal morphology of the Golgi stacks may serve to enhance the efficiency of the glycosylating enzymes localized here by increasing the membrane surface-to-lumenal volume ratio (Mellman and Simons, 1992).

A significant feature of the Golgi apparatus is its localization at the center of the cell near the MTOC (Rambourg and Clermont, 1990). Why the Golgi complex is localized here is not known but is probably related to its tight association with microtubules (Thyberg and Moskalewski, 1985). Disruption of microtubules with agents such as nocodazole and colchicine, which bind tubulin and inhibit their polymerization, results in the reversible fragmentation and dispersal of the Golgi complex (Turner and Tartakoff, 1989; Ho, *et al.*, 1989). Dispersed Golgi fragments in nocodazole-treated cells can still function in the processing and secretion of proteins. However, secretory vesicles insert in the plasma membrane randomly under these conditions. This is in contrast to cells with an intact microtubular system where membrane moving through the secretory pathway inserts in the plasma membrane in a polarized fashion at sites closest to the central Golgi complex (Rogalski and Singer, 1984). This type of polarized secretion has been proposed to play a role in cell-cell signaling and in directed cell migration (Kupfer, *et al.*, 1982).

3. TRANSPORT AND TARGETING FROM THE ER TO THE GOLGI APPARATUS

3.1. Control of Export from the ER

Early studies of protein export from the ER showed that different membrane and secretory proteins exit the ER at distinct rates, ranging from minutes to several hours (Fitting and Kabat, 1982; Lodish *et al.*, 1983; Williams *et al.*, 1985). This indicated that export is a selective process and led to the proposal that transport of molecules out of the ER required a signal or a recognition motif on the exported protein that enabled it to interact with a transport apparatus (Rose and Bergmann, 1983). Efforts to identify discrete signals on proteins required for export out of the ER have thus far been unsuccessful. Moreover, when an "inert" molecule within the ER lumen, an *N*-acyl glycotripeptide (which presumably lacks any "export" signal), was followed in cells to measure the rate of intracellular bulk flow transport, the half-time for its secretion was significantly faster than that for transport of most proteins (Wieland *et al.*, 1987). This suggested that transport out of the ER is a nonselective process, occurring by default.

The nonselective or "bulk flow" model for egress of protein and lipid from the ER is now widely accepted. According to this view, lumenal and membrane proteins are free to flow out of the ER into transport structures unless retarded in some way (Pfeffer and Rothman, 1987). The lumenal space of the ER, according to this model, functions like a two-phase system: a mobile or aqueous phase and a relatively immobile gel or matrix. The permanent resident proteins of the ER, including chaperone proteins and folding enzymes, would make up the gel/matrix. The degree of adsorption of newly synthesized proteins to this matrix by electrostatic and/or hydrophobic interactions would determine the rate at which these proteins enter the fluid phase which leads out of the ER, analogous to the process of adsorption chromatography. Thus, conditions promoting interaction with the gel/matrix, including aggregation, binding to BiP, or exposure of free sulfhydryl groups, result in the protein being excluded from the mobile phase and its retention within the ER lumen (Helenius *et al.*, 1993). Membrane proteins, in addition to lumenal ER proteins, could be retained by virtue of their interaction with the ER matrix, although additional retention mechanisms for membrane proteins can be envisioned (Poruchynsky and Atkinson, 1988; Klausner, 1989; Jackson *et al.*, 1990).

How are lumenal ER matrix components themselves retained in the ER? The most likely explanation is that they are unable to enter the mobile phase leading out of the ER because of extensive low-affinity interactions among themselves (Rothman and Orci, 1992). Interestingly, many ER resident proteins that are abundant components of the ER lumen and envisioned as components of the ER matrix contain the sequence KDEL at their C-terminus. Deletion of this KDEL sequence results in enhanced secretion of these proteins, while addition of this sequence to lysozyme results in their retention in the ER (Munro and Pelham, 1987). This suggested that an additional retention mechanism involving the KDEL sequence might be operating to retain this class of proteins within the ER (Pelham, 1990).

Since the number of proteins with KDEL sequences retained in the ER is far too numerous for there to be a specific receptor for this sequence in the ER,

Pelham (1990) suggested that KDEL receptors might be acting downstream, functioning to retrieve KDEL-containing proteins from a post-ER site. Consistent with this, biochemical evidence of Pelham (1988) showed that a fraction of KDEL-lysozyme cycles from the Golgi back into the ER, and putative KDEL receptors have been identified and localized to pre-Golgi and Golgi compartments (Lewis *et al.,* 1990; Vaux *et al.,* 1990).

Retrieval of KDEL-containing proteins from post-ER sites, however, may only play a backup role in the mechanism by which the cell retains KDEL proteins in the ER (Rothman and Orci, 1992). KDEL-containing proteins have never been morphologically detected in pre-Golgi and Golgi compartments (Bole *et al.,* 1989), even when recycling was slowed by either low temperature or nocodazole treatment, which results in a significant accumulation of other cycling proteins in the Golgi apparatus (Hsu *et al.,* 1991). Furthermore, removing KDEL sequences from ER-retained proteins only results in their slow secretion (Zagouras and Rose, 1989). Thus, it is more likely that multiple low-affinity interactions among KDEL-containing proteins within the ER rather than a post-ER retrieval process is the primary mechanism of retention of these molecules in the ER.

3.2. Genetic and *in Vitro* Reconstitution Studies of ER-to-Golgi Transport

There is a growing consensus among cell biologists that transport from the ER to the Golgi apparatus is a highly regulated process involving interactions of multiple gene products. Diverse approaches to studying the cellular machinery responsible for conveying proteins from the ER to the Golgi apparatus have ruled out the possibility that secretory proteins directly traverse from the ER to the Golgi apparatus, and support the view that this process occurs through topologically distinct transport intermediates. *In vitro* reconstitution studies following protein transport between the ER and Golgi complex in both mammalian and yeast cells have revealed that this process shares many components with intra-Golgi transport (Beckers *et al.,* 1987; Baker *et al.,* 1988; Ruohola *et al.,* 1988; Groesch *et al.,* 1990; Plutner *et al.,* 1991).

A genetic approach for dissecting the underlying mechanisms involved in ER-to-Golgi transport has produced many temperature-sensitive mutants in yeast *Saccharomyces cerevisiae* defective in this process. The secretory mutants sec 7, 12, 13, 16, 17, 18, 19, 20, 21, 22, and 23, bos 1, bet 1, sar 1, and ypt 1 mutants are all implicated in transport processes between the ER and the Golgi apparatus (Schekman, 1992; Kaiser and Schekman, 1990; Nakano and Muramatsu, 1989; Segev *et al.,* 1988; Newman and Ferro-Novick, 1987; Shim *et al.,* 1991). In most of these mutants, upon incubation at the nonpermissive temperature newly synthesized proteins accumulate in the ER and an extensive network of ER membrane is formed (Hicke and Schekman, 1990). A subclass of these mutants, including sec 17, 18, 20, and 22, also accumulate numerous 50-nm vesicles at the restrictive temperature. These vesicles are believed to be ER-to-Golgi intermediates because their generation requires the action of sec 12, 13, 16, and 23 (Kaiser and Schekman, 1990).

Functionally active transport vesicle intermediates can be produced from the ER in crude yeast lysates that fuse with Golgi *in vitro* (Baker *et al.*, 1988; Rexach and Schekman, 1991; Groesch *et al.*, 1990). Formation of the transport intermediate requires ATP and GTP hydrolysis, cytosol, the 21-kDa GTP binding protein, sar 1p, and involves the action of sec 12p, 13p, 16p, and 23p (d'Enfert *et al.*, 1991; Rexach and Schekman, 1991). Targeted fusion of the intermediate requires ATP, cytosol, Ca^{2+}, the GTP binding protein rab 1/ypt as well as sec 17p, 18p, 22p (Segev, 1988; Rexach and Schekman, 1991). The *in vivo* accumulation of transport vesicles in sec 17, 18, and 22 cells at the nonpermissive temperature is consistent with these biochemical observations.

Recent studies have demonstrated the conservation of the biochemical machinery involved in protein transport between the ER and Golgi apparatus. sec 18 and sec 17 gene products have been shown to be homologous to mammalian factors NSF (NEM-sensitive fusion protein) and SNAP (soluble NSF attachment protein) identified biochemically in a mammalian *in vitro* transport assay (Clary *et al.*, 1990; Wilson *et al.*, 1989; Kaiser and Schekman, 1990). Whereas NSF (homologous to sec 18p) is proposed to be involved in transport vesicle fusion, SNAP (homologous to sec 17p) is thought to be involved in NSF binding to membranes (Malhotra *et al.*, 1989; Wilson *et al.*, 1989; Clary *et al.*, 1990).

Cell fractionation and DNA sequence analysis have begun to define the biochemical characteristics of the yeast gene products necessary for ER-to-Golgi transport. Two sec mutant gene products shown to be integral membrane proteins are sec 12p and sec 20p. sec 12p has been proposed to have a role in promoting vesicle assembly/budding through the interaction of its cytosolic domain with sar 1p (d'Enfert *et al.*, 1991). sec 20p (50 kDa) is the only transmembrane protein known to contain the sequence HDEL at its C-terminus (Sweet and Pelham, 1992). In addition, sec 20p shows genetic interaction with an allele of the sac 1 gene, which in turn shows interactions with actin mutants. This, together with the observation that depletion of sec 20p from cells results in an elaboration of ER and clusters of small vesicles, has led to the suggestion that sec 20p might provide the connection between transport vesicles and some component of the cytoskeleton (Sweet and Pelham, 1992).

At least three sec gene products implicated in ER-to-Golgi traffic (sec 13p, sec 23p, and sec 7p) are present in the cytosol in high-molecular-weight complexes that can loosely associate with the cytoplasmic surfaces of membranes. sec 23p (85 kDa) is bound to a 105-kDa protein (Hicke *et al.*, 1992), and sec 13p (34 kDa) is part of a 400-kDa complex (Pryer *et al.*, 1990). Ultrastructural localization of the sec 23p mammalian homologue in exocrine and endocrine pancreatic cells showed a specific distribution to the cytoplasmic face of the ER near the Golgi apparatus (Orci *et al.*, 1991a). sec 7p has been observed by immunoelectron microscopy to be associated with the cytoplasmic surface of ER-to-Golgi transport vesicles (Franzusoff *et al.*, 1992) and is a large (230 kDa) cytosolic phosphoprotein recovered in both soluble and sedimentable forms. It is not yet known if any of these sec gene products are similar to components of the above-mentioned coatomer complexes observed in mammalian cells.

3.3. Role of the Pre-Golgi Intermediate Compartment in ER-to-Golgi Transport

The properties and characteristics of the transport intermediates that deliver membrane and protein from the ER to the Golgi apparatus have recently been examined. In mammalian cells, these intermediates take the form of small (90 nm) vesicles in addition to larger tubulovesicular structures (up to 200–500 nm in diameter) (Saraste and Svensson, 1991), and are largely devoid of resident components of the ER and Golgi apparatus (Hauri and Schweizer, 1992). At 16°C, however, newly synthesized viral glycoproteins accumulate in these structures indicating that they represent an intermediate compartment (IC) through which proteins moving from the ER to the Golgi apparatus must pass (Saraste and Kuismanen, 1984).

A number of cellular processes have been postulated to occur in the IC including: fatty acylation of proteins (Rizzolo and Kornfeld, 1988; Bonatti *et al.*, 1989), the first step of *O*-glycosylation (Tooze *et al.*, 1988), and the first enzyme step in the generation of the mannose-6-phosphate signal for lysosomal protein targeting (Kornfeld and Mellman, 1989). In addition, budding of coronavirus and murine leukemia virus is believed to occur at this site (Tooze *et al.*, 1988; Ulmer and Palade, 1991).

Lodish *et al.* (1987) and Paulik *et al.* (1988) were among the first to isolate low-density membrane fractions with characteristics expected for the IC. Further characterization and purification of the IC has been made possible with the availability of antibodies that preferentially recognize the IC in mammalian cells. Antibodies to two integral membrane proteins of 53 kDa (p53) and 58 kDa (p58) (Schweizer *et al.*, 1988; Saraste *et al.*, 1987) label a tubulovesicular membrane system extending from the *cis* side of the Golgi apparatus. At 16°C, structures containing p53 and p58 colocalize with sites of accumulation of newly synthesized membrane viral proteins (Schweizer *et al.*, 1990; Saraste and Svensson, 1991). These structures are scattered throughout the cytoplasm (Schweizer *et al.*, 1990) and have βCOP bound to their cytoplasmic surfaces (Duden *et al.*, 1991).

The biochemical characteristics of the IC have been pursued by Schweizer *et al.* (1991) who have isolated subcellular fractions enriched 40-fold for p53.

The purified IC fractions displayed a unique polypeptide pattern distinct from rough ER and *cis*-Golgi. Thus, they did not contain the ER proteins ribophorin I/II, BiP, and protein disulfide isomerase nor the *cis*-Golgi enzymes involved in generating the lysosomal targeting signal mannose-6-phosphate.

The observation that the structures comprising the IC at 16°C are scattered throughout the cytoplasm (Schweizer *et al.*, 1990; Saraste and Svensson, 1991) strongly suggests that sites of exit from the ER are not restricted to regions positioned near the Golgi apparatus as originally envisioned. The question of how membrane and protein within these peripheral structures are delivered to the centrally located Golgi complex is therefore significant. This could be accomplished by two different mechanisms. One possibility is that vesicles shuttle back and forth between the ER, the IC, and the Golgi complex with the IC acting as a stable intermediate. The alternative possibility is that the tubulovesicular structures and surrounding vesicles comprising the IC translocate as a unit through the cytoplasm and fuse with the Golgi complex. Under this mechanism, the IC would be only a transient structure continuously fusing with the Golgi complex and being re-formed from the ER.

Morphologic evidence favoring the latter model has been provided by Saraste and Svensson (1991) (but see also Bonatti and Torrisi, this volume). These authors demonstrated that p58 accumulates in large peripheral IC structures at 16°C. Rather than remaining peripherally distributed, however, these structures redistribute into the Golgi region when cells are briefly warmed to 37°C. This suggests that peripheral p58-containing structures are only transient structures that fuse with the Golgi complex upon transport into the region of the cell near the MTOC.

What directs the movement of IC structures from peripheral sites in the cytoplasm toward the cell center where the Golgi apparatus resides? Saraste and Svensson (1991) have proposed that microtubules might facilitate this movement. These authors showed that in addition to being inhibited at temperatures below 16°C, movement of p58-containing structures into the central Golgi region at 37°C requires intact microtubules (Saraste and Svensson, 1991). This conclusion is supported by studies of Lippincott-Schwartz *et al.* (1990) who followed the relocation of Golgi membrane proteins out of the ER in cells washed free of the drug brefeldin A (BFA), which redistributes Golgi membrane proteins into the ER (Lippincott-Schwartz *et al.*, 1989; Doms *et al.*, 1989). Within 5 min of washout of this drug, Golgi proteins appeared in punctate structures widely distributed within the cytoplasm. These punctate structures appeared to aggregate into larger structures by 10–15 min after washout but were still peripherally distributed in the cell. At 30 min after washout the structures became larger and remarkably had a distribution very similar to that of Golgi fragments in nocodazole-treated cells. After longer periods the Golgi aggregates moved into a perinuclear location resembling the distribution of the Golgi complex in control cells. This step could be inhibited by nocodazole (Lippincott-Schwartz, unpublished observations). These studies suggest, therefore, that sites of protein exit from the ER are widely distributed within this membrane network and that the subsequent long intracellular distances traveled by intermediate structures en route to the Golgi apparatus may involve an interaction with microtubules.

4. SORTING AND RECYCLING AT THE CGN

At some point in its movement toward the MTOC or cell center, the membranes comprising the IC fuse with the Golgi complex, delivering their content and membrane to this organelle. This is believed to occur at the CGN, which by electron microscopy appears as an array of vesicles and tubules that connect to the first one or two cisternae at the *cis* face of the Golgi stack (Rambourg and Clermont, 1990).

4.1. Distinction between the IC and CGN

Because the IC migrates toward and eventually fuses with the CGN, the two compartments have overlapping membrane constituents. p53, p58, and rab 2, for instance, are concentrated in both subcompartments (Schweizer *et al.*, 1990; Saraste and Svensson, 1991; Chavrier *et al.*, 1990). This has led to some confusion regarding the distinction between these compartments. At least one protein, the E1 glycoprotein of avian coronavirus, however, appears to be specifically localized to the CGN (Machamer *et al.*, 1990).

There are several reasons for conceptualizing the IC and CGN as distinct compartments. The IC consists of structures into which protein leaving the ER first accumulates. In many cells these structures are peripherally located and translocate toward the cell center where the Golgi complex resides (Saraste and Svensson, 1991; Lippincott-Schwartz *et al.*, 1990). They are therefore only transiently existing structures. The CGN, by contrast, is stably localized at the cell center where it interacts with the rest of the Golgi complex (Mellman and Simons, 1992). The CGN is involved not only in receiving but also in sorting components arriving from the ER. Sorting at the CGN involves the decision either to transfer molecules deeper into the Golgi stacks or to recycle them back into the ER.

4.2. Evidence for Golgi-to-ER Recycling

It has long been hypothesized that some mechanism must exist for returning lipid to the ER in the face of the continuous flow of membrane out of this organelle. Indeed, attempts at estimating the rate of lipid loss from the ER caused by bulk flow into the secretory pathway suggested that the rate of loss of lipid from the ER vastly exceeds the rate of lipid biosynthesis (Wieland et al., 1987). How might lipid be retrieved to the ER? Two possibilities have been most widely considered: cytosolic transfer of nonvesicular lipid back into the ER, and membrane recycling from the Golgi apparatus into the ER. While there is at present no evidence for lipid retrieval into the ER by monomeric diffusion, compelling evidence for a membrane transport pathway from the Golgi complex to the ER has come from studies of cells treated with the fungal metabolite brefeldin A, BFA.

Within minutes of adding BFA to cells, all membrane and content of the CGN and Golgi stacks are transported into the ER and export out of the ER is inhibited (Lippincott-Schwartz et al., 1989; Fujiwara et al., 1989; Doms et al., 1989; Young et al., 1990; Strous et al., 1991). That the pathway followed by Golgi components in the presence of BFA operates at some level in untreated cells for the selective recycling of molecules is supported by several studies. The normal steady-state distributions of p53 and p58 include the ER, IC, and CGN. To investigate whether these molecules constitutively cycle between these compartments, Lippincott-Schwartz et al. (1990) cooled cells to 16°C whereupon p53 became concentrated in the IC. A synchronized population of p53 molecules could then be followed upon warming. Brief warming to 37°C resulted in the movement of p53 first into the CGN (1 to 2 min) and then into the ER (10 min) where it colocalized with ER resident markers. Continued incubation at 37°C (30 min) returned p53 to its steady-state distributions in the ER, IC, and CGN. These temperature manipulations had no effect on the distributions of resident ER or Golgi proteins.

Similar results have been obtained from morphologic experiments following the intracellular dynamics of p58 (Saraste and Svensson, 1991), and MHC class I molecules in a mutant cell line (Hsu et al., 1991), suggesting that these molecules also constitutively cycle between the ER and Golgi apparatus. Ultrastructural studies showed that MHC class I in the mutant cell line moved only as far as the CGN and no farther into the Golgi stacks before recycling to the ER (Hsu et al., 1991). Consistent with this, biochemical experiments in the same study showed no processing to endo H resistance of the N-linked oligosaccharides of MHC class I.

Evidence for the recycling of lipids from the Golgi complex to the ER has recently been provided by Hoffman and Pagano (1993). These authors examined the fate of nonexchangeable fluorescent lipid analogues after delivery to the Golgi apparatus in semi-intact cells. The lipids were shown to redistribute from the Golgi to the ER within 30 min, with no obvious changes in the overall morphology of the Golgi or ER. Lipid redistribution into the ER was temperature and energy dependent, NEM sensitive, and required cytosol. Since these results were obtained in the absence of added drugs (i.e., BFA) and without temperature perturbations, they add further support to the idea that retrograde flow of membrane from Golgi to ER is a normal process. Future studies need to address the question of the extent of membrane cycling along this pathway.

4.3. Possible Roles of Membrane Cycling between the ER and Golgi Complex

A variety of functions could be served by a constitutive Golgi-to-ER recycling pathway. In addition to maintaining the membrane surface area of the ER in the face of continuous bulk flow of membrane out, recycling has been proposed to act as a "second line of defense" in preventing the loss of KDEL-containing ER resident proteins (Rothman and Orci, 1992; Pelham, 1991). According to this model, interaction of KDEL-containing proteins that have escaped the ER retention system with KDEL receptors in the Golgi complex would direct the escaped proteins into the retrograde retrieval pathway leading them back into the ER (Pelham, 1991; Rothman and Orci, 1992).

A different function of the Golgi-to-ER retrograde pathway could be to provide the cell with an additional quality control system to that of the ER for preventing molecules that have failed to assemble properly from being transported to the cell surface. An example of this is the MHC class I complex in mutant cells (Hsu *et al.*, 1991), where failure of the heavy and light chains to assemble properly leads to their failure to be transported to the cell surface. Rather than being retained in the ER, MHC class I molecules remain trapped within the ER/Golgi system futilely cycling between the ER and Golgi.

In addition to these functions, the Golgi-to-ER retrograde pathway may be utilized for remodeling of the Golgi apparatus during specific periods of a cell's lifetime, including the dispersal of the Golgi during mitosis (Lucocq *et al.*, 1989), alterations in Golgi location during myogenesis (Tassin *et al.*, 1985), and during microtubule disruption. Finally, use of the retrograde pathway might be one way the cell regulates net membrane inflow into the Golgi stacks.

4.4. Characteristics of the Golgi-to-ER Retrograde Transport Pathway

Insight into the characteristics of the retrograde recycling pathway into the ER has come from detailed studies using the drug BFA. Within minutes of adding BFA, the Golgi apparatus undergoes major structural alterations with the result that the CGN and Golgi stacks fuse into larger structures and give rise to an extensive array of long tubular processes that extend along microtubules to the cell periphery where they eventually fuse and mix with the ER (Lippincott-Schwartz *et al.*, 1990). Concomitant with these morphologic effects, anterograde transport including protein secretion is blocked (Misumi *et al.*, 1986). Remarkably, upon washing out BFA, Golgi proteins along with secretory proteins move out of the ER into IC-like structures, subsequently re-forming the Golgi apparatus and refilling the rest of the biosynthetic pathway (Lippincott-Schwartz *et al.*, 1989).

The membrane tubules that mediate "retrograde" transport of Golgi components into the ER in BFA-treated cells have distinct properties (Lippincott-Schwartz *et al.*, 1990; Donaldson *et al.*, 1991a). Fusing uniquely with the ER, these Golgi-derived tubules are about 90 nm in diameter and can extend up to 10–20 μm in length. Unlike membrane structures mediating anterograde transport, which are coated with COPs, retrograde tubules are devoid of coatomer or "uncoated." The tubules appear to utilize microtubules (moving toward the plus poles) and are not observed when cells are treated with microtubule-disrupting agents like nocodazole. In addition, the tubules are not formed at temperatures below 16°C, during ATP depletion, or in the presence of aluminum fluoride or GTPγS when added before BFA.

Do membrane tubules arising from the CGN mediate retrograde transport from Golgi to ER in cells under normal conditions? Transient tubules emanating from the Golgi are observed in living cells at 37°C (Cooper *et al.*, 1990). Evidence that such tubules might mediate retrograde transport comes from studies of Lippincott-Schwartz *et al.* (1990). These authors showed using immunofluorescence that at 16°C and after brief warm-up periods, the recycling molecule p53 appears in tubular processes emanating out of the CGN. Golgi resident proteins (including galactosyl transferase) were normally excluded from the p53-containing tubular processes, but upon BFA treatment at 16°C became colocalized with p53 in the tubular processes. This suggested that the tubular processes containing p53 were normally utilized for retrograde traffic since they were observed in the presence or absence of BFA.

The idea that "noncoated" membrane tubules mediating retrograde transport provide an alternative transport mechanism to "coated" structures mediating anterograde transport has emerged from studies revealing the biochemical basis of BFA's effects. Donaldson *et al.* (1990) observed that within 30 s of adding BFA to cells, the coatomer protein βCOP comes off Golgi and Golgi-associated membranes. This proximal effect results from the inability of coatomer to bind to membrane in the presence of BFA (Donaldson *et al.*, 1991a; Orci *et al.*, 1991b). As mentioned above, association of coatomer with membranes is a controlled process utilizing GTP and is believed to be essential for both ER-to-Golgi and intra-Golgi anterograde traffic (Klausner *et al.*, 1992; Rothman and Orci, 1992). That BFA prevents membrane binding of coatomer provides a simple explanation for the ability of BFA to block anterograde movement into the Golgi. That retrograde tubule trafficking continues in the presence of BFA is consistent with this pathway not requiring coatomer binding.

The relationship between "coat"-mediated anterograde traffic and tubule-mediated retrograde traffic was examined using BFA in a cell-free intra-Golgi transport system (Orci *et al.*, 1991b). When isolated Golgi membranes were treated with BFA, coatomer (detected by βCOP binding) was no longer associated with these membranes. Nevertheless, transport of protein from donor to acceptor Golgi fractions occurred with normal kinetics. This indicated that transport is not absolutely dependent on coatomer binding. To obtain an explanation for the *in vitro* "transport" under BFA conditions, ultrastructural examination of BFA-treated Golgi membrane preparations was performed. An extensive network of "uncoated" tubules that connected previously separate cisternae and stacks was observed. These tubule networks were proposed to represent the *in vitro* equivalents of membrane tubules mediating retrograde transport *in vivo*, allowing the mixing of contents of Golgi cisternae with each other. The parallels between tubule network formation of purified Golgi stacks and of tubules mediating retrograde transport *in vivo* were consistent with this interpretation: both required ATP to form and could be prevented from forming with pretreatment with GTPγS; and both mediated transport or mixing of molecules (Orci *et al.*, 1991b; Lippincott-Schwartz *et al.*, 1990; Donaldson *et al.*, 1991a).

These results give strong support to the idea that there are two alternative pathways of membrane traffic operating within the ER/Golgi system: one mediated via "coated" structures and the other mediated via "uncoated" tubules (Orci *et al.*, 1991b) (see Figure 2). BFA's effects on the ER Golgi system can then be rationalized in terms of its disruption of a normal balance between these two pathways that is regulated at the level of coatomer binding (Klausner *et al.*,

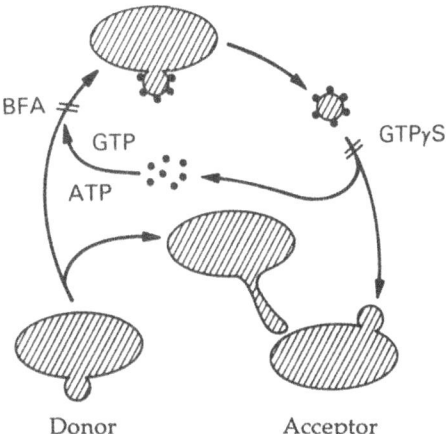

FIGURE 2. Two alternative membrane transport pathways for membrane trafficking. Two pathways, one mediated by discontinuous "coated" structures and the other by continuous tubules, are proposed to operate within the ER/Golgi system. Both pathways require the functioning of a constitutive budding process. Binding of coatomer and other cytosolic factors (i.e., ARF) to the bud would result in transport by the coat-mediated pathway. In the absence of coatomer binding, budding would give rise to tubule structures. Both pathways would require additional components for the recognition and targeting events necessary for the final fusion of donor and acceptor membranes. Regulation of transport along the "coated" vesicle versus noncoated tubular pathway would depend on the status of coatomer binding, with conditions favoring coatomer binding (i.e., GTP) resulting in enhanced "coat-mediated" transport and conditions favoring coatomer dissociation from membranes (i.e., BFA) resulting in enhanced tubular transport. GTPγS, which prevents coatomer dissociation from membranes, would inhibit both pathways, since coatomer-free buds are required for tubular traffic, and coatomer dissociation is required for vesicle fusion along the coat-mediated pathway.

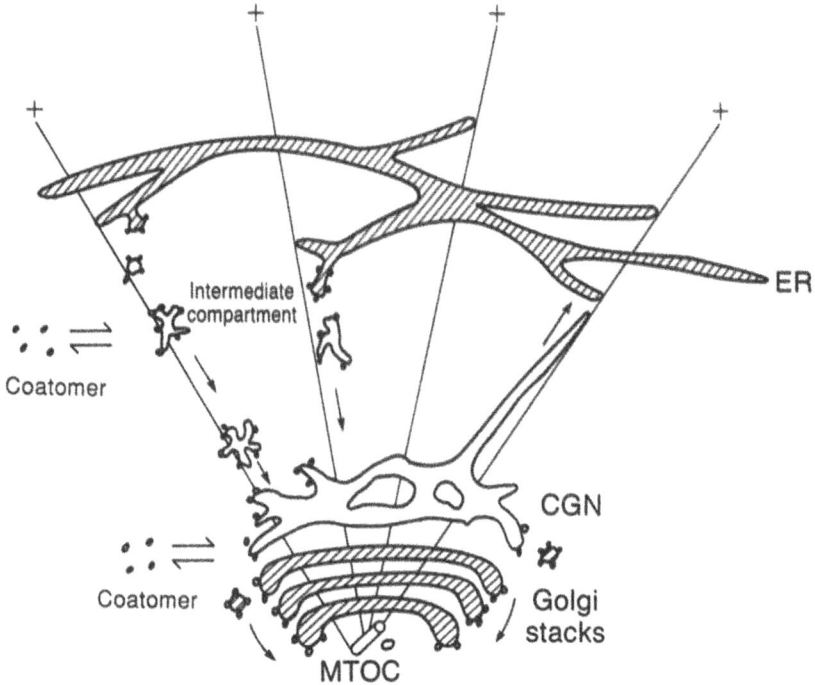

FIGURE 3. Hypothetical model for ER-to-Golgi membrane trafficking. Newly synthesized proteins leave the ER in membrane structures, which can arise from multiple sites throughout the cytoplasm. These peripheral transport intermediates (intermediate compartment) move toward the MTOC in a microtubule-facilitated manner. In addition to being relatively large (200–500 nm wide), the intermediates have coatomer bound to their cytoplasmic surfaces. As the intermediates reach the Golgi region they fuse with the CGN. Dissociation of coatomer would allow membrane buds to form tubular structures which utilize microtubules to move toward the cell periphery where they fuse with the ER. Sorting mechanisms for localizing membrane and protein into tubules versus coated structures would operate throughout this cycling system.

1992) (see Figure 3). Anterograde movement into the Golgi complex, which requires coatomer binding onto membrane, would be inhibited by BFA. Release of coatomer by BFA would be associated with Golgi tubulation and retrograde transport into the ER. The additional profound effect of BFA on Golgi structure suggests that maintenance of this organelle depends on a controlled balance of membrane input and output through these pathways.

What is the advantage to the cell of having two pathways—one leading into the Golgi and the other leading back to the ER—that exhibit such distinct structural characteristics? Such a system may provide the physical basis for sorting at the site(s) of recycling by the segregation of structures that maximize volume-to-surface area ratio (vacuoles) from structures that minimize volume-to-surface area ratio (tubules). This would be analogous to sorting of fluid-phase molecules from recycling membrane receptors in the endocytic system (Geuze *et al.*, 1983). At the level of the ER/Golgi system, anterograde-moving vesicles from the ER (which maximize volume content) would carry bulk flow lumenal constituents forward through the biosynthetic pathway while tubules (which minimize volume) would recycle membrane components back to the ER. This process coupled with the selective movement of membrane components into different transport structures would allow for efficient sorting of bulk flow from recycling components early in the biosynthetic pathway.

5. MECHANISMS FOR REGULATING MEMBRANE CYCLING BETWEEN THE ER AND GOLGI

5.1. Role of Coatomer Binding in ER-to-Golgi Trafficking

Assuming that ER-to-Golgi traffic involves two distinct transport mechanisms (i.e., coatomer-mediated anterograde transport and tubule-mediated retrograde transport), what normally regulates the balance between the two? An important clue to this question has come from the observation that under conditions that favor membrane accumulation of coatomer (incubation with GTPγS), retrograde tubules do not form, whereas when association of coatomer is prevented (by BFA), tubules predominate (Donaldson *et al.*, 1991a; Orci *et al.*, 1991b). This reciprocal relationship between BFA-sensitive coatomer binding and tubule production suggests that both pathways share a common regulatory system tied to the membrane assembly/disassembly of coatomer.

How might the status of coatomer binding result in different transport processes? Since budding is required for the formation of both tubules and "coated" vesicular structures, coatomer binding may not be required for budding *per se* as previously proposed (Rothman and Orci, 1990), but be superimposed on a budding process to serve other roles (Klausner *et al.*, 1992). For instance, binding of coatomer to membrane could inhibit or slow bud growth into a tubule by masking sites on microtubule motor proteins required for the extension of the tubule along microtubules. Extending this thinking to the level of the ER, if one considers the ER as a dynamic tubulating system utilizing microtubule motor proteins to move to the cell periphery, then blocking motor activity at specific sites (perhaps by coatomer binding) would inhibit tubule growth. Fission of the coated tubules would then result in detached structures free to be transported toward the cell center where the Golgi complex resides. Another role of coatomer binding might be to provide specificity to the membrane content in the regions of binding (Klausner *et al.*, 1992).

5.2.   Role of GTP-Binding Proteins in Regulating Coatomer Binding

Insight into the biochemical machinery that regulates coatomer binding is only just beginning. The observation that aluminum fluoride and GTPγS both promote association of βCOP with Golgi membranes implicates one or more GTP-binding proteins in initiating the association of coatomer with membranes (Donaldson *et al.*, 1991b). Both the low-molecular-weight GTP-binding protein, ARF, and a heterotrimeric G protein have been proposed to be involved in this process (Donaldson *et al.*, 1991b).

ARF is an abundant cytosolic protein that reversibly associates with Golgi membranes (Donaldson *et al.*, 1991b). Like βCOP, association of ARF with Golgi membranes can be enhanced in the presence of GTPγS and is inhibited by BFA. Current evidence suggests ARF plays an essential role in regulating βCOP (i.e., coatomer) binding (Donaldson *et al.*, 1992). Not only does Golgi membrane binding of cytosolic βCOP require the presence of ARF, but interaction of recombinant, myristoylated ARF, GTPγS, and Golgi membrane is sufficient to make the Golgi membrane fully competent for βCOP binding at a subsequent step (in the absence of free ARF and GTPγS). This initial ARF–membrane interaction step has been proposed to be the site of BFA action (Donaldson *et al.*, 1992).

Like other GTP-binding proteins, ARF can be envisioned to cycle between GDP- and GTP-bound forms. During activation, cytosolic ARF–GDP would interact with a membrane binding site, whereupon GDP/GTP exchange would convert it into the active GTP-bound form. This would initiate the process of coatomer binding. Hydrolysis of GTP would reverse activation with release of coatomer from the membrane. Hydrolysis-resistant GTP analogues like GTPγS would persistently activate ARF, resulting in persistent coatomer binding and a resistance to BFA action.

A role of heterotrimeric G protein(s) in coatomer binding to Golgi membrane was first suggested by Donaldson *et al.* (1991b) based on the effect of aluminum fluoride in promoting βCOP association with membrane. Aluminum fluoride is a potent activator of G proteins, but not of ARF and other low-molecular-weight GTP-binding proteins (Kahn, 1991). Support for this hypothesis was provided by experiments using βγ subunits of G proteins. Many effects of G proteins are mediated via free Gα subunits and addition of βγ subunits may inhibit these effects. When Donaldson *et al.* (1991b) added βγ subunits of a G protein to Golgi membranes and cytosol, GTPγS-dependent association of both βCOP and ARF to Golgi membranes was inhibited. Using a permeabilized cell system, Ktistakis *et al.* (1992) have provided further support for the role of a G protein in coatomer binding, showing that coatomer binding to membrane can be modulated by the G protein stimulator, mastoparin. The precise role that G proteins might play in the membrane association/dissociation cycle of coatomer is unknown, but Donaldson *et al.* (1992) suggested that they might enhance the formation or stability of active ARF–GTP complexes. One G protein, Giα3, has been localized to the Golgi apparatus and its overexpression results in the inhibition of proteoglycan secretion (Stow *et al.*, 1991).

Since the paradigm for G protein function on the plasma membrane is that they act as signal transducers, it is possible that the functioning of G proteins in regulating binding of coatomer to the Golgi complex is coupled to receptors that respond to signals within the cell. One molecule that might serve as such a receptor is ERD2 (the putative receptor for KDEL-containing proteins) (Semenza *et al.*, 1990; Hsu *et al.*, 1992). Lowering ERD2 levels in cells results in a block in normal transport through the secretory pathway and an accumulation of Golgi membrane (Semenza *et al.*, 1990). By contrast, overexpression of ERD2 or an ERD2 homologue in mammalian cells results in: the redistribution of Golgi membrane into the ER, loss of coatomer binding to membrane, and a block in secretion, all similar to the effects of BFA (Hsu *et al.*, 1992). Based on these observations, Hsu *et al.* (1992) have proposed that one function of KDEL receptors in the Golgi apparatus might be to regulate membrane traffic between the ER and Golgi apparatus. Interaction of KDEL receptors in the Golgi complex with KDEL proteins that have leaked out of the ER, according to their model, could provide a detection mechanism for the overall regulation of retrograde transport or recycling of proteins from the Golgi back into the ER. Overexpression of these receptors might lead to a constitutively "on" signal, resulting in an imbalance in membrane traffic with retrograde transport now favored, analogous to the effects of BFA.

6. SUMMARY

Membrane traffic between the ER and Golgi is now recognized as a carefully regulated process controlled by distinct anterograde (to the Golgi) and retrograde (to the ER) pathways. These pathways link two organelles with different morphologies, structures, and localizations within the cell. The ER, which is involved in multiple cellular functions including protein biosynthesis and folding, extends to the cell periphery and forms a dynamic tubule reticulum. By contrast, the Golgi apparatus, which functions in membrane sorting and recycling events, is localized at the center of the cell near the MTOC and is comprised of compact cisternal units. The required transport into the Golgi apparatus of newly synthesized proteins exported from the ER offers a twofold advantage to the cell. First, the rate of movement of membrane and protein through the biosynthetic pathway can be controlled by the selective use of a recycling pathway. Second, membrane moving through the biosynthetic pathway enters a structure specialized for sorting of membrane to different final destinations in the cell.

Control of biosynthetic transport within the ER/Golgi system involves the utilization of two alternative transport pathways: anterograde (ER to Golgi) and retrograde (Golgi to ER). These two pathways share a common regulatory system involving membrane assembly/disassembly of cytosolic coatomer proteins. Thus, conditions that favor irreversible coatomer binding (i.e., GTPγS) inhibit retrograde transport while producing anterograde transport intermediates. Conditions that prevent coatomer binding (i.e., BFA) inhibit anterograde transport and enhance retrograde transport.

The underlying biochemical machinery that normally balances anterograde and retrograde membrane traffic between the ER and Golgi is only just beginning to be understood. Any model to explain this system, however, must account for the morphologic characteristics of the membranes involved. Whereas anterograde traffic involves discontinuous "coated" structures moving from peripheral sites in the ER toward the central Golgi, retrograde traffic utilizes continuous "noncoated" tubule structures that move from a central site (i.e., the CGN) to the peripheral ER (see Figure 3). Such a system maximizes volume transport (utilizing vacuolar structures) in the anterograde direction and membrane transport (utilizing tubules) in the retrograde direction. It is therefore ideal for sorting of bulk flow lumenal contents from recycling membrane early in the biosynthetic pathway.

ACKNOWLEDGMENTS. I am most grateful for the comments and suggestions of Drs. Richard Klausner, Danny Cassel, Alex Franzusoff, Julie Donaldson, Peter Peters, Mark Terasaki, and Victor Hsu.

7. REFERENCES

Allan, V. J., and Kreis, T. E., 1986, A microtubule-binding protein associated with membranes of the Golgi apparatus, *J. Cell Biol.* **103**:2229–2239.

Baker, D., Hicke, L., Rexach, M., Schleyer, M., and Schekman, R., 1988, Reconstitution of SEC gene product-dependent intercompartmental protein transport, *Cell* **54**:335–344.

Beckers, C. J., Keller, D. S., and Balch, W. E., 1987, Semi-intact cells permeable to macromolecules: Use in reconstitution of protein transport from the endoplasmic reticulum to the Golgi complex, *Cell* **50**:523–534.

Bole, D. G., Dowin, R., Doriaux, M., and Jamieson, J. D., 1989, Immunocytochemical localization of BiP to the rough endoplasmic reticulum: Evidence for protein sorting by selective retention, *J. Histochem. Cytochem.* **37**:1817–1823.

Bonatti, S., Migliaccio, G., and Simons, K., 1989, Palmitylation of viral membrane glycoproteins takes place after exit from the ER, *J. Biol. Chem.* **264**:12590–12595.

Bonifacino, J. S., and Lippincott-Schwartz, J., 1991, Degradation of proteins within the endoplasmic reticulum, *Curr. Opin. Cell Biol.* **3**:592–600.

Booth, C., and Koch, L. E., 1990, Perturbation of cellular calcium induces secretion of luminal ER proteins, *Cell* **59**:729–737.

Chavrier, P., Parton, R. G., Hauri, H.-P., Simons, K., and Zerial, M., 1990, Localization of low molecular weight GTP binding proteins to exocytic and endocytic compartments. Cell 62:317–329.

Clary, D. O., Griff, I. C., and Rothman, J. E., 1990, SNAPs, a family of NSF attachment proteins involved in intracellular membrane fusion in animals and yeast. Cell 61:709–721.

Cooper, M. S., Cornell-Bell, A. H., Chernjavsky, A., Dani, J. W., and Smith, S. J., 1990, Tubulovesicular processes emerge from trans-Golgi cisternae, extend along microtubules, and interlink adjacent trans-Golgi elements into a reticulum. Cell 61:135–145.

Dabora, S. L., and Sheetz, M. P., 1988, The microtubule-dependent formation of a tubulovesicular network with characteristics of the ER from cultured cell extracts. Cell 54:27–35.

d'Enfert, C., Wuestehube, L. J., Lila, T., and Schekman, R., 1991, Sec12p-dependent membrane binding of the small GTP-binding protein sar1p promotes formation of transport vesicles from the ER. J. Cell Biol. 114:663–670.

Doms, R. W., Russ, G., and Yewdell, J. W., 1989, Brefeldin A redistributes resident and itinerant Golgi proteins to the endoplasmic reticulum. J. Cell Biol. 109:61–72.

Donaldson, J. G., Lippincott-Schwartz, J., Bloom, G. S., Kreis, T. E., and Klausner, R. D., 1990, Dissociation of a 110 kD peripheral membrane protein from the Golgi apparatus is an early event in brefeldin A action. J. Cell Biol. 111:2295–2306.

Donaldson, J. G., Lippincott-Schwartz, J., and Klausner, R. D., 1991a, Guanine nucleotides modulate the effects of brefeldin A in semipermeable cells: Regulation of the association of a 110 kD peripheral membrane protein with the Golgi apparatus. J. Cell Biol. 112:579–588.

Donaldson, J. G., Kahn, R. A., Lippincott-Schwartz, J., and Klausner, R. D., 1991b, Binding of ARF and βCOP to Golgi membranes: Possible regulation by a trimeric G protein. Science 254:1197–1199.

Donaldson, J. G., Cassel, D., Kahn, R. A., and Klausner, R. D., 1992, ADP-ribosylation factor, a small GTP-binding protein is required for binding of the coatomer protein βCop to Golgi membranes. Proc. Nat'l Acad. Sci. 89:6408–6412.

Duden, R., Griffiths, G., Frank, R., Argos, P., and Kreis, T. E., 1991, β-COP, a 110 kD protein associated with non-clathrin-coated vesicles and the Golgi complex, shows homology to β-adaptin. Cell 64:649–665.

Fitting, T., and Kabat, D., 1982, Evidence for a glycoprotein "signal" involved in transport between subcellular organelles. Two membrane glycoproteins encoded by murine leukemia virus reach the cell surface at different rates. J. Biol. Chem. 257:14011–14017.

Franzusoff, A., Lauze, E., and Howell, K. E., 1992, Immuno-isolation of sec 7p-coated transport vesicles from the yeast secretory pathway. Nature 355:173–175.

Fujiwara, T., Oda, K., and Ikehara, Y., 1989, Dynamic distribution of the Golgi marker thiamine pyrophosphatase is modulated by brefeldin A in rat hepatoma cells. Cell Struct. Funct. 14:605–616.

Geuze, H. J., Slot, J. W., Strous, G. J., Lodish, H. F., and Schwartz, A. L., 1983, Intracellular site of asialoglycoprotein receptor-ligand uncoupling: Double label immuno-electron microscopy during receptor-mediated endocytosis. Cell 32:277–287.

Griffiths, G., and Simons, K., 1986, The trans Golgi network: Sorting at the exit site of the Golgi complex. Science 234:438–443.

Groesch, M., Ruohola, H., Bacon, R., Rossi, G., and Ferro-Novick, S., 1990, Isolation of a functional vesicular intermediate that mediates ER to Golgi transport in yeast. J. Cell Biol. 111:45–53.

Hicke, L., Yoshihisa, T., and Schekman, R., 1992, Sec 23p and a novel 105-kDa protein function as a multimeric complex to promote vesicle budding and protein transport from the endoplasmic reticulum. Mol. Biol. Cell 3:667–676.

Helenius, A., Tatu, U., Marquardt, T., and Braakman, I., 1993, Protein folding in the endoplasmic reticulum. Serono Symposium (in press).

Hicke, L., and Schekman, R., 1990, Molecular machinery required for protein transport from the endoplasmic reticulum to the Golgi complex, *Bioessays* **12**:253–258.

Hauri, H-P., and Schweizer, A., 1992, The endoplasmic reticulum-Golgi intermediate compartment, *Curr. Opin. Cell Biol.* **4**:600–608.

Ho, W. C., Allan, V. J., van Meer, G., and Kreis, T. E., 1989, Redistribution of scattered Golgi elements occurs along microtubules, *Eur. J. Cell Biol.* **48**:250–263.

Hoffman, P. M., and Pagano, R. E., 1993, Retrograde movement of membrane lipids from the Golgi apparatus to the endoplasmic reticulum of perforated cells: Evidence for lipid recycling, *Eur. J. Cell Biol.* (in Press).

Hsu, V. W., Yuan, L. C., Nuchtern, J. G., Lippincott-Schwartz, J., Hammerling, G. J., and Klausner, R. D., 1991, A recycling pathway between the endoplasmic reticulum and the Golgi apparatus for retention of unassembled MHC class 1 molecules, *Nature* **352**:441–444.

Hsu, V. W., Shah, N., and Klausner, R. D., 1992, A brefeldin A-like phenotype is induced by the overexpression of a human ERD-2-like protein, ELP-1, *Cell* **69**:625–635.

Jackson, M. R., Nilsson, T., and Peterson, P. A., 1990, Identification of a consensus motif for retention of transmembrane proteins in the endoplasmic reticulum, *EMBO J.* **9**:3153–3162.

Kachar, J., and Reese, T., 1988, The mechanism of cytoplasmic streaming in characean algal cells: Sliding of endoplasmic reticulum along actin filaments, *J. Cell Biol.* **106**:1545–1552.

Kahn, R. A., 1991, Fluoride is not an activator of the smaller (20–25KDa) GTP-binding proteins, *J. Biol. Chem.* **266**:15595–15597.

Kaiser, C. A., and Schekman, R., 1990, Distinct sets of SEC genes govern transport vesicle formation and fusion in the secretory pathway, *Cell* **61**:723–733.

Klausner, R. D., 1989, Architectural editing: Determining the fate of newly synthesized membrane proteins, *New Biol.* **1**:3–8.

Klausner, R. D., Donaldson, J. G., and Lippincott-Schwartz, J., 1992, Brefeldin A: Insights into the control of membrane traffic and organelle structure, *J. Cell Biol.* **116**:1071–1080.

Kornfeld, S., and Mellman, I., 1989, The biogenesis of lysosomes, *Annu. Rev. Cell Biol.* **5**:483–525.

Kupfer, A. D., and Singer, S. J., 1982, Polarization of the Golgi apparatus and the microtubule organizing center in cultured fivroblasts at the edge of an experimental wound. *Proc. Nat'l Acad. Sci. USA* **79**:2603–2607.

Ktistakis, N. T., Linder, M. E., and Roth, M. G., 1992, Action of brefeldin A blocked by activation of pertussis-toxin-sensitive G protein, *Nature* **356**:344–346.

Lewis, M. J., Sweet, D. J., and Pelham, H.R.B., 1990, The ERD2 gene determines the specificity of the luminal ER protein retention system, *Cell* **61**:1359–1363.

Lippincott-Schwartz, J., Yuan, L. C., Bonifacino, J. S., and Klausner, R. D., 1989, Rapid redistribution of Golgi proteins into the ER in cells treated with brefeldin A: evidence for membrane cycling from Golgi to ER, *Cell* **56**:801–813.

Lippincott-Schwartz, J., Donaldson, J. G., Schweizer, A., Berger, E. G., Hauri, H.-P., Yuan, L. C., and Klausner, R. D., 1990, Microtubule-dependent retrograde transport of proteins into the ER in the presence of brefeldin A suggests an ER recycling pathway, *Cell* **60**:821–836.

Lippincott-Schwartz, J., Yuan, L., Tipper, C., Amherdt, M., Orci, L., and Klausner, R. D., 1991, Brefeldin A's effects on endosomes, lysosomes and the TGN suggest a general mechanism for regulating organelle structure and membrane traffic, *Cell* **67**:601–616.

Lodish, H. F., Kong, N., Snider, M., and Strous, G. J., 1983, Hepatoma secretory proteins migrate from the rough endoplasmic reticulum to Golgi at characteristic rates, *Nature* **304**:80–93.

Lodish, H. F., Kong, N., Hirani, S., and Rasmussen, J., 1987, A vesicular intermediate in the transport of hepatoma secretory proteins from the rough ER to the Golgi complex, *J. Cell Biol.* **104**:221–230.

Lucocq, J. M., Berger, E. G., and Warren, G., 1989. Mitotic Golgi fragments in HeLa cells and their role in the reassembly pathway, *J. Cell Biol.* **109**:463–474.

Machamer, C. E., Mentone, S. A., Rose, J. K., and Farquhar, M. G., 1990, The E1 glycoprotein of an avian coronavirus is targeted to the cis Golgi complex, *Proc. Natl. Acad. Sci. USA* **87**:6944–6948.

Malhotra, V., Serafini, T., Orci, L., Shepherd, J. C., and Rothman, J. E., 1989, Purification of a novel class of coated vesicles mediating biosynthetic protein transport through the Golgi stack, *Cell* **58**:329–336.

Mellman, I., and Simons, K., 1992, The Golgi complex: In vitro veritas? *Cell* **68**:829–840.

Misumi, Y., Miki, K., Takatsuki, A., Tamura, G., and Ikehara, Y., 1986, Novel blockade by brefeldin A of intracellular transport of secretory proteins in cultured rat hepatocytes, *J. Biol. Chem.* **261**:11398–11403.

Munro, S., and Pelham, H.R.B., 1987, A C-terminal signal prevents secretion of luminal ER proteins, *Cell* **48**:899–907.

Nakano, A., and Muramatsu, M., 1989, A novel GTP-binding protein Sar1p is involved in transport from the ER to the Golgi apparatus, *J. Cell Biol.* **109**:2677–2691.

Newman, A. P., and Ferro-Novick, S., 1987, Characterization of new mutants in the early part of the yeast secretory pathway isolated by [³H]mannose suicide selection, *J. Cell Biol.* **105**:1587–1594.

Orci, L., Ravazzola, M., Meda, P., Holcomb, C., Moore, H.-P., Hicke, L., and Schekman, R., 1991a, Mammalian sec 23p homologue is restricted to the endoplasmic reticulum transitional cytoplasm, *Proc. Natl. Acad. Sci. USA* **88**:8611–8615.

Orci, L., Tagaya, M., Amherdt, M., Perrelet, A., Donaldson, J. G., Lippincott-Schwartz, J., Klausner, R. D., and Rothman, J. E., 1991b, Brefeldin A, a drug that blocks secretion, prevents the assembly of non-clathrin-coated buds on Golgi cisternae, *Cell* **64**:1183–1195.

Palade, G. E., 1975, Intracellular aspects of the process of protein secretion, *Science* **189**:347–358.

Palade, G. E., and Siekevitz, P., 1956, Liver microsomes. An integrated morphological and biochemical study, *J. Biophys. Biochem. Cytol.* **2**:171–200.

Paulik, M., Nowack, D. D., and Morre, D. J., 1988, Isolation of a vesicular intermediate in the cell-free transfer of membrane from transitional elements of the endoplasmic reticulum to Golgi apparatus cisternae of rat liver, *J. Biol. Chem.* **263**:17738–17748.

Pelham, H.R.B., 1988, Evidence that luminal ER proteins are sorted from secreted proteins in a post-ER compartment, *EMBO J.* **7**:913–918.

Pelham, H.R.B., 1990, The retention signal for luminal ER proteins, *Trends Biochem. Sci.* **15**:483–486.

Pelham, H.R.B., 1991, Recycling of proteins between the endoplasmic reticulum and Golgi complex, *Curr. Opin. Cell Biol.* **3**:585–591.

Pfeffer, S. R., and Rothman, J. E., 1987, Biosynthetic protein transport and sorting by the endoplasmic reticulum and Golgi, *Annu. Rev. Biochem.* **56**:829–852.

Plutner, H., Cox, A. D., Pind, S., Khosravi-Far, R., Bourne, J. R., Schwaninger, R., Der, C. J., and Balch, W. E., 1991, Rab1b regulates vesicular transport between the endoplasmic reticulum and successive Golgi compartments, *J. Cell Biol.* **115**:31–43.

Poruchynsky, M. S., and Atkinson, P. H., 1988, Primary sequence domains required for the retention of rotovirus VP7 in the endoplasmic reticulum, *J. Cell Biol.* **107**:1697–1706.

Pryer, N. K., Salama, N. R., Kaiser, C. A., and Schekman, R., 1990, Yeast sec 13p is required in cytoplasmic form for ER to Golgi transport in vitro, *J. Cell Biol.* **111**:325a.

Rambourg, A., and Clermont, Y., 1990, Three-dimensional electron microscopy: Structure of the Golgi apparatus, *Eur. J. Cell Biol.* **51**:189–200.

Rexach, M. F., and Schekman, R. W., 1991, Distinct biochemical requirements for the budding, targeting, and fusion of ER-derived transport vesicles, *J. Cell Biol.* **114**:219–229.

Rizzolo, L. J., and Kornfeld, R., 1988, Post-translational protein modification in the endoplasmic reticulum. Demonstration of fatty acylase and doxymannoiirimycin-sensitive alpha-mannosidase activities. *J. Biol. Chem.* **263**:9520–9525.

Rogalski, A., Bergmann, J., and Singer, S. J., 1984, Effect of microtubule assembly status on the intracellular processing and expression of an integral protein of the plasma membrane. *J. Cell Biol.* **99**:1101–1109.

Rose, J. K., and Bergmann, J. E., 1983, Altered cytoplasmic domains affect intracellular transport of vesicular stomatitis virus glycoprotein, *Cell* **34**:513–524.

Rothman, J. E., and Orci, L., 1990, Movement of proteins through the Golgi stack: A molecular dissection of vesicular transport, *FASEB J.* **4**:1460–1468.

Rothman, J. E., and Orci, L., 1992, Molecular dissection of the secretory pathway, *Nature* **355**:409–415.

Ruohola, H., Kabcenell, A. K., and Ferro-Novick, S., 1988, Reconstitution of protein transport from the endoplasmic reticulum to the Golgi complex in yeast: The acceptor Golgi compartment is defective in sec 23 mutant, *J. Cell Biol.* **107**:1465–1476.

Saraste, J., and Kuismanen, E., 1984, Pre- and post-Golgi vacuoles operate in the transport of Semliki forest virus membrane glycoproteins to the cell surface. *Cell* **38**:535–549.

Saraste, J., and Svensson, K., 1991, Distribution of the intermediate elements operating in Er to Golgi transport. *J. Cell Sci.* **100**:415–430.

Saraste, J., Palade, G. E., and Farquhar, M. G., 1987, Antibodies to rat pancrease Golgi subfractions: Identification of a 58-kD cis-Golgi protein, *J. Cell Biol.* **105**:2021–2029.

Schekman, R., 1992, Genetic and biochemical analysis of vesicular traffic in yeast, *Curr. Opin. Cell Biol.* **4**:587–592.

Schweizer, A., Fransen, J.A.M., Bachchi, T., Ginsel, L., and Hauri, H. P., 1988, Identification, by a monoclonal antibody, of a 53 kD protein associated with a tubulo-vesicular compartment at the cis-side of the Golgi apparatus, *J. Cell Biol.* **107**:1643–1653.

Schweizer, A., Fransen, J., Matter, K., Kreis, T. E., Ginsel, L., and Hauri, H. P., 1990, Identification of an intermediate compartment involved in protein transport from ER to Golgi apparatus, *Eur. J. Cell Biol.* **53**:185–196.

Schweizer, A., Matter, K., Ketcham, C. M., and Hauri, H. P., 1991, The isolated ER–Golgi intermediate compartment exhibits properties that are different from ER and cis-Golgi, *J. Cell Biol.* **113**:45–54.

Segev, N., Mulholland, J., and Botstein, D., 1988, The yeast GTP-binding YPT1 protein and a mammalian counterpart are associated with the secretion machinery, *Cell* **52**:915–924.

Semenza, J. C., Hardwick, K. G., Dean, N., and Pelham, H.R.B., 1990, ERD 2, a gene required for the receptor-mediated retrieval of luminal ER proteins from the secretory pathway, *Cell* **61**:1349–1357.

Serafini, T., Stenbeck, G., Brecht, A., Lottspeich, F., Orci, L., Rothman, J. E., and Wieland, F. T., 1991, A coat subunit of Golgi-derived nonclathrin-coated vesicles with homology to the clathrin-coated vesicle coat protein β-adaptin, *Nature* **349**:215–220.

Shim, J., Newman, A. P., and Ferro-Novick, S., 1991, The BOS1 gene encodes an essential 27 kD putative membrane protein that is required for vesicle transport from the ER to the Golgi complex in yeast, *J. Cell Biol.* **113**:55–64.

Stow, J. L., de Almeida, J. B., Narula, N., Holtzman, E. J., Ercolani, L., Ausiello, D. A., 1991, A heterotrimeric G protein, G alphai-3, on Golgi membranes regulates the secretion of a heparan proteoglycan in LLC-PKI epithelial cells. *J. Cell Biol.* **114**:1113–1124.

Strous, G. J., Berger, E. G., van Kerkhof, P., Bosshart, H., Berger, B., and Geuze, H. J., 1991, Brefeldin A induces a microtubule-dependent fusion of galactosyltransferase-containing vesicles with the rough endoplasmic reticulum, *J. Biol. Cell.* **71**:25–31.

Sweet, D. J., and Pelham, H.R.B., 1992, The Saccharomyces cerevisiae SEC 20 gene encodes a membrane glycoprotein which is sorted by the HDEL retrieval system, *EMBO. J.* **11**:423–432.

Tassin, A. M., Paintrand, M., Berger, E. G., and Bornens, M., 1985, The Golgi apparatus remains associated with the microtubule organizing center during myogenesis, *J. Cell Biol.* **101**:630–638.

Terasaki, M., and Jaffe, L. A., 1991, Organization of the sea urchin egg endoplasmic reticulum and its reorganization at fertilization, *J. Cell Biol.* **114**:929–940.

Terasaki, M., and Sardet, 1991, Demonstration of calcium uptake and release by sea urchin egg cortical endoplasmic reticulum, *J. Cell Biol.* **115**:1031–1037.

Terasaki, M., Chen, L. B., and Fujiwara, K., 1986, Microtubules and the endoplasmic reticulum are highly interdependent structures, *J. Cell Biol.* **103**:1557–1568.

Thyberg, J., and Moskalewski, S., 1985, Microtubules and the organization of the Golgi complex. *Exp. Cell Res.* **159**:1–16.

Tooze, J., Tooze, S. A., and Warren, G., 1988, Site of addition of N-acetyl-galactosamine to the E1 glycoprotein of mouse hepatitis virus-A59, *J. Cell Biol.* **106**:1475–1487.

Turner, J. R., and Tartakoff, A. M., 1989, The response of the Golgi complex to microtubule alterations: The roles of metabolic energy and membrane traffic in Golgi complex organization, *J. Cell Biol.* **109**:2081–2088.

Ulmer, J. B., and Palade, G. E., 1991, Effects of brefeldin A on the Golgi complex, endoplasmic reticulum and viral envelope glycoproteins in murine erythroleukemia cells, *Eur. J. Cell Biol.* **54**:38–54.

Vaux, D., Tooze, J., and Fuller, S. D., 1990, Identification by anti-idiotypic antibodies of an intracellular membrane protein that recognizes a mammalian endoplasmic reticulum signal, *Nature* **345**:495–501.

Wieland, F. T., Gleason, M. L., Serafinik, T. A., and Rothman, J. E., 1987, The rate of bulk flow from the endoplasmic reticulum to the cell surface, *Cell* **50**:289–300.

Wilgram, G. F., and Kennedy, E. P., 1963, Intracellular distribution of some enzymes catalyzing reactions in the biosynthesis of complex lipids, *J. Biol. Chem.* **238**:2615–2619.

Williams, D. B., Swiedler, S. J., and Hart, G. W., 1985, Intracellular transport of membrane glycoproteins: Two closely related histocompatibility antigens differ in their rates of transit to the cell surface, *J. Cell Biol.* **101**:725–734.

Wilson, D. W., Wilcox, C. A., Flynn, G. C., Chen, E., Kuang, W.-J., Henzel, W. J., Block, M. R., Ullrich, A., and Rothman, J. E., 1989, A fusion protein required for vesicle-mediated transport in both mammalian cells and yeast, *Nature* **339**:355–359.

Young, W. W., Lutz, M. S., Mills, S. E., and Lechler-Osborn, S., 1990, Use of brefeldin A to define sites of glycosphingolipid synthesis: GA2/GM2/GD2 synthase is trans to the brefeldin A block, *Proc. Natl. Acad. Sci. USA* **87**:6838–6842.

Zagouras, P., and Rose, J. K., 1989, Carboxy-terminal SEKDEL sequences retard but do not retain two secretory proteins in the ER, *J. Cell Biol.* **109**:2633–2640.

Chapter 6

The Intermediate Compartment between Endoplasmic Reticulum and Golgi Complex in Mammalian Cells

Stefano Bonatti and Maria Rosaria Torrisi

1. INTRODUCTION

Several lines of evidence strongly support the view that the traffic between ER and Golgi complex involves one (or more) previously unidentified membranous structure. This fascinating new finding is based on data obtained in several different experimental systems with different methodology. Unavoidably, many terms have been suggested for this structure: pre-Golgi vacuoles (Saraste and Kuismanen, 1984), budding compartment (Tooze et al., 1988), intermediate compartment (Schweizer et al., 1990), intermediate elements (Saraste and Svensson, 1991), salvage compartment (Munro and Pelham, 1987), cis-Golgi network (Mellman and Simons, 1992). Throughout this review, we will use only the term "intermediate compartment." Many different findings and opinions on

Abbreviations used in this chapter: CHO, Chinese hamster ovary; DNP, dinitrophenol; EM, electron microscopy; ER, endoplasmic reticulum; SFV, Semliki Forest virus; VSV, vesicular stomatitis virus.

Stefano Bonatti   Department of Biochemistry and Medical Biotechnology, University of Naples "Federico II," Naples, Italy.   **Maria Rosaria Torrisi**   Department of Experimental Medicine, University of Rome "La Sapienza," Rome, Italy.

Subcellular Biochemistry, Volume 21: Endoplasmic Reticulum, edited by N. Borgese and J. R. Harris. Plenum Press, New York, 1993.

the structure and function of the intermediate compartment have been reported and will be considered throughout this review: thus, we will present our own view from the beginning, to give the readers a guideline in this intricate pathway.

This review does not pretend to be exhaustive; however, we made an effort to recall and discuss together new and old findings. Emphasis is placed on results obtained by EM immunocytochemistry. Numerous other recent reviews have addressed some aspects of the intermediate compartment: Hurtley and Helenius (1989), Pelham (1989), Balch (1990), Rose and Doms (1988), Rothman and Orci (1992), Klausner et al. (1992), Mellman and Simons (1992).

2.   CURRENT VIEW OF THE GOLGI COMPLEX

The definition of the overall structure of the Golgi complex has been a very difficult task for cell biologists. Recently, Rambourg and Clermont (1990) have summarized their work on the three-dimensional structure of the Golgi complex. They think that the Golgi complex consists of a continuous ribbon assembled by three portions: the *cis* element, formed by interconnected tubules with few vesicles and saccules (the *cis*-Golgi tubular network); the midcompartment, formed by few aligned saccules and showing compact as well as fenestrated zones; the *trans* element, formed by discontinuous saccules that yeld progressively to a *trans*-tubular network. Besides these kinds of structures, smaller "varicose" tubules have been visualized in the *cis*-Golgi region of neurons from spinal ganglia (Lindsey and Ellisman, 1985a). These tubules were sometimes seen bridging Golgi elements with the ER, separate ER cisternae, or just protruding from both organelles with blind endings. In addition, various forms of the *cis* element were detected in the same cell or in adjacent cells of the same type (vesicle form, vesicle-network form, network-saccule form, etc.; Lindsey and Ellisman, 1985b). This observation led to the suggestion that the *cis* elements of the Golgi complex "undergo morphological transformations as part of their normal function" (Lindsey and Ellisman, 1985b).

All three-dimensional reconstructions of the *cis*-Golgi region rely on a harsh staining procedure like osmification, and therefore they must be critically considered. However, we think that at least two observations are very relevant for the structure and function of the intermediate compartment: (1) The *cis*-Golgi region is not anatomically well defined. Thus, it is difficult to identify elements that belong to an intermediate compartment located in close proximity to the Golgi complex from elements that belong to the *cis*-Golgi itself. (2) It is possible that transient tubular connections play a role in the communications between ER and Golgi complex. This hypothesis challenges the traditional view that interorganelle traffic would depend on vesicular transport (Palade, 1975), and goes along with the new concept of organelle tubulation derived from the use of brefeldin A (Klausner et al., 1992).

3. STRUCTURE OF THE INTERMEDIATE COMPARTMENT

3.1. A Model for the Traffic between ER and Golgi Complex

The cartoon shown in Figure 1 summarizes our view of the membrane structures most likely involved in the traffic between ER and Golgi complex. The drawing is strictly derived from all of the evidence that will be discussed in this review, and it gives a view of the anterograde and retrograde transport in a cell not specialized for secretion, for which more data are available. We think that five compartments (or subcompartments, see below) are involved. These are: smooth ER cisternae and tubules (region 1); ER exit sites (region 2); peripheral tubulovesicular structures (region 3); perinuclear, pre-Golgi vacuoles (region 4); *cis*-Golgi tubular network (region 5).

According to the drawing, newly made secretory and membrane proteins may reach region 1, but not region 2, if not competent for ER export. An alternative view of this subcompartmentalization is that noncompetent proteins may also get access at a low rate to region 2 and reach the Golgi area, but would

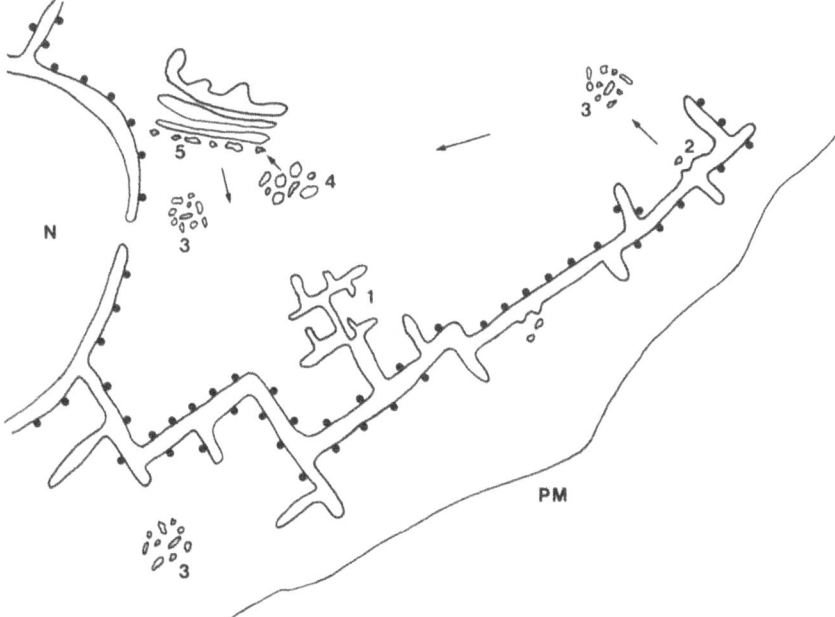

FIGURE 1. Schematic drawing of our model for ER-to-Golgi traffic. Newly synthesized secretory and membrane proteins, not competent for export, may reach smooth ER cisternae and tubules (region 1). Export from ER may occur at exit sites (region 2). Tubulovesicular structures (region 3) and pre-Golgi vacuoles (region 4) represent intermediate structures involved in the export from ER to the *cis*-Golgi (region 5).

then quickly recycle to the ER. The latter hypothesis would extend the current view of the "quality control" process for ER export (Helenius *et al.*, 1992). In several instances, region 1 will expand and occupy a significant cytoplasmic area.

The export from the ER is seen occurring from exit sites (region 2), which are randomly distributed throughout the rough ER. The exit would take place by means of transport vesicles, tubular buds, or connecting tubules. Under normal metabolic conditions the proteins move to the Golgi stacks very quickly, with no detectable accumulation in intermediate structures (regions 3 and 4). However, current evidence shows the involvement of peripheral tubulovesicular structures (region 3), and suggests the existence of perinuclear, pre-Golgi vacuoles (region 4). We envision that both structures are constitutively engaged in the export from the rough ER to the Golgi (they are not artifactually generated by blocking the transport at low temperature), and are distinct from both the exit sites and the *cis*-Golgi tubular network. We think it is likely that the transfer of the proteins from the peripheral elements to the perinuclear region is mediated by carrier vesicles or tubular buds. The access to region 3 may not be totally precluded to proteins not competent for the export or incompletely folded and oligomerized.

The little evidence available on the retrograde transport to the ER indicates that it originates from region 4 and/or 5; no information is available about the involvement in this process of the peripheral tubulovesicular structures as well as of the exit sites. Conversely, it has been shown that this step of transport may be mediated by tubular buds.

Our view of the intermediate compartment is mainly based on the finding that these structures are not usually close to the Golgi complex, as originally suggested, but rather randomly distributed in the cytoplasm and frequently adjacent to the ER (Lotti *et al.*, 1992). For this reason, we disagree with the proposal that such intermediate structures should be considered as part of the *cis*-Golgi network (Mellman and Simons, 1992). Conversely, the pre-Golgi vacuoles (region 4) could represent a specialized region of the *cis*-Golgi (region 5), functionally overlapping with it. Although clear-cut evidence is lacking, we favor the view that distinct structures operate at the *cis*-Golgi.

3.2. Low Temperature Facilitates the Study of the Intermediate Compartment

The first proof of the existence of an intermediate compartment in the export from the ER to the Golgi complex was presented by Saraste and Kuismanen (1984). They used a temperature-sensitive strain of Semliki Forest virus, the spike glycoproteins of which accumulate in the ER of the infected cell at 39°C. Normal transport to the plasma membrane through the Golgi complex was restored when the temperature was shifted to 28°C. Incubation at 15°C led to a very different situation. The glycoproteins accumulated in vacuoles of about 300 nm average diameter, mostly located adjacent to or very close to the Golgi stacks. Smaller vesicles of about 100 nm diameter were also labeled in the same area.

The level of trimming of the oligosaccharide side chains confirmed that the SFV glycoproteins had not reached the Golgi complex. However, the Golgi complex itself appeared to be markedly altered at the end of the incubation at low temperature, even in uninfected cells. It consisted of smaller stacks of one or two cisternae, the rest being replaced by vesicular structures.

Saraste and Kuismanen proposed that between ER and Golgi complex are "pre-Golgi vacuoles," involved in the generation of new cisternae by progressive fusion and maturation events. This conclusion was partially confirmed by the finding that a temperature-sensitive transport step was detectable in rat pancreatic lobules *in vitro* (Saraste *et al.*, 1986), as shown by cell-fractionation procedures after pulse–chase labeling. Acinar cells incubated at low temperature were analyzed by EM. A large increase in the number of small vesicles in the region between the Golgi stacks and the transitional elements of the ER was observed; furthermore, the Golgi stacks appeared partially fragmented and disorganized (Saraste *et al.*, 1986).

The effect of low temperature on the secretory pathway in pancreatic lobules *in vitro* has also been studied by Tartakoff (1986). Cell-fractionation experiments after pulse–chase labeling showed retention of secretory proteins in the ER fraction at 15°C; however, the effect of incubation at 10°C was more carefully examined. A dramatic increase in the number, size, and complexity of the transitional elements of the ER and of their associated protrusions was observed by EM; conversely, incubation in the presence of DNP, which is known to lower the ATP level, led to the disappearance of such elements. More recently, the effects of incubation at 10°C have been confirmed on tissue-cultured cells secreting prolactin (Nasciutti *et al.*, 1992). Thus, it is likely that ATP depletion and incubation at 10°C affect exit from the ER at two separate stages. Noteworthy, proliferation of transitional elements at 15°C has not been reported.

The transport from the ER to the Golgi complex has been addressed recently using as marker the G glycoprotein coded by the ts-045 mutant strain of vesicular stomatitis virus. This protein is a very good temperature-sensitive transport mutant (Gallione and Rose, 1985); very little export from the ER is detectable at 39°C, and the reversion of the phenotype upon transition to the permissive temperature (31°C) is fast and almost complete (Kreis and Lodish, 1986). The immunofluorescence analysis of ts-045-infected Vero cells previously kept at 39°C and subsequently shifted to 15°C has been reported (Bonatti *et al.*, 1989; Schweizer *et al.*, 1990; Duden *et al.*, 1991; Hobman *et al.*, 1992). Under these conditions, G glycoprotein gave a peculiar staining pattern, clearly distinguishable from the ones shown at 39 or 31°C (ER and Golgi, respectively). Short incubation time (1 h) suggested that G glycoprotein was present in roughly rounded membranous structures randomly dispersed in the cytoplasm (Bonatti *et al.*, 1989); confocal fluorescence microscopy confirmed this view (Bonatti *et al.*, 1989). However, longer incubations at 15°C (3 h) resulted in a more compact signal in the perinuclear area, co-staining structures labeled by anti-p53 antibody (see Section 5; Schweizer *et al.*, 1990). EM immunocytochemistry on cryosections performed under this last condition showed labeling of tubulovesicular structures at the *cis* side of the Golgi complex (Schweizer *et al.*, 1990); however, low-level labeling in these structures was also detected at the nonpermissive temperature.

Interestingly, when SFV ts-1 glycoproteins were visualized by immunofluorescence microscopy after 2 h incubation at 15°C, labeling was localized at the perinuclear region as well as at the periphery of the cell (Kuismanen and Saraste, 1989); thus, it is likely that incubation at this temperature induces a progressive compaction of the signal in the perinuclear region of the cell.

The morphology of the structures containing G glycoprotein in infected Vero cells after 1 h incubation at 15°C, and the transfer of this protein to the Golgi complex upon temperature shift-up have been extensively analyzed by postembedding EM immunocytochemistry (Lotti *et al.*, 1992). The intermediate compartment appeared to be formed by about 30–40 separate units of clustered small vesicles and short tubules, having an average diameter of 80 nm 2a). No direct continuity of these units with either ER or Golgi cisternae was detected. Interestingly, these units were not confined to the perinuclear region of the cell, where the great majority of the Golgi complex was localized. The majority of the tubulovesicular units were found in the peripheral region of the cell, far from the Golgi complex; they were also near Golgi stacks, but the organelle more frequently found in close proximity with the units was the ER. Rab2 protein, a small GTP binding protein previously associated with vesicles in between ER and Golgi complex (Chavrier *et al.*, 1990), was present in these structures (Figure 2b,c). However, at variance with G glycoprotein, it was also present in the *cis* region of the Golgi complex. Because of the peculiar morphology, the tubulovesicular units could also be localized in uninfected cells: this allowed establishing that the general morphology and intracellular distribution of the units were not affected by low temperature or viral infection (Figure 3). Time-course experiments were performed to follow the transport of G glycoprotein to the Golgi complex when the permissive temperature was restored (Figure 4). A progressive and parallel transfer was detected from either peripheral or perinuclear units; moreover, the intracellular distribution of the units did not change. These results suggested that the intermediate compartment is a station in the secretory pathway, connected through a transport step to the Golgi complex. The alternative explanation, that the tubulovesicular units themselves move to the *cis*-Golgi, was not supported by these results (Lotti *et al.*, 1992).

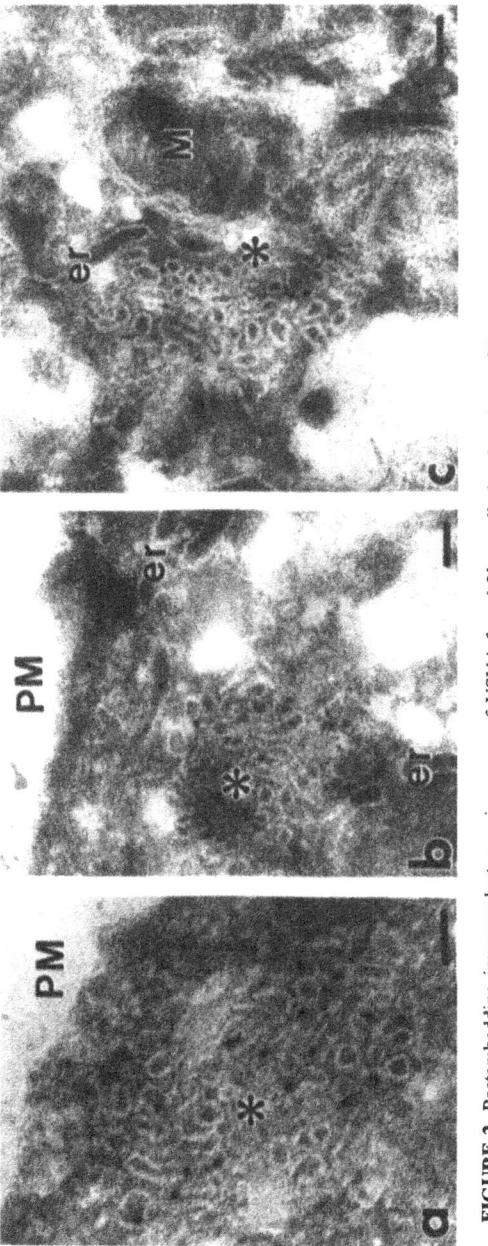

FIGURE 2. Postembedding immunoelectron microscopy of VSV-infected Vero cells incubated at 15°C: clusters of small vesicles and short tubules are immunolabeled with anti-G protein antibody (a) and anti-Rab2 antibody (b, c). The tubulovesicular units (asterisks) are mostly located at the cell periphery (a, b) and in proximity to ER cisternae. (Reproduced from *J. Cell Biol.* **118**:43–50, 1992.)

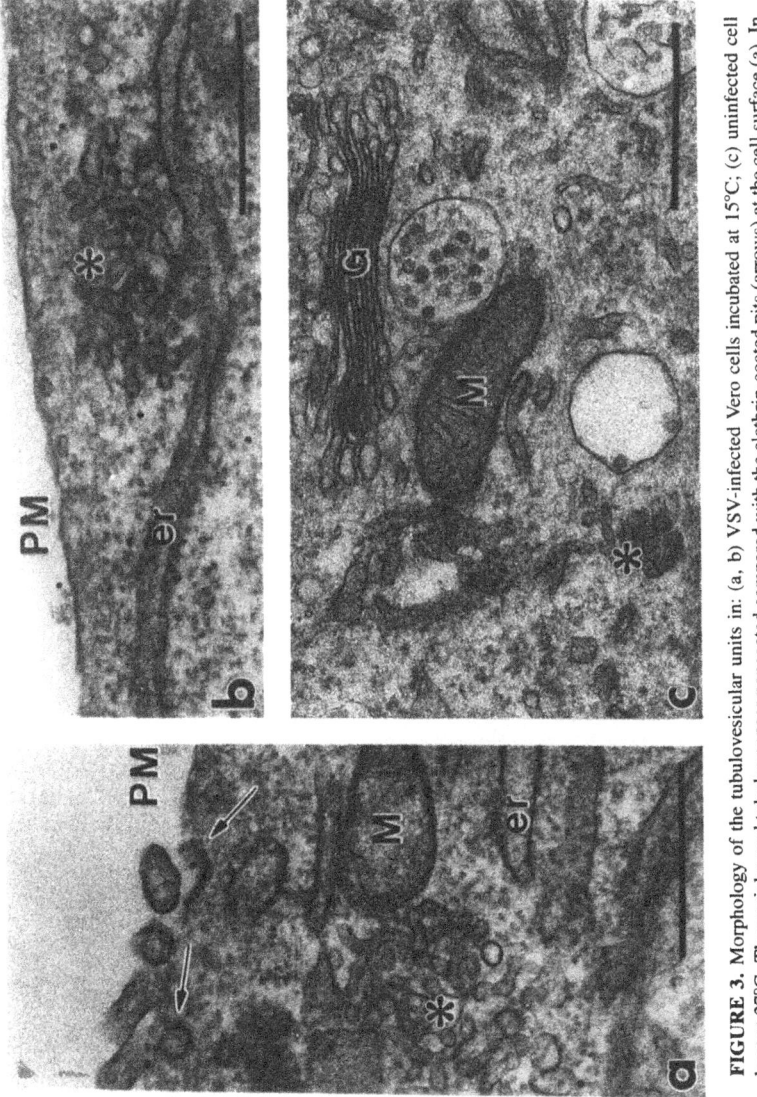

FIGURE 3. Morphology of the tubulovesicular units in: (a, b) VSV-infected Vero cells incubated at 15°C; (c) uninfected cell kept at 37°C. The vesicles and tubules appear noncoated compared with the clathrin-coated pits (arrows) at the cell surface (a). In b, a tubulovesicular unit faces transitional elements of the rough ER. The units can also be found in proximity to Golgi cisternae (c). (a, b: reproduced from *J. Cell Biol.* **118**:43–50, 1992. G, Golgi complex. Bars: a, b, 0.5 μm; c, 1μm.

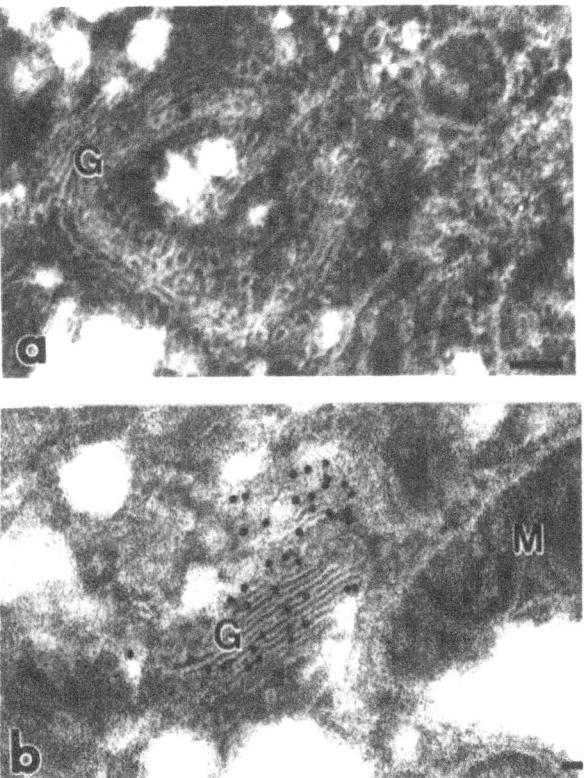

FIGURE 4. Immunolabeling with anti-G protein antibody of infected cells incubated at (a) 15°C and (b) 31°C. Golgi cisternae appear almost unlabeled at 15°C, whereas at 31°C all Golgi stacks are heavily labeled. (a: reproduced from *J. Cell Biol.* **118**:43–50, 1992.) Bars: 0.1 μm.

3.3.   Marker Proteins for the Intermediate Compartment

To date, immunocytochemical studies have indicated few proteins as candidate markers for the intermediate compartment: p58 (Saraste *et al.*, 1987; Saraste and Svensson, 1991), p53 (Schweizer *et al.*, 1988, 1990), Rab2 protein (Chavrier *et al.*, 1990, Lotti *et al.*, 1992), Rab1b protein (Plutner *et al.*, 1991). Furthermore, it cannot be excluded that p58 and p53 are the rat and human equivalent of the same membrane protein.

p58 was identified by means of a polyclonal antibody raised against a rat pancreas light Golgi fraction. Immunofluorescence microscopy showed that in pancreatic acinar cells, p58 was mainly localized at the plasma membrane (both apical and basolateral domains) and in the Golgi complex region (Saraste *et al.*, 1987). In mouse myeloma and NRK cells, the same antibod labeled the perinuclear Golgi region in immunofluorescence assays, and the fenestrated *cis*-most Golgi cisterna and vesicles or tubules in continuity with it in immunoperoxidase staining (Saraste *et al.*, 1987). Weak peroxidase reaction was also observed in the penultimate *cis*-Golgi cisterna, in tubules or vesicles not close to the Golgi complex, in the ER, and in the transitional elements of the ER (Saraste *et al.*, 1987). When the cells were infected at 39°C with the SFV ts-1 mutant, shifted for 1 h at 15°C, and then double labeled for immunofluorescence microscopy, only part of the p58-positive structures co-stained with the viral glycoproteins (Saraste and Svensson, 1991). The more peripheral p58-positive structures were seen accumulating in response to the incubation at 15°C, and rapidly reclustering when the temperature was shifted to 37°C (Saraste and Svensson, 1991). This last finding is intriguing, considering the opposite behavior shown by p53 protein (Lippincott-Schwartz *et al.*, 1990; Schweizer *et al.*, 1990; see below and Section 5). Considering the limitations of immunofluorescence light microscopy (Schweizer *et al.*, 1990; Lotti *et al.*, 1992), we think that quantitative EM immunocytochemistry should be used to prove this point.

The identification of human p53 occurred similarly to p58: a cell fraction enriched for Golgi membranes from Caco-2 cells was used as immunogen, and one of the isolated mAbs clearly lighted the Golgi complex in immunofluorescence assays (Schweizer *et al.*, 1988). Immunogold labeling of ultrathin cryosections demonstrated that p58 was mostly localized in tubulovesicular structures at the *cis* side of the Golgi complex, but also in a fenestrated *cis* cisterna (Schweizer *et al.*, 1988). In a subsequent study (Schweizer *et al.*, 1990), immunofluorescence microscopy and immunogold labeling of cryosections were performed in VSV ts-045-infected cells shifted for 3 h at 15°C. It was shown that VSV ts-045 G glycoprotein and p53 partially colocalized in tubulovesicular structures at the *cis* side of the Golgi complex. Noteworthy, these structures were seen more dispersed throughout the cytoplasm of uninfected cells at 37°C than after 3 h incubation at 15°C; low temperature and viral infection together induced the strongest perinuclear compaction of p53-containing elements (Schweizer *et al.*, 1990).

The third putative marker of the intermediate compartment, Rab2 protein, was identified by screening a cDNA library for Rab-family protein members, and by immunizing rabbits with a short specific peptide corresponding to the identified sequence (Chavrier *et al.*, 1990). Immunofluorescence microscopy and immunogold labeling of cryosections also showed that Rab2 protein was localized in tubulovesicular structures at the *cis* side of the Golgi complex, with a minor but significant localization in the first *cis*-Golgi cisternae (Chavrier *et al.*, 1990). Recently, this protein has been colocalized by EM immunocytochemistry in the tubulovesicular structures containing VSV ts-045 G glycoprotein after incubation for 1 h at 15°C (Lotti *et al.*, 1992).

Another member of the Rab family, Rab1b protein, is a potential marker of the intermediate compartment because it partially colocalizes with Rab2 protein in immunofluorescence microscopy (Plutner *et al.*, 1991). In addition, *in vitro* studies showed that this protein is required in the transport from the ER to the medial-Golgi region (Plutner *et al.*, 1991).

It is remarkable that all of the putative marker proteins of the intermediate compartment are present at all stages of the secretory pathway from the ER to the Golgi complex. This finding alone may suggest the existence of a recycling pathway from the Golgi to the ER. Unfortunately, these are not abundant cellular proteins, thus impairing a quantitative immunocytochemical approach, and for none of them is the physiological function known. It is likely that Rab2 and Rab1b proteins, members of the ras superfamily, associate with the cytosolic side of the intermediate compartment membranes, and that their function involves a GTP-binding/hydrolysis cycle (Balch, 1990).

All attempts at identifying enzymatic activity that may serve as markers of the intermediate compartment have given inconclusive results. A number of posttranslational modifications have been reported to occur in a nondefined post-ER pre-Golgi, *cis*-Golgi area: palmitylation (Bonatti *et al.*, 1989), mannosidase I activity (Dunphy and Rothman, 1983), initiation of *O*-glycosylation (Roth, 1991; Pascale *et al.*, 1992), phosphorylation of lysosomal proteins (Kornfeld and Kornfeld, 1985). On the other hand, no marker proteins have been localized in the *cis*-Golgi by EM immunocytochemistry. Therefore, the present picture points to a large undefined area, which extends from the rough ER to the medial/*trans*-Golgi, where several membrane and lumenal protein markers have been localized. An important contribution in this area may derive from the cell-free system developed by Balch and collegues for the study of ER-to-Golgi transport (Beckers *et al.*, 1987). Indeed, several biochemical requirements for the transport have already been identified (Beckers *et al.*, 1990).

3.4. Purification of the Intermediate Compartment

Several studies have been reported concerning the purification of membrane structures involved in ER-to-Golgi transport but distinct from both organelles. Lodish *et al.* (1987) showed that pulse–chase-labeled hepatoma secretory proteins are transiently present in light-density vesicles (lighter than the Golgi-derived vesicles) before they reach the Golgi complex. Saraste *et al.* (1987) have isolated a membrane fraction from pancreatic lobules pulse-labeled at 37°C, and subsequently incubated 1 h at 16°C, which apparently contained the front of the secretory proteins in transit to the Golgi complex. Morré and colleagues have reported the purification of putative transitional vesicles generated in a cell-free transport system derived from rat liver (Paulik *et al.*, 1988). However, only partial purification and biochemical characterization of the fractions were achieved in these studies.

Schweizer *et al.* (1991) have reported the isolation from Vero cells of a membrane fraction enriched about 40-fold in p53 protein. This fraction formed a broad band in a 10–30% metrizamide gradient, and was not enriched for heavy-chain binding protein, protein disulfide isomerase; ribophorin I and II (all markers of the rough ER), galactosyltransferase (marker of the *trans*-Golgi region), or N-acetylglucosamine-1-phosphodiester-δ-N-acetylglucosaminidase (the uncovering enzyme, a putative *cis*-Golgi region marker; Goldberg and Kornfeld, 1983). Some enrichment (6-fold) was found for the N-acetylglucosaminylphosphotransferase, the enzyme phosphorylating lysosomal hydrolases, for which a post-ER/pre-Golgi localization was previously suggested (Goldberg and Kornfeld, 1983). The polypeptide composition of the isolated fraction was different from both ER and Golgi complex, and transmission EM showed almost exclusively small, smooth uncoated vesicles, having an average diameter of about 90 nm (Schweizer *et al.*, 1991). The vesiculation, and the absence of a coat, could in principle be related to the methodology used; however, there is a good morphological correspondence between these vesicles and the intermediate compartment structures observed *in vivo* (Schweizer *et al.*, 1990; Lotti *et al.*, 1992).

4. ER–GOLGI TRANSPORT INVOLVES SPECIALIZED REGIONS OF THE ER

4.1. Dynamic Behavior of the ER

The structure of the ER has been studied by several morphological approaches (Buckley and Porter, 1975). It consists of a complex tubular network of cisternae; often, this network constitutes a large fraction of intracellular membranes and shows high surface-to-volume ratio. Two morphologically defined domains of the ER are ubiquitous constituents of mammalian cells: the rough ER, composed of cisternae with membrane-bound ribosomes, and the smooth ER, characterized by more dilated cisternae devoid of bound ribosomes. Most relevant to this review is the finding that the ER is associated with microtubules and has an extremely dynamic behavior (Franke, 1971; Terasaki *et al.*, 1986; Lee and Chen, 1988; Dabora and Sheetz, 1988). It has been shown that the formation of the reticular network depends on the movement of tubular membranes on intersecting microtubules (Lee and Chen, 1988; Dabora and Sheetz, 1988). These networks are in a continuous remodeling process; thus, ER membranes may be quickly reorganized, redistributed, and perhaps compartmentalized in the cell cytoplasm, in response to different needs.

4.2. The Transitional Elements of the ER

The first suggestion of the existence of intermediate structures between rough and smooth ER, and the Golgi complex, was made several years ago by

Palade and co-workers. These structures were named transitional elements (Jamieson and Palade, 1967; Palade, 1975), and were thought to be involved in the export from the ER of secretory products as well as membrane proteins and lipids. Transitional elements were easily observed in tissues and cultured cells specialized for secretion as asymmetric cisternae, partially rough and partially smooth. The smooth portion of the transitional elements shows typical small protrusions, probably corresponding to vesicles budding from or fusing with the ER, and free vesicles of average diameter 50–80 nm; frequently, these vesicles appeared coated by fibrillar material, reminiscent (but distinguishable) of the clathrin coat of endocytic vesicles (Jamieson and Palade, 1967).

In cell types where the Golgi complex is highly polarized, as in pancreatic cells where secretion is regulated and the secretory products are stored at the *trans* side of the Golgi, transitional elements are characterized by a clear sidedness with the smooth portion and the protrusions oriented toward the *cis*-Golgi (Merisko *et al.*, 1986; Tartakoff, 1986; Orci *et al.*, 1991). A similar sidedness has been observed in cells actively involved in constitutive secretion, such as myeloma cells (Ottosen *et al.*, 1980).

However, in cells not specialized for secretion, where the Golgi complex is less polarized, it is possible to find regions of the ER that share only some features with the transitional elements as defined above. Partly smooth/partly rough structures and protrusions appear not to be oriented, with their smooth portions and protrusions and associated vesicles facing both the cell center and the cell periphery, and only occasionally localized in close apposition to the Golgi complex (Lotti *et al.*, 1992; Torrisi *et al.*, unpublished).

Only indirect evidence supports the view that export from the ER is mediated by the transitional elements. Conditions that prevent the traffic between ER and Golgi complex, such as ATP depletion or incubation at 10°C, led to the disappearance or to the proliferation of the protrusions and associated vesicles, respectively (Merisko *et al.*, 1986; Tartakoff, 1986; Orci *et al.*, 1991; Nasciutti *et al.*, 1992). The mammalian homologue of yeast Sec23p, a protein required for ER–Golgi traffic in yeast, is restricted to the cytoplasm around the transitional elements in pancreatic cells (Orci *et al.*, 1991). Therefore, it is possible that the transitional elements described in cells specialized for secretion do not represent the main starting point of the export from the ER; indeed, they could be the site where the retrograde transport from the Golgi to the ER is directed. As illustrated in Figure 1, we predict that the anterograde transport to the Golgi complex originates from several exit sites randomly distributed throughout the rough ER.

4.3. Smooth ER Expansion and Viral Budding Compartment

Several laboratories have documented interesting cases of membrane and secretory proteins that interrupt their normal export from the ER to the plasma membrane before the Golgi complex. A chimeric protein formed by growth hormone linked to the C-terminal portion of influenza hemagglutinin was first shown by immunolabeled cryosections to accumulate in modified smooth ER cisternae (Rizzolo et al., 1985). These cisternae surrounded the Golgi complex and were continuous with the rough ER. They were clearly detectable by conventional EM, but only in the cells expressing this mutant protein. A naturally occurring mutant form of LDL receptor also accumulates in smooth extension of rough ER cisternae (Pathak et al., 1988); however, these structures are not a peculiar feature of the mutant cells (Pathak et al., 1988). It was suggested that smooth ER elements are normal constituents of the secretory pathway and may represent the equivalent of the transitional elements described in pancreatic cells; they would dramatically expand only when proteins not competent for the export are synthesized in high amount (Rizzolo et al., 1985; Pathak et al., 1988).

Other examples of very large increase of tubular smooth ER have been described in hepathocytes of phenobarbital-treated rats (Remmer and Merker, 1963; Orrenius et al., 1965) and in compactin-resistant CHO cells adapted to grow in the absence of cholesterol (Chin et al., 1982). In both cases, the need to accommodate a very high level of either detoxifying enzymes or HMG-CoA reductase is thought to be the trigger for the generation of new smooth ER. At variance, chicken chondrocytes grown in scorbutic monolayer cultures accumulate procollagen in a dilated rough ER (Pacifici and Iozzo, 1988). More recently, smooth tubular expansion of the rough ER has been shown to contain the accumulated, precursors of the large chondroitin sulfate proteoglycan (Vertel et al., 1989), and rubella virus glycoprotein E1 expressed by permanently transformed CHO cells (Hobman et al., 1992). In both instances, an anti-ER polyclonal antibody (Louvard et al., 1982) did not label these convoluted structures, thus showing that they represent a subcompartment within the ER.

In the case of rubella E1 protein accumulated in CHO cells, the relation of these structures to the intracellular transport of VSV ts-045 G glycoprotein has been studied by immunofluorescence microscopy (Hobman et al., 1992). At the nonpermissive temperature, some colabeling of G and E1 glycoprotein was detected; shifting the temperature to 32°C, G glycoprotein entered the Golgi complex and no labeling in the tubular smooth ER was evident any longer. Furthermore, when the temperature was shifted to 15°C, G clearly entered the intermediate compartment, leaving the ER but not gaining access to the Golgi. This finding was interpreted as indicating that the smooth tubular network is proximal to the intermediate compartment. However, there is no evidence in favor of the idea that the tubular smooth ER is an intermediate station between ER and Golgi complex.

In the absence of a clear picture about the site where ER export takes place, the colocalization of rubella virus E1 and VSV ts-045 G glycoproteins does not prove the point: indeed, newly synthesized viral membrane proteins move freely to and from the inner nuclear membrane and the smooth ER (Torrisi et al., 1987;

Bergmann and Fusco, 1990). Noteworthy, large tubular smooth membrane networks that extend the rough ER are generated in cells infected with simian virus 40 (Kartenbeck *et al.*, 1989); these networks are morphologically very similar to the one described above and contain much of the incoming virus but not its progeny (Kartenbeck *et al.*, 1989). It is conceivable that the generation of expanded networks of smooth ER reflects the need of the cell to segregate large amount of products that cannot be quickly removed by export to the Golgi or degradation *in situ* (Bonifacino and Lippincott-Schwartz, 1991).

A different situation has been described in the case of E1 glycoprotein coded by mouse hepatitis virus-A59 (Tooze *et al.*, 1988). This glycoprotein is detectable immunocytochemically in the ER and throughout the Golgi stacks of the infected cells; however, early during the infection, virus budding is first observed in pre-Golgi structures called budding compartment. Moreover, lowering the temperature to 31°C blocks the transport of the lumenal viral particle to the Golgi stacks, thus offering a suitable marker for the morphological analysis. The budding compartment appeared to be formed by smooth vesiculotubular membranes irregularly organized; it was always detected close to the ER and particularly to putative transitional elements of the ER, but no evidence of direct continuity with those structures was found (Tooze *et al.*, 1988). Often, the cytoplasmic side of the budding compartment was coated, although differently from the clathrin-coated vesicles (Tooze *et al.*, 1988). This description of the budding compartment, obtained early postinfection as well as from uninfected cells, is in several aspects similar to that for the intermediate compartment defined by means of incubation at 15°C (Saraste and Kuismanen, 1984; Schweizer *et al.*, 1990; Lotti *et al.*, 1992). A budding compartment for the retroviruses produced in Friend erythroleukemia cells has been described by conventional EM (Walter and Tandler, 1989; Ulmer and Palade, 1991). This secondary site of virus production consists of an irregular network of smooth membranes continuous to the rough ER: therefore, it probably represents another example of ER compartmentalization.

5. INTERMEDIATE COMPARTMENT, SALVAGE FUNCTION, AND GOLGI–ER RETROGRADE TRANSPORT

The discovery that ER lumenal proteins maintain their topological location by means of a retrieval process, which involves a C-terminal retention sequence and occurs downstream in the secretory pathway, suggested that the intermediate compartment could be the site of such "salvage" function (Munro and Pelham, 1987; Warren, 1987). Indirect evidence in favor of this hypothesis is the finding that one putative receptor for such retention sequence gave an immunofluorescence pattern reminiscent of the one described for the intermediate compartment

(Vaux *et al.*, 1990). However, a secretory protein tagged with the retention signal accumulated in the ER but underwent posttranslational modifications thought to occur in the *cis*-Golgi (Pelham, 1988); and a resident ER protein was shown to be galactosylated, a typical *trans*-Golgi modification (Nguyen Van *et al.*, 1989). Furthermore, other more characterized receptors have been shown by immu- nofluorescence to be normally localized throughout the Golgi complex (Lewis and Pelham, 1992; Hsu *et al.*, 1992). Thus, it appears likely that animal cells may perform the salvage function in several compartments along the secretory pathway (Lewis and Pelham, 1992).

The retrograde pathway from the Golgi to the ER has been described first by studying the effect of brefeldin A in tissue-cultured cells (Doms *et al.*, 1989; Ulmer and Palade, 1989; Lippincott-Schwartz *et al.*, 1989, 1990). More recent- ly, brefeldin A-like effects have been shown in a temperature-sensitive CHO mutant (Zuber *et al.*, 1991), and in cells overexpressing retrieval receptors for ER lumenal proteins (Hsu *et al.*, 1992). In all of these cases, dramatic mor- phological and biochemical effects were clearly detected: disassembly and disap- pearance of the Golgi complex, swelling of the ER, interruption of the secretory pathway in the ER, and relocation in the ER of Golgi enzymes. However, ER– Golgi recycling of membrane proteins may also occur without destroying the integrity of the secretory pathway. Lippincott-Schwartz *et al.* (1989) have shown by immunofluorescence that p53 moves from a more dispersed cytoplasmic location to the Golgi region of the cell when the temperature is shifted from 37 to 15°C; upon restoring the physiological temperature, p53 goes first in the ER and then returns to its normal post-ER location. The same cycling was shown to occur in a mutant cell line in which class I MHC molecules failed to assemble (Hsu *et al.*, 1991): as a result, these proteins are stored in the ER, but apparently passing through the *cis* cisternae of the Golgi complex. Therefore, the current evidence may suggest the existence of a fast recycling pathway from the ER to the Golgi area (passing through the intermediate compartment) and then back to the ER; a direct recycling between ER and peripheral elements of the intermedi- ate compartment has yet to be proven.

A very important finding shown in the studies described above is that the retrograde traffic may occur by tubular buds rather than by carrier vesicles (Klausner *et al.*, 1992). That the anterograde transport may be based on tubular buds has not been demonstrated, but it represents a fascinating new hypothesis.

6. MAIN UNANSWERED QUESTIONS ABOUT THE INTERMEDIATE COMPARTMENT

The drawing shown in Figure 1 describes the transfer of newly synthesized proteins from the ER to the Golgi complex as may be visualized by EM immunocytochemistry. Obviously, this picture has to be translated in a three-dimensional view, as in the case of the Golgi complex (see Section 2). In addition, this morphological view has to be confronted with the evidence obtained from the biochemical approach to the study of ER-to-Golgi transport.

Concerning the first problem, EM analysis of ultrathin sections, after either conventional or immunocytochemical staining, results in poor detection of tubules of small average diameter: these will in general appear as small vesicles. In principle, examination of serial sections should circumvent this problem, but this approach is almost unfeasible in immunocytochemistry. These limitations complicate a three-dimensional evaluation of the intermediate compartment. Two hypotheses may be suggested concerning the relation between ER and peripheral tubulovesicular structures, as well as between perinuclear pre-Golgi vacuoles and cis-Golgi tubular network (regions 2–3 and 4–5 in Figure 1, respectively): (1) convoluted membranous tubules permanently join these structures. In this case membrane and secretory proteins may move by diffusion within these tubules. (2) No physical connection exists, the transport being mediated by carrier vesicles or tubular buds. Conversely, we do not consider the same alternative for the transport from the peripheral to the perinuclear structures. As discussed above, it is very likely that this transport is mediated by carrier vesicles or tubular buds.

ER-to-Golgi transport has been studied in a cell-free system following the remodeling of the carbohydrate side chains of VSV ts-045 G glycoprotein as a marker for arrival in the Golgi complex (Beckers et al., 1990). This approach suggests that at least one round of vesicular transport is involved in the export. Given that yeast-derived cell-free systems are yielding the same indication (Groesh et al., 1990; Rexach and Schekman, 1991), we think it is almost certain that one dissociative transport step is involved, and tentatively suggest placing it in between the peripheral and the perinuclear structures of the intermediate compartment (regions 3 and 4 in Figure 1). This location would better agree with the quantitative immunocytochemical analysis already described (Lotti et al., 1992), but it does not preclude the possibility that other dissociative steps are also involved, as discussed above.

The most obvious unanswered question about the intermediate compartment concerns its function. It may play a role in the folding/oligomerization of newly synthesized proteins, and several posttranslational modifications have been ascribed to this compartment; however, the more common hypothesis is the constitutive participation in anterograde and retrograde transport between ER and Golgi complex, as well as the role in the sorting/retrieval process of ER proteins. Ironically, one could say that there have been more suggestions about the function of the intermediate compartment than solid evidence to prove its existence. At the moment, the only demonstrated function is the involvement in the anterograde transport; however, most likely the intermediate compartment is endowed with multiple functions, as all other organelles.

7. FUTURE PERSPECTIVES

It is clear that the full understanding of ER–Golgi communications requires a multidisciplinary and coordinated effort. In line with the main theme of this review, we will briefly indicate some areas that have to be addressed mainly by EM immunocytochemistry: (1) The characterization of the exit sites of the ER. (2) The combination of semithin sectioning and enzymatic staining to investigate the structure of the tubulovesicular elements. Recently, this approach has given very clear results in the visualization of the tubular endosomes (Tooze and Hollinshead, 1991). (3) The mapping in the *cis*-Golgi region of the several putative marker enzymes that have been localized there by indirect evidence. (4) The search for other marker proteins of the intermediate compartment. Membrane fractions enriched in such elements are the obvious starting material to isolate new markers, but the case of the human homologue of yeast sec23p (Orci *et al.*, 1991) indicates that this search should not be limited to the lumenal and membrane content of these vesicles.

The unanswered questions on the intermediate compartment are more numerous than those solved. However, the impressive progress made in the last few years, and the number of laboratories actively engaged in this area of research, make it likely that soon we will know more about the structure and function of the intermediate compartment.

ACKNOWLEDGMENTS. We thank Drs. Ari Helenius and Pietro De Camilli for a critical reading of the manuscript and for many helpful suggestions.

8. REFERENCES

Balch, W. E., 1990, Small GTP-binding proteins in vesicular transport, *Trends Biochem. Sci.* **15:**473–477.

Beckers, C. J. M., Keller, D. S., and Balch, W. E., 1987, Semi-intact cells permeable to macromolecules: Use in reconstitution of protein transport from the endoplasmic reticulum to the Golgi complex, *Cell* **50:**523–534.

Beckers, C. J. M., Plutner, H., Davidson, H. W., and Balch, W. E, 1990, Sequential intermediates in the transport of protein between the endoplasmic reticulum and the Golgi, *J. Biol. Chem.* **265:**18298–18310.

Bergmann, J. E., and Fusco, P. J., 1990, The G protein of vesicular stomatitis virus has free access into and egress from the smooth endoplasmic reticulum of UT-1 cells, *J. Cell Biol.* **110:**625–635.

Bonatti, S., Migliaccio, G., and Simons, K., 1989, Palmitylation of viral membrane glycoproteins takes place after exit from the endoplasmic reticulum, *J. Biol. Chem.* **264:**12590–12595.

Bonifacino, J. S., and Lippincott-Schwartz, J., 1991, Degradation of proteins within the endoplasmic reticulum, *Curr. Opin. Cell Biol.* **3:**592–600.

Buckley, I. K., and Porter, K. R., 1975, Electron microscopy of critical point dried whole cultured cells, *J. Microsc. (Oxford)* **104**:107–120.

Chavrier, P., Parton, R. G., Hauri, H. P., Simons, K., and Zerial, M. 1990, Localization of low molecular weight GTP binding proteins to exocytic and endocytic compartments, *Cell* **62**:317–329.

Chin, D. J., Luskey, K. L., Anderson, R. G. W., Faust, J. R., Goldstein, J. L., and Brown, M. S., 1982, Appearance of crystalloid endoplasmic reticulum in compactin-resistant Chinese hamster cells with a 500-fold increase in 3-hydroxy-3-methylglutaryl-coenzyme A reductase, *Proc. Natl. Acad. Sci. USA* **79**:1185–1189.

Dabora, S. L., and Sheetz, M. P., 1988, The microtubule-dependent formation of a tubulovesicular network with characteristics of the ER from cultured cells, *Cell* **54**:27–35.

Doms, R. W., Russ, G., and Yewdell, J. W., 1989, Brefeldin A redistributes resident and itinerant Golgi proteins to the endoplasmic reticulum, *J. Cell Biol.* **109**:61–72.

Duben, R., Griffiths, G., Frank, R., Argos, P., and Kreis, T. K., 1991, β-COP, a 110 kD protein associated with non-clathrin-coated vesicles and the Golgi complex, shows homology with β-adaptin, *Cell* **64**:649–665.

Dunphy, W. G., and Rothman, J. E., 1983, Compartmentation of asparagine-linked oligosaccharide processing in the Golgi apparatus, *J. Cell Biol.* **97**:270–275.

Franke, W. W., 1971, Cytoplasmic microtubules linked to endoplasmic reticulum with cross-bridges, *Exp. Cell Res.* **66**:486–489.

Gallione, C. J., and Rose, J. K., 1985, A single amino acid substitution in a hydrophobic domain causes temperature-sensitive cell-surface transport of a mutant viral glycoprotein, *J. Virol.* **54**:374–382.

Goldberg, D. E., and Kornfeld, S., 1983, Evidence for extensive subcellular organization of asparagine-linked oligosaccharide processing and lysosomal enzymes phosphorylation, *J. Biol. Chem.* **258**:3159–3165.

Groesch, M., Ruohola, H., Bacon, R., Rossi, G., and Ferro-Novick, S., 1990, Isolation of a functional vesicular intermediate that mediates ER to Golgi transport in yeast, *J. Cell Biol.* **111**:45–53.

Helenius, A., Marquardt, T., and Braakman, I., 1992, The endoplasmic reticulum as a protein-folding compartment, *Trends Cell Biol.* **2**:227–231.

Hobman, T. C., Woodward, L., and Farquhar, M. G., 1992, The rubella virus E1 glycoprotein is arrested in a novel post-ER, pre-Golgi compartment, *J. Cell Biol.* **118**:795–812.

Hsu, V. W., Yuan, L. C., Nuchtern, J. G., Lippincott-Schwartz, J., Hammerling, G. J., and Klausner, R. D., 1991, A recycling pathway between the endoplasmic reticulum and the Golgi apparatus for retention of unassembled MHC class I molecules, *Nature* **352**:441–444.

Hsu, V. W., Shah, N., and Klausner, R. D., 1992, A brefeldin A-like phenotype is induced by the overexpression of a human ERD-2-like protein, ELP-1, *Cell* **69**:625–635.

Hurtley, S. M., and Helenius, A., 1989, Protein oligomerization in the endoplasmic reticulum, *Annu. Rev. Cell Biol.* **5**:277–307.

Jamieson, J. D., and Palade, G. E., 1967, Intracellular transport of secretory proteins in the pancreatic exocrine cell. I: Role of the peripheral elements of the Golgi complex, *J. Cell Biol.* **34**:577–596.

Kartenbeck, J., Stukenbrok, H., and Helenius, H., 1989, Endocytosis of simian virus 40 into the endoplasmic reticulum, *J. Cell Biol.* **109**:2721–2729.

Klausner, R. D., Donaldson, J. G., and Lippincott-Schwartz, J., 1992, Brefeldin A: Insights into the control of membrane traffic and organelle structure, *J. Cell Biol.* **116**:1071–1080.

Kornfeld, R., and Kornfeld, S., 1985, Assembly of asparagine linked oligosaccharides, *Annu. Rev. Biochem.* **54**:631–664.

Kreis, T. E., and Lodish, H. F., 1986, Oligomerization is essential for transport of vesicular stomatitis viral glycoprotein to the cell surface, *Cell* **46**:929–937.

Kuismanen, E., and Saraste, J., 1989, Low temperature-induced transport blocks as tools to manipulate membrane traffic, *Methods Cell Biol.* **32**:257–274.

Lee, C., and Chen, L. B., 1988, Dynamic behavior of endoplasmic reticulum in living cells, *Cell* **54**:37–46.

Lewis, M. J., and Pelham, H. R. B., 1992, Ligand induced redistribution of a human KDEL receptor from the Golgi complex to the endoplasmic reticulum, *Cell* **68**:353–364.

Lindsey, J. D., and Ellisman, M. H., 1985a, The neuronal endomembrane system. I: Direct links between rough endoplasmic reticulum and the cis element of the Golgi apparathus, *J. Neurosci.* **5**:3111–3123.

Lindsey, J. D., and Ellisman, M. H., 1985b, The neuronal endomembrane system. II: The multiple forms of the Golgi apparatus cis element, *J. Neurosci.* **5**:3124–3134.

Lippincott-Schwartz, J., Yuan, L. C., Bonifacino, J. S., and Klausner, R. D., 1989, Rapid redistribution of Golgi proteins into the ER in cells treated with brefeldin A: Evidence for a membrane cycling from Golgi to ER, *Cell* **56**:801–813.

Lippincott-Schwartz, J., Donaldson, J. G., Schweizer, A., Berger, E. G., Hauri, H. P., Yuan, L. C., and Klausner, R. D., 1990, Microtubule-dependent retrograde transport of proteins into the ER in the presence of brefeldin A suggests an ER recycling pathway, *Cell* **60**:821–836.

Lodish, H. F., Kong, N., Hirani, S., and Rasmussen, J., 1987, A vesicular intermediate in the transport of hepatoma secretory proteins from the rough endoplasmic reticulum to the Golgi complex, *J. Cell Biol.* **104**:221–230.

Lotti, L. V., Torrisi, M. R., Pascale, M. C., and Bonatti, S., 1992, Immunocytochemical analysis of the transfer of vesicular stomatitis virus G glycoprotein from the intermediate compartment to the Golgi complex, *J. Cell Biol.* **118**:43–50.

Louvard, D., Reggio, H., and Warren, G., 1982, Antibodies to the Golgi complex and the rough endoplasmic reticulum, *J. Cell Biol.* **92**:92–106.

Mellman, I., and Simons, K., 1992, The Golgi complex: In vitro veritas? *Cell* **68**:829–840.

Merisko, E. M., Fletcher, M., and Palade, G. E., 1986, The reorganization of the Golgi complex in anoxic pancreatic acinar cells, *Pancreas* **1**:95–109.

Munro, S., and Pelham, H. R. B., 1987, A C-terminal signal prevents secretion of luminal ER proteins, *Cell* **48**:899–907.

Nasciutti, L. E., Picart, R., Rosenbaum, E., Tixier-Vidal, A., and Tougard, C., 1992, Effect of reduced temperatures and brefeldin A on prolactin secretion and on subcellular distribution of the secretory product and membrane antigens in GH3 pituitary cells, *Biol. Cell* **72**:25–35.

Nguyen Van, P., Peter, F., and Soling, H. D., 1989, Four intracisternal calcium-binding glycoproteins from rat liver microsomes with high affinity for calcium. No indication for calsequestrin-like proteins in inositol 1,4,5-trisphosphate-sensitive calcium sequestering rat liver vesicles, *J. Biol. Chem.* **264**:17494–17501.

Orci, L., Ravazzola, M., Meda, P., Holcomb, C., Moore, H. P., Hicke, L., and Schekman, R. S., 1991, Mammalian sec23p homologue is restricted to the endoplasmic reticulum transitional cytoplasm, *Proc. Natl. Acad. Sci. USA* **88**:8611–8615.

Orrenius, S., Ericsson, J. L. E., and Ernster, L., 1965, Phenobarbital-induced synthesis of the microsomal drug-metabolizing enzyme system and its relationship to the proliferation of endoplasmic membranes, *J. Cell Biol.* **25**:627–639.

Ottosen, P. D., Courtoy, P. J., and Farquhar, M. G., 1980, Pathways followed by membrane recovered from the surface of plasma cells and myeloma cells, *J. Exp. Med.* **152**:1–19.

Pacifici, M., and Iozzo, R., 1988, Remodeling of the rough endoplasmic reticulum during stimulation of procollagen secretion by ascorbic acid in cultured chondrocytes, *J. Biol. Chem.* **263**:2483–2492.

Palade, G. E., 1975, Intracellular aspects of the process of protein synthesis, *Science* **189**:347–358.

Pascale, M. C., Erra, M. C., Malagolini, N., Serafini-Cessi, F., Leone, A., and Bonatti, S., 1992,

Post-translational processing of an O-glycosylated protein, the human CD8 glycoprotein, during the intracellular transport pathway to the plasma membrane, *J. Biol. Chem.* **267**:25196–25201.

Pathak, R. K., Merkle, R. K., Cummings, R. D., Goldstein, J. L., Brown, M. S., and Anderson, R. G. W., 1988, Immunocytochemical localization of mutant low density lipoprotein receptors that fail to reach the Golgi complex, *J. Cell Biol.* **106**:1831–1841.

Paulike, M., Nowack, D. D., and Morré, J. D., 1988, Isolation of a vesicular intermediate in the cell-free transfer of membrane from transitional elements of the endoplasmic reticulum to Golgi apparatus cisternae of rat liver, *J. Biol. Chem.* **263**:17738–17748.

Pelham, H. R. B., 1988, Evidence that luminal ER proteins are sorted from secreted proteins in a post-ER compartment, *EMBO J.* **7**:913–918.

Pelham, H. R. B., 1989, Control of protein exit from the endoplasmic reticulum, *Annu. Rev. Cell Biol.* **5**:1–23.

Plutner, H., Cox, A. D., Pind, S., Khosravi-Far, R., Bourne, J. R., Schwaninger, R., Der, C. J., and Balch, W. E., 1991, Rab1b regulates vesicular transport between the endoplasmic reticulum and successive Golgi compartments, *J. Cell Biol.* **115**:31–43.

Rambourg, A., and Clermont, Y., 1990, Three-dimensional electron microscopy: Structure of the Golgi apparatus, *Eur. J. Cell Biol.* **51**:189–200.

Remmer, H., and Merker, H. J., 1963, Drug induced changes in liver endoplasmic reticulum: Association with drug metabolizing enzymes, *Science* **142**:1657–1658.

Rexach, M. F., and Schekman, R. W., 1991, Distinct biochemical requirements for the budding, targeting, and fusion of ER-derived transport vesicles, *J. Cell Biol.* **114**:219–229.

Rizzolo, L. I., Finidori, J., Gonzalez, A., Arpin, M., Ivanov, I. E., Adesnik, M., and Sabatini, D. D., 1985, Biosynthesis and intracellular sorting of growth hormone–viral envelope glycoprotein hybrids, *J. Cell Biol.* **101**:1351–1362.

Rose, J. K., and Doms, R. W., 1988, Regulation of protein export from the endoplasmic reticulum, *Annu. Rev. Cell Biol.* **4**:257–288.

Roth, J., 1991, Localization of glycosylation sites in the Golgi apparatus using immunolabeling and cytochemistry, *J. Electron Microsc. Tech.* **17**:121–131.

Rothman, J. E., and Orci, L., 1992, Molecular dissection of the secretory pathway, *Nature* **355**:409–415.

Saraste, J., and Kuismanen, E., 1984, Pre- and post-Golgi vacuoles operate in the transport of Semliki Forest virus membrane glycoproteins to the cell surface, *Cell* **38**:535–549.

Saraste, J., and Svensson, K., 1991, Distribution of the intermediate elements operating in ER to Golgi transport, *J. Cell Sci.* **100**:415–430.

Saraste, J., Palade, G. E., and Farquhar, M. G., 1986, Temperature sensitive steps in the transport of secretory proteins through the Golgi complex in exocrine pancreatic cells, *Proc. Natl. Acad. Sci. USA* **83**:6425–6429.

Saraste, J., Palade, G. E., and Farquhar, M. G., 1987, Antibodies to rat pancreas Golgi subfractions: Identification of a 58-kD cis-Golgi protein, *J. Cell Biol.* **105**:2021–2029.

Schweizer, A., Fransen, J. A. M., Bachi, T., Ginsel, L., and Hauri, H. P., 1988, Identification, by a monoclonal antibody, of a 53-kD protein associated with a tubulo-vesicular compartment at the cis-side of the Golgi apparatus, *J. Cell Biol.* **107**:1643–1653.

Schweizer, A., Fransen, J. A. M., Matter, K., Kreis, T. K., Ginsel, L., and Hauri, H. P., 1990, Identification of an intermediate compartment involved in protein transport from endoplasmic reticulum to Golgi apparatus, *Eur. J. Cell Biol.* **53**:185–196.

Schweizer, A., Matter, K., Ketcham, C. M., and Hauri, H. P., 1991, The isolated ER–Golgi intermediate compartment exhibits properties that are different from ER and cis-Golgi, *J. Cell Biol.* **113**:45–54.

Tartakoff, A. M., 1986, Temperature and energy dependence of secretory protein transport in the exocrine pancreas, *EMBO J.* **5**:1477–1482.

Terasaki, M., Chen, L. B., and Fujikawa, K., 1986, Microtubules and the endoplasmic reticulum are highly interdependent structures, *J. Cell. Biol.* **103**:1557–1568.

Tooze, J., and Hollinshead, M., 1991, Tubular endosomal networks in AtT20 and other cells, *J. Cell Biol.* **115**:635–653.

Tooze, S. A., Tooze, J., and Warren, G., 1988, Site of addition of N-acetyl-galactosamine to the E1 glycoprotein of mouse hepatitis virus-A59, *J. Cell Biol.* **106**:1475–1487.

Torrisi, M. R., Lotti, L. V., Pavan, A., Migliaccio, G., and Bonatti, S., 1987, Free diffusion to and from the inner nuclear membrane of newly synthesized plasma membrane glycoproteins, *J. Cell Biol.* **104**:733–737.

Ulmer, J. B., and Palade, G. E., 1989, Targeting and processing of glycophorins in murine erythroleukemia cells: Use of brefeldin A as perturbant of intracellular traffic, *Proc. Natl. Acad. Sci. USA* **86**:6992–6996.

Ulmer, J. B., and Palade, G. E., 1991, Effect of brefeldin A on the Golgi complex, endoplasmic reticulum and viral envelope glycoproteins in murine erythroleukemia cells, *Eur. J. Cell Biol.* **54**:38–54.

Vaux, D., Tooze, J., and Fuller, S., 1990, Identification by anti-idiotype antibodies of an intracellular membrane protein that recognizes a mammalian endoplasmic reticulum retention signal, *Nature* **345**:495–501.

Vertel, B. M., Velasco, A., LaFrance, S., Walters, L., and Kaczam-Daniel, K., 1989, Precursors of chondroitin sulfate proteoglycan are segregated within a subcompartment of the chondrocyte endoplasmic reticulum, *J. Cell Biol.* **109**:1827–1836.

Walter, R. J., and Tandler, B., 1989, Viruses and annulate lamellae in Friend erythroleukemia cells, *J. Submicrosc. Cytol. Pathol.* **21**:93–101.

Warren, G., 1987, Signals and salvage sequences, *Nature* **327**:17–18.

Zuber, C., Roth, J., Misteli, T., Nakano, A., and Moremen, K., 1991, DS28-6, a temperature-sensitive mutant of Chinese hamster ovary cells, expresses key phenotypic changes associated with brefeldin A treatment, *Proc. Natl. Acad. Sci. USA* **88**:9818–9822.

Chapter 7

The Endoplasmic Reticulum as a Site of Protein Degradation

AnnaMaria Fra and Roberto Sitia

1. INTRODUCTION

Proteins that are not secreted or otherwise released by cells must be eventually degraded. Cytosolic degradation cannot cope with proteins that have been delivered unidirectionally to membrane-bound compartments. Hence, proteolytic systems are likely to be present in all compartments where proteins tend to accumulate (e.g., mitochondria, peroxisomes). In this respect, the endoplasmic reticulum (ER) appears to be no exception, since an independent degradative

Abbreviations used in this chapter: ALLM, *N*-acetyl-leucyl-leucyl-methioninal; ALLN, *N*-acetyl-leucyl-leucyl-norleucinal; apo B-100, apolipoportein B-100; ASGR, asialoglycoprotein receptor; BiP, heavy-chain binding protein; BFA, brefeldin A; CCCP, carbonyl-cyanide-chlorophenyl hydrazone; CFTR, cystic fibrosis transmembrane conductance regulator; CHO, Chinese hamster ovary; ER, endoplasmic reticulum; GC, Golgi complex; HMG-CoA, 3-hydroxy-3-methylglutaryl coenzyme A; Ig, immunogloobulin; LDL, low-density lipoprotein; PDI, protein disulfide isomerase; TCR, T cell receptor; TLCK, *N*-tosyl-L-lysine chloromethyl ketone; TPCK, *N*-tosyl-L-phenylalanine chloromethyl ketone; VLDL, very-low-density lipoprotein; VSV, vesicular stomatitis virus; ZPCK, *N*-carbobenzoxy-L-phenylalanine chloromethyl ketone.

AnnaMaria Fra   Department of Biology and Technology Research, San Raffaele, Milan, Italy.   **Roberto Sitia**   Department of Biology and Technology Research, San Raffaele, Milan, and National Institute for Cancer Research, Genoa, Italy.
Subcellular Biochemistry, Volume 21: Endoplasmic Reticulum, edited by N. Borgese and J. R. Harris. Plenum Press, New York, 1993.

system has been identified in this organelle (Lippincott-Schwartz *et al.,* 1988; Klausner and Sitia, 1990).

In this chapter, we shall present some facts and speculations about the process of ER degradation. Although degradation in the ER of many membrane and lumenal proteins has been described in considerable detail, it is still premature to draw unifying models. Thus, after having summarized the data available on some well-characterized experimental systems, we shall try to focus on the most relevant open questions. Owing to their strict relationship with the problem of degradation in the ER, it is, however, necessary to revisit some of the key aspects of protein transport and quality control within the exocytic pathway.

1.1. The ER as Part of the Central Vacuolar System: Retention and Retrieval of Resident Proteins

The ER is the port of entry of newly synthesized proteins destined to the central vacuolar system, which includes the organelles of the exocytic and endocytic pathways (reviewed by Palade, 1975; Klausner, 1989). Following translocation into the ER, transport of proteins to the distal organelles of the secretory pathway is thought to occur by "bulk flow" (Wieland *et al.,* 1987). Thus, specific mechanisms must exist that prevent resident ER proteins from being transported farther by the bulk flow. This can be mediated either by true *retention* (transport incompetency) or by *retrieval* from some distal compartment. Transport incompetency may result from the formation of aggregates that are excluded from the vesicular flow. This appears to be the case for ribophorins and other proteins involved in ribosome binding and membrane translocation, which are thought to form large supramolecular complexes (Hortsch and Meyer, 1985). In the case of soluble proteins, a C-terminal tetrapeptide motif, KDEL (and a few variants), is sufficient to confer ER localization (Munro and Pelham, 1987), via a receptor-mediated retrieval system from the Golgi stacks (Semenza *et al.,* 1990; Lewis *et al.,* 1990; Lewis and Pelham, 1992; Hsu *et al.,* 1992). However, the rate of transport to the Golgi complex (GC) differs considerably among individual ER lumenal proteins from which the KDEL sequence has been deleted (Haugejorden *et al.,* 1991), suggesting that other mechanisms might be at work to determine their ER residency. Several transmembrane ER proteins have been shown to contain a retention motif in their cytoplasmic tails, which consists of two lysines, one positioned at -3 and the other at -4 or -5 from the C-terminus. Also in this case, ER residency seems to depend on retrieval from some distal compartment, rather than on true retention (Jackson *et al.,* 1990). Thus, not only lipid homeostasis, but also compartment identity require efficient retrograde transport systems that selectively balance anterograde transport steps. Compelling evidence for the existence of a retrograde Golgi–ER membrane flow has been obtained (Lippincott-Schwartz *et al.*, 1990; see also Lippincott-Schwartz, this volume).

1.2. Quality Control of Newly Synthesized Proteins in the ER

Also, the proteins destined to the distal organelles of the central vacuolar system spend considerable time in the ER where they undergo folding, assembly, and several other posttranslational modifications (see Rowling and Freedman, this volume). Thus, the ER functions not only as a maternity department for a vast group of proteins, but also as a school where proteins mature and are thoroughly checked (quality control), before proceeding to the GC (Rose and Doms, 1988; Hurtley and Helenius, 1989; Pelham, 1989). Although this quality control activity is essential when newly synthesized proteins first acquire their three-dimensional structure, in the case of resident proteins it may well continue for their entire life span, and regulate their activity and turnover. As a result of quality control, unassembled or misfolded proteins are unable to reach the GC, either because they form aggregates or because they interact with ER resident proteins such as BiP or protein disulfide isomerase (PDI). These interactions might serve the dual role of preventing premature transport and facilitating folding and assembly.

2. CHARACTERISTICS OF DEGRADATION IN THE ER

Degradation in the ER was originally defined by (1) its insensitivity to lysosomotropic drugs, such as NH_4Cl or chloroquine, (2) immunolocalization experiments, which showed accumulation of the degradation substrate in the ER, and (3) lack of maturation of the substrate oligosaccharide moieties, characteristic of ER retained proteins (Lippincott-Schwartz *et al.*, 1988). In general, degradation in the ER is not inhibited by a number of drugs that perturb transport at various steps along the exocytic pathway, e.g., brefeldin A (BFA), monensin, carbonyl-cyanide-chlorophenyl-hydrazone (CCCP). In addition, with the exception of the asialoglycoprotein receptor (ASGR), degradation intermediates cannot be detected. In the case of transmembrane proteins, this implies the coordinated degradation of the lumenal, transmembrane, and cytosolic portions. A list of proteins degraded in the ER is reported in Table I. Among membrane proteins, the most studied examples include subunits of the T cell receptor (TCR) and ASGR. Immunoglobulin (Ig) chains offer a nice model for comparing the fate of membrane and lumenal proteins, and so do certain ribophorin mutants. HMG-CoA reductase and apolipoprotein B-100 (apo B-100) (see Davis, and Ericsson and Dallner, this volume) are of interest for their involvement in important

Table I
Substrates of Degradation in the ER

Proteins	References[a]
TCR subunits	Lippincott-Schwartz et al. (1988), Chen et al. (1988), Bonifacino et al. (1989), Stafford and Bonifacino (1991), Wileman et al. (1991)
Immunoglobulin subunits	Sitia et al. (1987), Weiss and Bogen (1991), Bachhawat and Pillai (1991), Amitay et al. (1992)
Asialoglycoprotein receptor	Lederkremer and Lodish (1991), Amara et al. (1989), Wikström and Lodish (1991, 1992)
Apoliproprotein B	Sato et al. (1990), Davis et al. (1990)
HMG-CoA reductase	Roitelman et al. (1992), Chun et al. (1990), Chuck et al. (1990)
Ribophorin mutants	Tsao et al. (1992)
α_1-Antitrypsin mutants	Le et al. (1990, 1992)
Cystic fibrosis transmembrane conductance regulator	Cheng et al. (1990)
LDL receptors	Esser and Russell (1988)
β-Lactamase–α-globin chimera	Stoller and Shields (1989)
Acetylcholine receptors	Blount and Merlie (1990)
Retinol binding protein	Tosetti et al. (1992)
β-Hexosaminidase mutant	Lau and Neufeld (1989)
Acetylcholinesterase	Rotundo (1988), Rotundo et al. (1989)

[a]For space limitation, only selected references are included.

metabolic pathways, while the behavior of cystic fibrosis transmembrane conductance regulator (CFTR) and of α_1-antitrypsin mutants is relevant in disease.

2.1. T Cell Receptor Subunits

The functional form of antigen receptors expressed on the surface of T lymphocytes is composed of at least seven subunits: α, β, γ, δ, ϵ, and ζ_2. The assembly pathway has been characterized in detail: formation of a pentameric structure comprising α, β, γ, δ, and ϵ chains is required for exiting the ER. Further assembly with a ζ_2 homodimer, or a $\zeta\eta$ heterodimer, is essential for surface expression. The availability of cell lines expressing various combinations of TCR subunits has allowed a detailed molecular dissection of the posttranslational events regulating assembly, transport, and degradation. Thus, in either T cells or fibroblasts, individual chains or partial complexes fail to reach the GC, as indicated by the absence of oligosaccharide maturation, and are selectively degraded in the ER. Following an initial lag period (ranging from 10 to 30 min) α, β, and δ chains are degraded with half lives of about 30 min, while ϵ and ζ chains are more stable. The fate of γ chains has been shown to depend on the state of assembly: thus, free γ chains are rapidly degraded, whereas ϵ–γ complexes are rather stable, even though they are unable to reach the Golgi. These observations point to the existence of structural motifs that target individual chains to degradation, and that can be masked upon oligomerization. When the last 28 amino acids of the TCRα chain, comprising the transmembrane and cytosolic portions, are deleted, the truncated soluble protein is retained in the lumen of the ER and displays a long half-life. This observation indicates that retention *per se* is not sufficient to induce degradation in the ER and suggests that the deleted sequences contain information for rapid degradation. Indeed, transplanting the C-terminal portion of TCRα onto CD4 or Tac results in rapid, BFA-insensitive degradation of the chimeric proteins. Further site-directed mutagenesis showed that the presence of a single charged residue within the transmembrane region is sufficient to cause rapid degradation in the ER. The position and the side chain of the charged residue determine the ultimate fate of the chimeric protein, degradation being maximal when an arginine is placed ten residues below the lumen–membrane boundary. While covalent association with TCRβ chains does not inhibit the degradation of TCRα, assembly with CD3-δ chains results in the stabilization of both TCRα and Tac-TCRα chimeric proteins (Bonifacino *et al.*, 1990a, b, 1991). Similarly, assembly with CD3-εγ dimers inhibits, albeit partially, the degradation of TCR α–β complexes (Wileman *et al.*, 1990).

In T cells lacking ζ chains, [α, β, γ, δ, ε] pentamers are routed to lysosomes, where all individual chains are degraded with identical kinetics (Letourneur and Klausner, 1992); this observation demonstrates once more the independence of the ER and lysosomal degradative pathways.

2.2. HMG-CoA Reductase

In the biosynthetic pathway of sterols and isoprenoids, HMG-CoA reductase catalyzes the conversion of HMG-CoA into mevalonic acid. HMG-CoA reductase is a high-mannose glycoprotein of 97 kDa that is thought to span the ER membrane eight times (see Ericsson and Dallner, this volume). The enzyme is regulated at different levels, including the degradation of the protein. In the presence of mevalonate, the half-life of the protein is reduced severalfold. The transmembrane domain is essential for the sterol-regulated degradation: when transferred to a reporter protein (β-galactosidase), it is sufficient to confer rapid, sterol-sensitive degradation in the ER. Replacement of the seventh transmembrane stretch abolishes regulated degradation (Roitelman *et al.*, 1992, and references therein). When UT-1 cells are grown in the presence of lovostatin, an inhibitor of mevalonate biosynthesis, drastic morphological modifications are induced, including the enlargement of the ER and the formation of the so-called crystalloid ER. The membranes of the latter are highly enriched in HMG-CoA reductase. Withdrawal from lovostatin leads to the disappearance of the crystalloid ER, a phenomenon involving selective autophagy. However, the mevalonate-induced HMG-CoA reductase degradation is not sensitive to 3-methyladenine, an inhibitor of autophagy, indicating that there are at least two pathways to dispose of the excess enzyme. Cycloheximide inhibits the regulated degradation, suggesting that a short-lived protein might be involved in the process (Chun et al., 1990).

2.3. Apolipoprotein B-100

apoB-100 is a major component of serum low- and very-low-density lipoproteins (LDL and VLDL, respectively). It is a large secretory protein rich in short hydrophobic sequences that bind lipids. This interaction is essential for making lipid soluble, as well as to prevent precipitation of hydrophobic apoB-100 in aqueous solutions. In liver cells about half of newly synthesized apoB-100 is translocated into the ER lumen and transported to the Golgi. The other half remains associated with the ER membrane and, as shown by protease sensitivity tests, is partially exposed to the cytosol. This pool is degraded by a BFA- and leupeptin-insensitive pathway (Davis et al., 1990; Sato et al., 1990). This two-step translocation pathway may be important in regulating the amount of apoB-100 secreted (see Davis, this volume).

2.4. Cystic Fibrosis Transmembrane Conductance Regulator

CFTR, the gene product of the cystic fibrosis-associated gene, consists of a protein of ±170 kDa with six membrane-spanning regions (Riordan et al., 1989). CFTR is a member of a class of related proteins involved in the ATP-dependent transport of molecules across cellular membranes. When transfected into COS-7 cells, constructs of CFTR engineered to mimic the mutations found in cystic fibrosis patients are unable to reach the cell surface. Immunolocalization experiments and analysis of the oligosaccharide processing suggest that CFTR mutants are retained and degraded in the ER. Interestingly, the relevant mutations are present in the cytosolic domain (Cheng et al., 1990). Also in the cases of VSV G protein and influenza virus hemagglutinin, mutations in the cytosolic domains were found to affect dramatically the transport competency (Rose and Doms, 1988, and references therein).

2.5. α_1-Antitrypsin Variants

α_1-Antitrypsin is a serine protease inhibitor secreted by hepatocytes. In the PiZ allele, a Glu-342 to Lys substitution is responsible for retention in the ER. The defective α_1-antitrypsin secretion predisposes homozygous individuals to pulmonary emphysema. In hepatocytes, part of PiZ accumulates in dilated ER cisternae in the form of stable insoluble aggregates, which can cause liver damage also in transgenic mice (Carlson *et al.*, 1989). The majority, however, is degraded in the ER by a mechanism that is sensitive to inhibitors of protein synthesis and pH modulators, but insensitive to BFA. Interestingly, attachment of KDEL at the C-terminus of PiZ prevents intracellular degradation (Le *et al.*, 1990, 1992).

2.6. Asialoglycoprotein Receptors

The ASGR is expressed on liver cells and binds serum glycoproteins bearing terminal galactose residues. It is composed of two type II membrane subunits, H1 and H2, which form oligomers. The H2 chain, which is less abundant than H1, is present in two forms arising by differential splicing, H2a and H2b, which differ for the presence or absence of a five-amino-acid peptide (Lederkremer and Lodish, 1991). When expressed alone in fibroblasts, the fate of the three subunits differs considerably. While H1 chains can be found, at least in part (about 50%), on the cell surface, all H2a is degraded intracellularly. Since a considerable number of H2b chains reach the cell surface, the miniexon present in H2a molecules might be important in determining the fate of this subunit. The degradation of both forms of H2 has been shown to occur in at least two steps. A rapid proteolytic event, sensitive to TLCK or TPCK, but insensitive to ATP depletion or cycloheximide, generates a soluble 35-kDa intermediate containing the C-terminal lumenal domain. The latter accumulates in the rough ER, and is degraded by a second mechanism, which is energy dependent and requires protein synthesis. The oligosaccharide moieties of the 35-kDa intermediates are Man_{6-9}-GlcNAc$_2$. Taken together, these observations indicate that the initial endoproteolytic cleavage occurs in the rough ER itself (Amara *et al.*, 1989; Wikström and Lodish, 1991, 1992).

2.7. Ribophorin Mutants

Ribophorin I is a type I membrane glycoprotein segregated into the rough portion of the ER, and it is part of the oligosaccharyl-transferase complex (Kelleher *et al.*, 1992). When expressed in HeLa cells the native protein is very stable. By contrast, two truncated variants, one constituted by most of the lumenal domain, the other including also the transmembrane region, are degraded without a detectable lag. While the degradation of the membrane-bound mutant is monophasic with a half-life of about 50 min, the soluble mutant shows a biphasic curve with a threefold increase in the rate of degradation after 50 min. In the presence of BFA, degradation of the soluble form becomes monophasic, with a half-life that is intermediate between the two; furthermore, CCCP inhibits the rapid phase of degradation, but only when added before this is initiated. Taken together, these results suggest the existence of two nonlysosomal degradation sites, one in the ER itself and the other requiring vesicular transport (Tsao *et al.*, 1992).

2.8. Immunoglobulins

Immunoglobulins are heteromultimeric glycoproteins composed of heavy (H) and light (L) chains. During B cell differentiation, different isotypes (e.g., IgM, IgG) are produced as the result of switch recombination. Furthermore, differential RNA processing events allow the generation of both membrane and soluble forms of H chains which constitute the B cell antigen receptor and the secreted antibodies, respectively. In the absence of L, both membrane and secretory H chains are retained in the ER by interactions involving BiP (which was originally defined as the "heavy-chain binding protein"; Haas and Wabl, 1983) and the first H constant domain (C_H1) (Bole *et al.*, 1986; Hendershot *et al.*, 1987). Expression of a functional L chain displaces BiP, and H chains are rapidly assembled into the basic H2L2 structure (Hendershot, 1990). This is sufficient to allow secretion of "monomeric" Ig, such as IgG, IgE, or IgD. By contrast, IgM and IgA are present in the serum and in secretions as disulfide-linked oligomers. Polymerization is essential for IgM secretion. A conserved cysteine residue (Cys-575) is present in the C-terminal extension found in μ, but not in γ, ϵ, or δ chains (the μ_s tailpiece). Cys-575 is not only responsible for the formation of the intersubunit disulfide bonds, but it also acts as an intracellular retention element for unpolymerized IgM (Sitia *et al.*, 1990). In plasma cells, free μ chains are rapidly degraded in the ER. When L chains are present, about one-third of secretory IgM is still degraded by a BFA-insensitive pathway, probably because of inefficient oligomerization. Replacement of Cys-575 with other amino acids inhibits degradation. Furthermore, attachment of the μ_s tailpiece to a myc-tagged cathepsin D (a lysosomal protease) is sufficient to cause retention and BFA-insensitive degradation of the chimeric protein (CDMμtp) only when the critical cysteine is present. Thus, Cys-575 appears to be responsible for assembly, retention, *and* degradation of IgM assembly intermediates (Fra *et al.*, unpublished data). For unknown reasons, B lymphocytes are unable to form oligomers and hence all secretory IgM are retained and degraded intracellularly (Sitia *et al.*, 1988; Amitay *et al.*, 1992; Fra *et al.*, unpublished data). However, at variance with the situation in plasma cells, in B cells the degradation of IgM subunits is partially inhibited by BFA (Amitay *et al.*, 1992).

That the mechanisms of retention and degradation might differ for membrane and lumenal proteins is suggested by studies on a pre-B lymphoma that produces both the membrane and the secretory forms of γ chains (γ_m and γ_s, respectively), but no L chains. Although neither γ_m nor γ_s chains reach the Golgi, the former is rapidly degraded, while γ_s forms a stable complex with BiP (Bachhawat and Pillai, 1991).

3. THE BIOCHEMISTRY OF DEGRADATION IN THE ER

Proteolytic enzymes are functionally grouped into four distinct classes: aspartine, serine, cysteine, and metal proteases. The precise nature of the proteases responsible for pre-Golgi degradation is still unknown. Indirect information has been gathered on the basis of inhibition studies. In the absence of a reproducible *in vitro* ER degradation assay, there are several obvious limitations in the use of inhibitors *in vivo,* as the drugs should be: (1) membrane permeant, (2) nontoxic, and (3) inactive on membrane traffic within the exocytic pathway. The results obtained in several systems are summarized in Table II.

3.1. Involvement of Cysteine and Metal Proteases

There is general agreement as to the involvement of cysteine proteases: indeed, sulfhydryl reagents, such as diamide or *N*-ethylmaleimide, completely inhibit the degradation of most substrates. Furthermore, ALLN, ALLM, TPCK, membrane-permeant E-64 derivatives, and other cysteine protease inhibitors significantly reduce the rate of degradation. Also, blocking the function of calpains,

Table II
Inhibitors of Degradation in the ER

Inhibitors	Main function	Substrates[a]
Diamide	Thiol oxidizer	HMG-CoA reductase, TCR subunits[b]
TPCK[c]	Cysteine-protease inhibitor	TCR subunits, ASGR (first step)
ALLN–ALLM[c]	Cysteine-protease inhibitor	HMG-CoA reductase
TLCK[c]	Serine-protease inhibitor	ASGR (first step)[d]
1,10-Phenanthroline	Metalloprotease inhibitor	Ribophorin I, TCR subunits
Co^{2+}	Ca^{2+} antagonist	HMG-CoA reductase, ribophorin I
Ionomycin–A23187	Ca^{2+} ionophores	HMG-CoA reductase, ribophorin I
Thapsigargin	Ca^{2+}-ATPase inhibitor	HMG-CoA reductase
Cycloheximide[e]	Protein synthesis inhibitor	HMG-CoA reductase, ASGR (second step)
Hexokinase + glucose NaCN + KF Deoxyglucose + NaN$_3$	ATP depletion[f]	HMG-CoA reductase

[a]References in the text.
[b]TCR subunits and derived chimeras.
[c]Abbreviations: ALLN: *N*-acetyl-leucyl-leucyl-norleucinal; ALLM: *N*-acetyl-leucyl-leucyl-methioninal; TPCK: *N*-tosyl-L-phenylalanine chloromethyl ketone; TLCK: *N*-tosyl-L-lysine chloromethyl ketone.
[d]Inactive on TCRα subunit.
[e]Inactive on TCR subunits, ribophorin I, and ASGR (first step).
[f]Inactive on ASGR (first step) and TCR subunits.

Ca^{2+}-dependent cysteine proteases (Pontremoli and Melloni, 1986; Melloni and Pontremoli, 1989), results in stabilization of TCR subunits.

The inhibition exerted by 1,10-phenanthroline, a metal chelator, points to an involvement of metalloproteases. Interestingly, Mn^{2+} and Mg^{2+}, but not Zn^{2+}, can reverse the activity of 1,10-phenanthroline (Wileman et al., 1991; Tsao et al., 1992).

3.2. The Role of Calcium

In view of the pleiotropic role of Ca^{2+} in cellular metabolic pathways, the results obtained by the use of Ca^{2+} ionophores should be interpreted with caution. Nonetheless, it is generally accepted that Ca^{2+} ions play an important role in degradation in the ER, and more in general in quality control mechanisms. The proposed role of calnexin in controlling the fate of membrane proteins is likely to depend on Ca^{2+} (Ahluwalia et al., 1992). In addition, the mechanisms of retention of some lumenal ER resident proteins may involve their Ca^{2+} binding activity (Booth and Koch, 1989). Changes in the Ca^{2+} levels might thus affect, directly or indirectly, the fate of other soluble proteins (Suzuki et al., 1991). Another possible link is represented by calpains (see above). Thus, it is not surprising that Ca^{2+} ionophores often have profound effects on ER degradation. Indeed, Ca^{2+} depletion has been shown to accelerate the degradation of TCR and CD3 subunits (Wileman et al., 1991), but to inhibit the degradation of truncated ribophorin mutants (Tsao et al., 1992) and of HMG-CoA reductase. In the latter case, oscillations in the Ca^{2+} concentrations, perhaps induced by mevalonate, might explain the regulated catabolism of HMG-CoA reductase (Roitelman et al., 1991).

How can the opposite effects of Ca^{2+} ions on different substrates be explained? In a recent study, Inoue and Simoni (1992) have compared the degradation of $TCR\alpha$ and HMG-CoA reductase expressed in CHO cells. The results, summarized in Table III, indicate that although the two proteins appear to be degraded in the same subcellular compartment, the mechanisms responsible for degradation differ. Thus, while ALLM and diamide inhibit the degradation of both molecules, TPCK and ZPCK are active only on $TCR\alpha$. The reverse is true for cycloheximide and energy blockers, which do not affect $TCR\alpha$ but inhibit both basal and mevalonate-induced degradation of HMG-CoA reductase.

3.3. Inhibitors of Protein Synthesis and Energy Production

The effects of cycloheximide point to a role for short-lived protein(s), perhaps a protease itself, in the regulated degradation of HMG-CoA reductase. By contrast, other substrates, such as $TCR\alpha$, are insensitive to inhibitors of protein synthesis. A selective requirement for protein synthesis was demonstrated also

Table III
Differential Stabilization of TCRα Subunit[a] and HMG-CoA Reductase[a]
by Pharmacological Treatments

Inhibitors	TCRα subunit	HMG-CoA reductase[b]
ALLM (100 μg/ml)	+	+
TPCK (10 μg/ml)	+	−
TLCK (100 μg/ml)	−	−
ZPCK[c] (10 μg/ml)	+	−
A23187 (1 μM)–ionomycin (1 μM)	−	+
Thapsigargin (500 μM)	−	+
Co^{2+} (1 mM)	−	+
Diamide (1 mM)	+	+
Cycloheximide (500 μM)	−	+
NaCN (1 mM) + KF (10 mM)	−	+
Deoxyglucose (25 mM) + NaN$_3$ (2.5 mM)	−	+

[a]Both proteins expressed in CHO cells (Inoue and Simoni, 1992).
[b]Basal and mevalonate-induced degradation were similarly affected by inhibitors with the only exception of Ca^{2+} ionophores which display a stronger effect on the latter.
[c]Abbreviation (see also Table II) ZPCK: N-carbobenzoxy-L-phenylalanine chloromethyl ketone.

for the second proteolytic step of ASGR (Wikström and Lodish, 1991). The requirements for ATP largely overlap with those for protein synthesis (see for an example Tables II and III): the effects of ATP depletion on degradation in the ER may thus be secondary to the inhibition of protein synthesis. Alternative possibilities are a requirement for substrate phosphorylation, the involvement of an energy-dependent vesicular transport step, or the release of substrates from BiP induced by ATP.

3.4. Is a Proteasome Involved?

Ubiquitin-dependent degradation is mediated by multicatalytic protease complexes, also referred to as proteasomes. These large macromolecular structures (20 S) are composed of numerous subunits, and may further associate with other polypeptides, i.e., CF-1 and CF-2, to form an even larger complex (26 S). In mice, rats, and humans, the genes encoding CF-1 and CF-2 are located in the MHC complex, suggesting a linkage with antigen presentation (see Cerundolo, this volume). Proteasomes are abundant in the cytosol and nucleus, but they have also been detected in the microsomal fraction, although their presence in the lumen of the ER has not been reported convincingly. They may thus be involved in the degradation of the cytosolic domains of transmembrane proteins.

4.   THE PROBLEM OF SELECTIVITY

The half lives of individual polypeptides range from a few minutes to several days. Thus, a critical question underlying the process of protein degradation is what determines its selectivity (Olson and Dice, 1989; Bachmair and Varshawsky, 1989). In principle, specificity can be obtained in two ways. Proteins can be selectively sorted to a specialized compartment, where proteases nonspecifically degrade them. Alternatively, short-lived proteins can be tagged so as to become specific substrates of noncompartmentalized proteases. Lysosomes are the best example of the first case. By contrast, one of the best characterized proteolytic systems in the cytosol relies on selective ubiquitin conjugation of the substrate, and subsequent degradation by the multicatalytic protease complex (Hershko, 1988). In both cases, however, specific structural features, hereafter referred to as "degradation targeting signals," must be present on short-lived proteins.

It is well known that folding, assembly, and certain posttranslational mod-

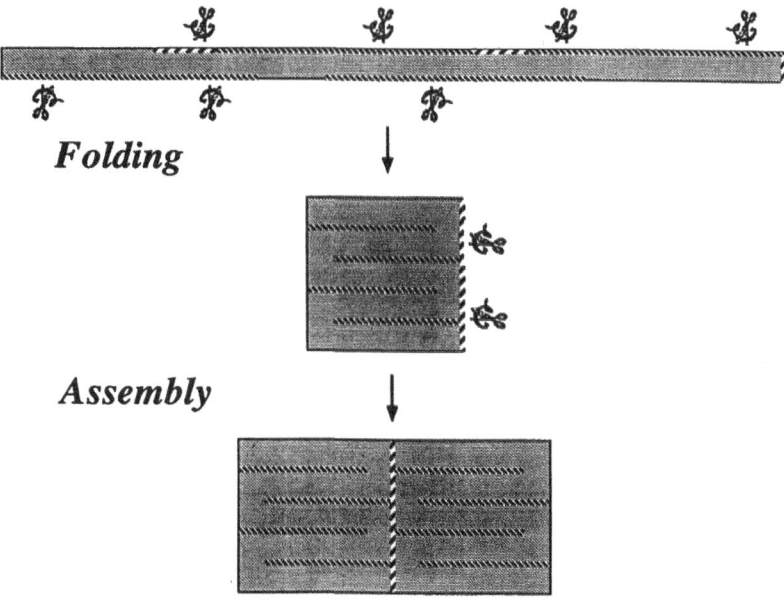

FIGURE 1. Folding and assembly protect proteins from degradation. The simplest way to explain the increased stability of folded and assembled molecules is the reduction in solvent-exposed surfaces. In the schematic drawing shown, the two unfolded precursors have a total surface of 104 units, the two unassembled subunits of 40 and the assembled homodimer of 30. In addition, proteases (symbolized here as scissors) could recognize selectively surfaces involved in folding and assembly.

ifications (e.g., glycosylation) increase protein stability. This may be the result of a decrease in the number of accessible degradation targeting signals (Figure 1).

4.1. Retention in the ER Is Not Sufficient to Cause Degradation

Defective folding, defective posttranslational modifications, and/or unbalanced subunit biosynthesis lead to the accumulation of transport-incompetent polypeptides in the ER, which become substrates for degradation. The easiest way to link degradation to transport incompetency would be to assume that retention in the ER increases the chances of encountering a protease. This simple model is untenable for several reasons. First, proteins retained in the ER are degraded at vastly different rates. Second, most ER resident proteins, such as BiP or PDI, are very stable. Third, in the case of a rapidly degraded PiZ mutant, addition of a KDEL motif makes the protein more stable (Le *et al.*, 1990). Thus, the failure of a protein to exit the ER is not sufficient to cause degradation; the latter must depend on the recognition of specific structural features. Since very little is known about the enzymology and the precise localization of the ER proteolytic system, it is not possible at present to determine whether suscep-

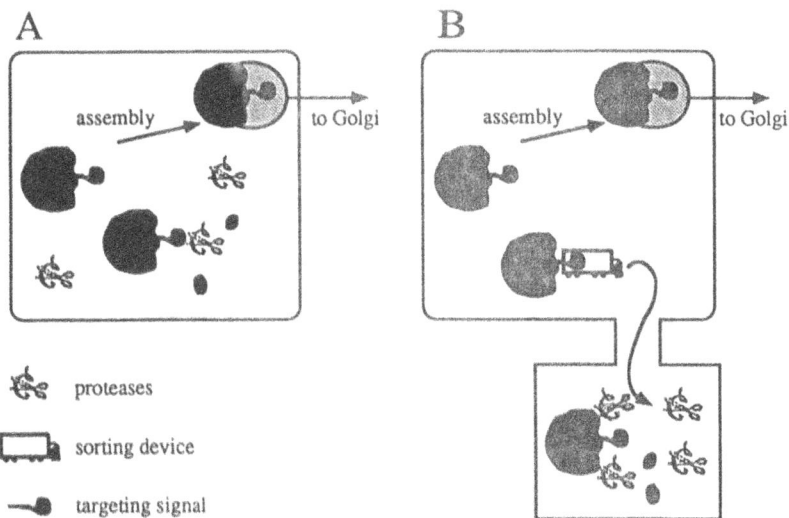

FIGURE 2. Specificity of degradation in the ER: sorting elements or protease attachment sites? A degradation targeting signal (the pipelike structure) might perform its duties in two ways: Either it acts as a protease attachment site (A) or as a sorting device to a degradative subcompartment (the little truck in B). Assembly would mask the degradation targeting signal, allowing transport to the Golgi complex (see also Figure 3).

tibility of a given protein reflects its quality as a substrate for protease(s) or the presence of sorting signals to a specific subcompartment (Figure 2).

4.2. Folding and Assembly Increase Protein Stability

Also within the ER, assembled complexes are generally more stable than isolated chains. However, not all subunit interactions lead to an increased resistance to proteolysis. For instance, while the association with CD3-ε chains stabilizes CD3-γ, TCRα chains are equally degraded alone or when complexed with TCRβ (Bonifacino *et al.*, 1989). Thus, assembly *per se* is not always sufficient to protect from degradation. Stabilization occurs only when oligomerization masks all degradation targeting signals.

When extended with two independent ER degradation targeting signals, the transmembrane and cytosolic regions of the TCRα chain and the μ_s C-terminal cysteine, both γ2b and μ heavy chains are retained efficiently in the ER, but the latter are much more rapidly degraded (Fra *et al.*, unpublished data). Thus, the presence of degradation targeting signal(s) is not the only element that determines the rate of degradation. γ2b heavy chains could be intrinsically more resistant to ER proteases than μ chains. It is possible that ER resident proteins have been selected for their intrinsic resistance to ER proteases. It would be of interest to determine whether the addition of an ER degradation signal is sufficient to cause their degradation.

4.3. Colocalized Determinants for Assembly, Retention, and Degradation

From a comparative analysis of the results obtained on TCR and Ig, a general model emerges that links assembly, retention, and degradation by the existence of colocalized determinants for the three alternative fates (Figure 3).

The data in Section 2.1 clearly indicate that sequences within the transmembrane region of both α and β TCR chains are responsible for both assembly and rapid degradation of unassembled chains. A similar situation is encountered in surface Ig, which also requires association with "CD3-like" molecules in order to be transported to the cell surface (Sitia *et al.*, 1987; Hombach *et al.*, 1988; Wienands *et al.*, 1990; Campbell and Cambier, 1990). There are no charged residues in the transmembrane portion of the membrane form of μ chains; however, four hydrophilic residues have been shown to mediate the retention of IgM in the absence of associated molecules (Williams *et al.*, 1990). A common feature of T and B cell receptors is thus the remarkable hydrophilicity of their transmembrane regions. Polar and charged residues would be out of place in the lipid bilayer, and look for hydrophilic amino acid(s) in the transmembrane regions of other proteins to interact with. Indeed, both TCR- and B cell receptor-associated molecules have charged amino acids within their transmembrane re-

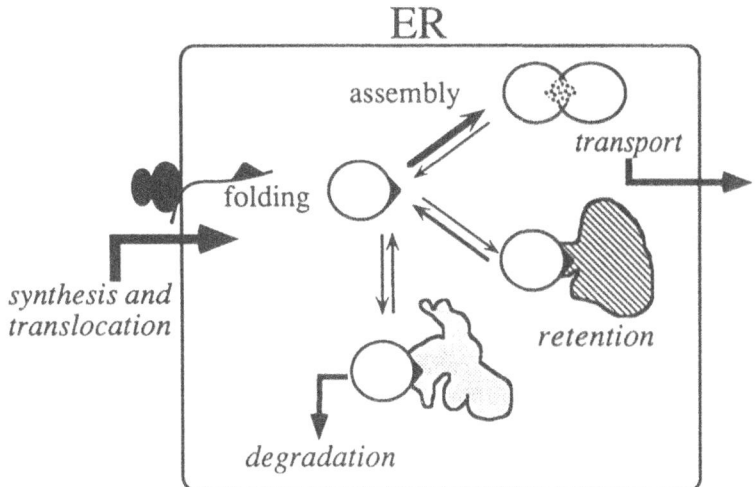

FIGURE 3. Colocalized determinants for assembly, retention, and degradation explain quality control. Cotranslational translocation into the ER, folding, assembly, and transport to the Golgi are sequential events in the life of newly synthesized proteins. Unassembled subunits interact transiently with BiP, PDI, and/or other resident proteins in the ER (striped) via the same element(s) that also mediate assembly and degradation (the dark triangle). Oligomerization will mask the signal (now in light gray) and allow transport. In the scheme, the degradative pathway is mediated by the molecular drawn in gray, which might be either a sorting device or a protease itself (see also Figure 2).

gions. The conservation of these residues in TCRα and δ from different species contrasts with the divergence of the surrounding sequences (Clevers *et al.*, 1988). In the absence of their cognate assembly interfaces, the interaction with yet unidentified proteins could lead to retention–degradation. Among the candidates for quality control of membrane protein assembly is calnexin, an 88-kDa abundant membrane protein that associates transiently with unassembled class I and TCR subunits (Wada *et al.*, 1991; Ahluwalia *et al.*, 1992; Hochstenbach *et al.*, 1992). Alternatively, the hydrophilicity of the transmembrane region might impair its function as a stop transfer sequence (Shin *et al.*, 1993).

Do soluble proteins obey the same general rule? That this is indeed the case is suggested by some of the studies on Ig referred to above, which imply the C-terminal cysteine of secretory μ chains in assembly, retention, and degradation of IgM subunits (Sitia *et al.*, 1990; Fra *et al.*, in preparation). For lumenal molecules, the situation would be obviously reversed, hydrophobic surfaces and thiol groups being the reactive equivalents to hydrophilic residues within lipid bilayers. Individual folding or assembly events will result in the masking of specific retention–degradation determinant(s). The biosynthesis of oligomeric IgM is a paradigm of the hierarchical organization of quality control events: first, the Cμ1–BiP interaction controls μ–L assembly. Only later will the role of the

C-terminal cysteine emerge in preventing secretion of unpolymerized $\mu 2L2$ subunits.

In the model outlined, it is a competition between the three reactions that explains quality control (Figure 3). An obvious prediction is that, when secretion is required, assembly reactions are favored compared with those leading to retention and degradation. Indeed, free light chains can dissociate BiP–H chain complexes (Hendershot, 1990). Several lines of evidence point to a role of BiP in the retention of soluble subunits by recognition of hydrophobic surfaces or peptide loops (Hendershot et al., 1987; Flynn et al., 1989, 1991) which are likely to become inaccessible upon oligomerization. Alternatively, hydrophobic interactions or disulfide interchange reactions may lead to the formation of aggregates, too big to enter transport vesicles. BiP (or ER-oxidoreductases such as PDI or ERp72) would then work as "true chaperones," preventing the irreversible denaturation or cross-linking of their substrates, being themselves retained while performing this function (Gething and Sambrook, 1990, 1992).

4.4. What Is the Role of BiP and Other Chaperones in ER Degradation?

If on the one hand the interaction with BiP might inhibit degradation, either by competing with substrate recognition by proteases or by preventing sorting to the degradative compartment, on the other hand it may keep the substrate in an accessible configuration. If indeed BiP and other chaperones can act as catalysts of protein folding and assembly, they would, under certain conditions, catalyze the reverse reactions. It is conceivable that partially assembled oligomers must be disassembled and unfolded for efficient degradation to occur. A similar requirement is probably shared by ribosome-inactivating toxins, such as Shiga or tetanus toxin (Sandvig et al., 1992). These molecules must in fact cross a membrane bilayer to exert their toxicity in the cytosol. Some of them might indeed utilize a recycling pathway (perhaps involving directly or indirectly the KDEL receptor) to reach the ER, and exploit the folding/unfolding activities to reach a translocation-competent state (Pelham et al., 1992). It remains to be seen whether the same chaperones catalyze these reactions, perhaps as the result of recycling into specialized ER subregions (Sitia and Meldolesi, 1992), or whether other proteins are involved.

4.5. Protein Sorting and Storage in the ER as the Result of Aggregation

In general, the overexpression of misfolded proteins in the ER results in an increase of BiP biosynthesis (Kassenbrook et al., 1988; Kozutsumi et al., 1988). This has been interpreted as a cellular response to prevent formation of aggregates that may be toxic, perhaps disturbing transport within the exocytic compartment. In plants, however, protein aggregation within the ER is part of certain developmental programs. In maize seeds, for instance, large aggregates of zeins accumulate in dilated ER cisternae (called protein bodies) as a food store. Similar aggregates are formed in *Xenopus* oocytes upon injection of zein mRNA, indicating that tissue- or species-specific factors are not required (Wallace *et al.*, 1988). Several examples of protein sorting within the ER as a result of protein condensation have been described in mammalian cells. These are referred to as "lumenal bodies," and include aggregates of PiZ in liver (Le *et al.*, 1990), procollagen in chondrocytes (Vertel *et al.*, 1989), zymogens in exocrine pancreas (Tooze *et al.*, 1989, 1990), and mutant Igs in myelomas (Valetti *et al.*, 1991). Intriguingly, although these aggregates represent a macroscopic accumulation of aberrant proteins in the ER, their presence does not always induce overexpression of BiP. Thus, these insoluble aggregates are somehow ignored by the quality control mechanisms. It remains to be established whether it is insufficient degradation that leads to precipitation, or *vice versa*. In chondrocytes and myelomas, removal of the conditions necessary for aggregate formation leads to the disappearance of lumenal bodies (Pacifici and Iozzo, 1988; Valetti *et al.*, 1991), suggesting that they might serve some physiological role.

5.   WHERE DOES DEGRADATION TAKE PLACE?

How do cells discriminate between proteins that have failed to assemble and proteins that have not yet had a chance to try? This represents a crucial problem, since from what we have learned on the nature of ER degradation targeting signals, nascent proteins are bound to be optimal targets. Nascent proteins must thus be somehow protected from proteases. In principle, this could be obtained by two main mechanisms. First, one can envisage newborn proteins as being shielded from proteases by a wrap of chaperones; alternatively, a specific subcompartment of the ER, distinct from that of synthesis, might be reserved for degradation. The peculiar temperature dependency of the ER degradation process, which suggests the involvement of a transport step, has been invoked in favor of a separate degradative subcompartment. On the other hand, evidence for degradation occurring in the ER itself has been obtained primarily from experiments on semipermeable cells, where degradation of TCR subunits and the first proteolytic event in ASGR have been shown to proceed in the absence of ATP or cytosol, a condition that prevents membrane fusion events (Stafford and Bonifacino, 1991; Wikström and Lodish, 1992). In these systems, a number of biochemical and morphological assays confirm that degradation does not require migration into the Golgi stack. However, since cells are first labeled and then made permeable, the requirement for a rapid transport step will go undetected even in this system. Thus, until the crucial experiment of coupling *in vitro* synthesis and degradation is performed, the existence of a distinct subcompartment, possibly within the ER network itself, cannot be formally excluded.

Does degradation take place in the sarcoplasmic reticulum? While continuities with the rough ER are easy to detect during myogenesis, in mature muscle cells the sarcoplasmic reticulum is thought to be an independent anatomical entity. Nonetheless, a subset of newly synthesized acetylcholinesterase is transported to the sarcoplasmic reticulum before undergoing nonlysosomal proteolysis. Thus, the possibility exists that a degradative pathway is present also in this structure (Rotundo *et al.*, 1989).

Are There Multiple ER Degradation Pathways?

Part of the contradictory data about the localization of degradation in the ER may be explained by the existence of more than one pathway, and perhaps more than one site, for pre-Golgi degradation. This could be the case for membrane and lumenal proteins, as suggested by the different behavior of γ_m and γ_s heavy chains in pre-B cells (Bachhawat and Pillai, 1991), and by the experiments of Tsao *et al.* (1992), where a characteristic biphasic degradation was observed only for soluble ribophorin mutants. A similar situation may also explain the two-step degradation of ASGR, where the soluble 35-kDa intermediate (generated by a CCCP- and cycloheximide-insensitive proteolytic event) requires energy and protein synthesis in order to be disposed of (Wikström and Lodish, 1991, 1992).

An example of cell-type specificity is perhaps found in the case of secretory IgM. In plasma cells, degradation of IgM subunits is insensitive to BFA. By contrast, this drug exerts an inhibitory effect in B cell lines (Amitay *et al.*, 1991, 1992; Fra *et al.*, unpublished data), suggesting that the same protein can be degraded following different pathways during B cell development (Rabinovitch *et al.*, personal communication).

6. THE FUNCTIONAL ROLES OF DEGRADATION IN THE ER

Cells respond to different environmental conditions by changing the concentration of specific intracellular proteins. This can be achieved by altering the rate of synthesis, the rate of degradation, or both. The rate of degradation can in turn be regulated by exposing or masking structural determinants on selected proteins, or by modulating the activity of a given proteolytic pathway. Among the many examples of important cellular functions being controlled by regulated proteolysis, suffice it to recall cyclin degradation by the ubiquitin pathway as one of the key elements in cell cycle control (Glotzer *et al.*, 1991).

6.1. Quality Control

The crucial role of ER degradation in the quality control of newly synthe-sized proteins has been amply dealt with in the preceding pages. Can the degra-dation process be regulated? The different responses of different substrates to Ca^{2+} ionophores suggest that this could indeed be the case. Thus, oscillations in the lumenal Ca^{2+} concentrations could modulate selectively the half lives of given (groups of) proteins. Such a model has been proposed to explain the mevalonate-regulated degradation of HMG-CoA reductase (Roitelman et al., 1991).

6.2. Controlling Certain Metabolic Pathways

Within the complex lipid metabolic pathways (Goldstein and Brown, 1990), HMG-CoA reductase is not the sole target of regulated ER degradation. As recalled above, the rate of catabolism of apoB-100 is dependent on exogenous LDL supply. Similarly, secretory retinol binding protein is degraded in the ab-sence of retinol or retinoic acid (Tosetti et al., 1992). As to the structural motifs involved in regulated degradation, it is tempting to assume that apoproteins are the structural equivalents to assembly subunits. Binding of the prosthetic group may thus result in the direct masking of the structural retention/degradation determinant, or change the overall conformation of the polypeptides.

6.3. ER Degradation and the Control of Gene Expression

The world of antigen receptors presents us with clear examples where degradation in the ER is exploited as a mechanism to execute certain develop-mental programs. In immature CD4+CD8+T lymphocytes, the fate of appar-ently normal TCR complexes is post translationally controlled: cross-linking of CD4 molecules results in increased expression of TCR, an event thought to be mediated by reduced degradation (Bonifacino et al., 1990c).

The differential handling of secretory IgM by B and plasma cells is intrigu-ing, as it suggests the existence of yet unidentified mechanism(s) by which cells can modulate quality control, and in turn the ultimate fate of a molecule (Sitia et al., 1990; Alberini et al., 1990). The failure to oligomerize IgM, and hence to secrete this isotype, has long been attributed to the absence of J chain in B cells (Koshland, 1985). However, it is quite clear that regardless of the presence or absence of this subunit, B lymphocytes are intrinsically lazy in the oligomeriza-tion process, with the subsequent victory of retention/degradation over secretion (Randall et al., 1992). As the word intrinsic falls short in satisfying the devoted scientist, the mechanisms underlying the regulated oligomerization of IgM will hopefully lead to new insights into the more general problem of quality control.

6.4. Antigen Presentation

As soon as the ER degradative pathway was discovered, its potential role in antigen presentation was appreciated (Nuchtern et al., 1989). In general, class I MHC present endogenous antigens (e.g., viral proteins, oncogenes) while exogenous antigens are presented by class II (Monaco, 1992; Neefjes and Ploegh, 1992). The sequences of several immunogenic peptides bound to class I molecules have been determined recently (Falk et al., 1990, 1991; Rötzschke et al., 1990), and common motifs are encountered in some of these nonamers. What is the role of ER degradation in the generation of such peptides (Weiss and Bogen, 1991)? For an updated review of the immunology and cell biology of these phenomena, the reader is referred to Cerundolo (this volume).

7. PERSPECTIVES

At this point, the reader will probably be uncertain as to whether the unresolved issues concerning degradation in the ER are indeed numerous, or whether the authors have been most unclear. In our opinion, among the most relevant unsolved issues are those concerning the precise location of the pre-Golgi proteolytic system(s), the nature of the proteases involved, and the mechanisms by which the latter recognize their substrates. Using the word *precise* may uncover a bias of the authors toward the existence of a (sub)compartment, perhaps of the ER itself, devoted to proteolysis. Such a compartment need not be surrounded by a continuous membrane. It is possible that the various activities underlying quality control are segregated into functional subcompartments, perhaps generated by unequal distribution of membrane and soluble proteins within the ER. Although several pieces of evidence are available in favor of the existence of distinct ER subcompartments (see Sitia and Meldolesi, 1992, for a review), the subcellular structure of the ER network is far from being completely understood.

Even assuming that a distinct degradative subcompartment exists, how is the degradation of lumenal, transmembrane, and cytosolic domains coordinated? Autophagy, a process by which a given membrane-bound compartment is fused to lysosomes, is responsible for the clearance of the crystalloid ER (Orci et al., 1984) and pancreatic intracisternal granules (Tooze et al., 1990). However, in view of the insensitivity to methyl-adenine and lysosomotropic agents, we can exclude an involvement of autophagy in the rapid and selective ER degradation. A cross talk between different classes of proteases must then be invoked.

Answers to all of these questions will require the success of several independent experimental approaches. A reproducible translation/degradation *in vitro* assay, as well as the availability of antibodies to some of the factors involved in ER proteolysis would clearly be welcome at this stage.

ACKNOWLEDGMENTS. We thank Cristina Alberini, Nica Borgese, Aldo Ceriotti, Claudio Fagioli, Dario Finazzi, Silvia Guenzi, Jacopo Meldolesi, César Milstein, Michael Neuberger, and Caterina Valetti for helpful discussions and suggestions, and Cristina DeMarchi for outstanding secretarial assistance. This work was made possible by grants from AIRC, CNR (PF-BTBS and PF-IG), Ministero della Sanità (AIDS Special Project), and Fondazione Monte Tabor. AMF is a recipient of a fellowship from AIRC.

8. REFERENCES

Alberini, C. M., Bet, P., Milstein, C., and Sitia, R., 1990, Secretion of immunoglobulin M assembly intermediates in the presence of reducing agents, *Nature* **347**:485–487.

Ahluwalia, N., Bergeron, J. J. M., Wada, I., Degen, E., and Williams, D. B., 1992, The p88 molecular chaperone is identical to the endoplasmic reticulum membrane protein, calnexin, *J. Biol. Chem.* **267**:10914–10918.

Amara, J. F., Lederkremer, G., and Lodish, H. F., 1989, Intracellular degradation of unassembled asialoglycoprotein receptor subunits: A pre-Golgi, nonlysosomal endoproteolytic cleavage, *J. Cell Biol.* **109**:3315–3324.

Amitay, R., Bar-Nun, S., Haimovich, J., Rabinovich, E., and Shachar, I., 1991, Post-translational regulation of IgM expression in B lymphocytes, *J. Biol. Chem.* **266**:12568–12573.

Amitay, R., Shachar, I., Rabinovich, E., Haimovich, J., and Bar-Nun, S., 1992, Degradation of secretory immunoglobulin M in B lymphocytes occurs in a postendoplasmic reticulum compartment and is mediated by a cysteine protease, *J. Biol. Chem.* **267**:20694–20700.

Bachhawat, A. K., and Pillai, S., 1991, Distinct intracellular fates of membrane and secretory immunoglobulin heavy chains in a pre-B cell line, *J. Cell Biol.* **115**:619–624.

Bachmair, A., and Varshawsky, A., 1989, The degradation signal in a short-lived protein, *Cell* **56**:1019–1032.

Blount, P., and Merlie, J. P., 1990, Mutational analysis of muscle nicotinic acetylcholine receptor subunit assembly, *J. Cell Biol.* **111**:2613–2622.

Bole, D. G., Hendershot, L. M., and Kearney, J. F., 1986, Posttranslational association of immunoglobulin heavy chain-binding protein with nascent heavy chains in nonsecreting and secreting hybridomas, *J. Cell Biol.* **102**:1558–1566.

Bonifacino, J. S., Suzuki, C. K., Lippincott-Schwartz, J., Weissman, A. M., and Klausner, R. D., 1989, Pre-Golgi degradation of newly synthesized T-cell antigen receptor chains: Intrinsic sensitivity and the role of subunit assembly, *J. Cell Biol.* **109**:73–83.

Bonifacino, J. S., Cosson, P., and Klausner, R. D., 1990a, Colocalized transmembrane determinants for ER degradation and subunit assembly explain the intracellular fate of TCR chains, *Cell* **63**:503–513.

Bonifacino, J. S., Suzuki, C. K., and Klausner, R. D., 1990b, A peptide sequence confers retention and rapid degradation within the endoplasmic reticulum, *Science* **247**:79–82.

Bonifacino, J. S., McCarthy, S. A., Maguire, J. E., Nakayama, T., Singer, D. S., Klausner, R. D., and Singer, A., 1990c, Novel post-translational regulation of TCR expression in CD4+CD8+ thymocytes influenced by CD4, *Nature* **344:**247–251.

Bonifacino, J. S., Cosson, P., Shah, N., and Klausner, R. D, 1991, Role of potentially charged transmembrane residues in targeting proteins for retention and degradation within the endoplasmic reticulum, *EMBO J.* **10:**2783–2793.

Booth, C., and Koch, G. L. E., 1989, Perturbation of cellular calcium induces secretion of luminal ER proteins, *Cell* **59:**729–737.

Campbell, K. S., and Cambier, J. C., 1990, B lymphocyte antigen receptors (mIg) are noncovalently associated with a disulfide linked, inducibly phosphorylated glycoprotein complex, *EMBO J.* **9:**441–448.

Carlson, J. A., Rogers, B. B., Sifers, R. N., Finegold, M. J., Clift, S. M., DeMayo, F. J., Bullock, D. W., and Woo, S. L. C., 1989, Accumulation of PiZ α1-antitrypsin causes liver damage in transgenic mice, *J. Clin. Invest.* **83:**1183–1190.

Chen, C., Bonifacino, J. S., Yuan, L. C., and Klausner, R. D., 1988, Selective degradation of T cell antigen receptor chains retained in a pre-Golgi compartment, *J. Cell Biol.* **107:**2149–2161.

Cheng, S. H., Gregory, R. J., Marshall, J., Paul, S., Souza, D. W., White, G. A., O'Riordan, C. R., and Smith, A. E., 1990, Defective intracellular transport and processing of CFTR is the molecular basis of most cystic fibrosis, *Cell* **63:**827–834.

Chuck, S. L., Yao, Z., Blackhart, B. D., McCarthy, B. J., and Lingappa, V. R., 1990, New variation on the translocation of proteins during early biogenesis of apolipoprotein B, *Nature* **346:**382–385.

Chun, K. T., Bar-Nun, S., and Simoni, R. D., 1990, The regulated degradation of 3-hydroxy-3-methylglutaryl-CoA reductase requires a short-lived protein and occurs in the endoplasmic reticulum, *J. Biol. Chem.* **265:**22004–22010.

Clevers, H., Alarcon, B., Wileman, T., and Terhorst, C., 1988, The T cell receptor/CD3 complex: A dynamic protein ensemble, *Annu. Rev. Immunol.* **6:**629–662.

Davis, R. A., Thrift, R. N., Wu, C. C., and Howell, K. E., 1990, Apolipoprotein B is both integrated into and translocated across the endoplasmic reticulum membrane. Evidence for two functionally distinct pools, *J. Biol. Chem.* **265:**10005–10011.

Esser, V., and Russell, D. W., 1988, Transport-deficient mutations in the low density lipoprotein receptor. Alterations in the cysteine-rich and cysteine-poor regions of the protein block intracellular transport, *J. Biol. Chem.* **263:**13276–13281.

Falk, K., Rötzschke, O., and Rammensee, H. G., 1990, Cellular peptide composition governed by major histocompatibility complex class I molecules, *Nature* **348:**248–251.

Falk, K., Rötzschke, O., Stevanovic, S., Jung, G., and Rammensee, H. G., 1991, Allelespecific motifs revealed by sequencing of self-peptides eluted from MHC molecules, *Nature* **351:**290–296.

Flynn, G. C., Chappell, T. G., and Rothman, J. E., 1989, Peptide binding and release by proteins implicated as catalysts of protein assembly, *Science* **245:**385–390.

Flynn, G. C., Pohl, J., Flocco, M. T., and Rothman, J. E., 1991, Peptide-binding specificity of the molecular chaperone BiP, *Nature* **353:**726–730.

Gething, M.-J., and Sambrook, J., 1990, Transport and assembly processes in the endoplasmic reticulum, *Semin. Cell Biol.* **1:**65–72.

Gething, M.-J., and Sambrook, J., 1992, Protein folding in the cell, *Nature* **355:**33–45.

Glotzer, M., Murray, A. W., and Kirschner, M. W., 1991, Cyclin is degraded by the ubiquitin pathway,*Nature* **349:**132–138.

Goldstein, J. L., and Brown, M. S., 1990, Regulation of the mevalonate pathway, *Nature* **343:**425–430.

Haas, I. G., and Wabl, M., 1983, Immunoglobulin heavy chain binding protein, *Nature* **306**:387–389.

Haugejorden, S. M., Srinivasan, M., and Green, M., 1991, Analysis of the retention signals of twoluminal endoplasmic reticulum proteins by in vitro mutagenesis, *J. Biol. Chem.* **266**:6015–6018.

Hendershot, L. M., 1990, Immunoglobulin heavy chain and binding protein complexes are dissociated in vivo by light chain addition, *J. Cell Biol.* **111**:829–837.

Hendershot, L., Bole, D., Köhler, G., and Kearney, J. F., 1987, Assembly and secretion of heavy chains that do not associate posttranslationally with immunoglobulin heavy chain-binding protein, *J. Cell Biol.* **104**:761–767.

Hershko, A., 1988, Ubiquitin-mediated protein degradation, *J. Biol. Chem.* **263**:14368–14373.

Hochstenbach, F., David, V., Watkins, S., and Brenner, M. B., 1992, Endoplasmic reticulum resident protein of 90 kilodaltons associates with the T- and B-cell antigen receptors and major histocompatibility complex antigens during their assembly, *Proc. Natl. Acad. Sci. USA* **89**:4734–4738.

Hombach, J., Sablitzky, F., Rajewsky, K., and Reth, M., 1988, Transfected plasmacytoma cells do not transport the membrane form of IgM to the cell surface, *J. Exp. Med.* **167**:652–657.

Hortsch, M., and Meyer, D. I., 1985, Immunochemical analysis of rough and smooth microsomes from rat liver. Segregation of docking protein in rough membranes, *Eur. J. Biochem.* **150**:559–564.

Hsu, V. W., Shah, N., and Klausner, R. D., 1992, A brefeldin A-like phenotype is induced by the overexpression of a human ERD-2-like protein, ELP-1, *Cell* **69**:625–635.

Hurtley, S. M., and Helenius, A., 1989, Protein oligomerization in the endoplasmic reticulum, *Annu. Rev. Cell Biol.* **5**:277–307.

Inoue, S., and Simoni, R. D., 1992, 3-Hydroxy-3-methylglutaryl-coenzyme A reductase and T cell receptor α subunit are differentially degraded in the endoplasmic reticulum, *J. Biol. Chem.* **267**:9080–9086.

Jackson, M. R., Nilsson, T., and Peterson, P. A., 1990, Identification of a consensus motif for retention of transmembrane proteins in the endoplasmic reticulum, *EMBO J.* **9**:3153–3162.

Kassenbrook, C. K., Garcia, P. D., Walter, P., and Kelly, R. B., 1988, Heavy-chain binding protein recognizes aberrant polypeptides translated in vitro, *Nature* **333**:90–93.

Kelleher, D. J., Kreibich, G., and Gilmore, R., 1992, Oligosaccharyltransfectase activity is associated with a protein complex composed of ribophorins I and II and a 48 kd protein, *Cell* **69**:55–65.

Klausner, R. D., 1989, Sorting and traffic in the central vacuolar system, *Cell* **57**:703–706.

Klausner, R. D., and Sitia, R., 1990, Protein degradation in the endoplasmic reticilum, *Cell* **62**:611–614.

Koshland, M. E., 1985, The coming of age of the immunoglobulin J chain, *Annu. Rev. Immunol.* **3**:425–453.

Kozutsumi, Y., Segal, M., Normingtom, K., Gething, M.-J., and Sambrook, J., 1988, The presence of malfolded proteins in the endoplasmic reticulum signals the induction of glucose regulated proteins, *Nature* **332**:462–464.

Lau, M. M. H., and Neufeld, E. F., 1989, A frameshift mutation in a patient with Tay-Sachs disease causes premature termination and defective intracellular transport of the α-subunit of β-hexosaminidase, *J. Biol. Chem.* **264**:21376–21380.

Le, A., Graham, K. S., and Sifers, R. N., 1990, Intracellular degradation of the transport-impaired human PiZ α1-antitrypsin variant, *J. Biol. Chem.* **265**:14001–14007.

Le, A., Ferrell, G. A., Dishon, D. S., Le, Q.-Q. A., and Sifers, R. N., 1992, Soluble aggregates of the human PiZ α1-antitrypsin variant are degraded within the endoplasmic reticulum by a mechanism sensitive to inhibitors of protein synthesis, *J. Biol. Chem.* **267**:1072–1080.

Lederkremer, G. Z., and Lodish, H. F., 1991, An alternatively spliced miniexon alters the subcellular fate of the human asialoglycoprotein receptor H2 subunit, *J. Biol. Chem.* **266:**1237–1244.

Letourneur, F., and Klausner, R. D., 1992, A novel di-leucine motif and a tyrosine-based motif independently mediate lysosomal targeting and endocytosis of CD3 chains, *Cell* **69:**1143–1157.

Lewis, M. J., and Pelham, H. R. B., 1992, Ligand-induced redistribution of a human KDEL-receptor from the Golgi complex to the endoplasmic reticulum, *Cell* **68:**353–364.

Lewis, M. J., Sweet, D. J., and Pelham, H. R. B., 1990, The ERD2 gene determines the specificity of the luminal ER protein retention system, *Cell* **61:**1359–1363.

Lippincott-Schwartz, J., Bonifacino, J. S., Yuan, L., and Klausner, R. D., 1988, Degradation from the endoplasmic reticulum: Disposing of newly synthesized proteins, *Cell* **54:**209–229.

Lippincott-Schwartz, J., Donaldson, J. G., Schweizer, A., Berger, E. G., Hauri, H. P., Yuan, L. C., and Klausner, R. D., 1990, Microtubule dependent retrograde transport of proteins into the ER in the presence of brefeldin A suggests an ER recycling pathway, *Cell* **60:**821–836.

Melloni, E., and Pontremoli, S., 1989, The calpains, *Trends Neurosci.* **12:**438–443.

Monaco, J. J., 1992, A molecular model of MHC class-I-restricted antigen processing, *Immunol. Today* **13:**173–178.

Munro, S., and Pelham, H. R. B., 1987, A C-terminal signal prevents secretion of luminal ER proteins, *Cell* **48:**899–907.

Neefjes, J. J., and Ploegh, H. L., 1992, Intracellular transport of MHC class II molecules, *Immunol. Today* **13:**179–184.

Nutchern, J. G., Bonifacino, J. S., Biddison, W. E., and Klausner, R. D., 1989, Brefeldin A implicates egress from endoplasmic reticulum in class I restricted antigen presentation, *Nature* **339:**223–226.

Olson, T. S., and Dice, J. F., 1989, Regulation of protein degradation rates in eukaryotes, *Curr. Opin. Cell Biol.* **1:**1194–1200.

Orci, L., Brown, M. S., Goldstein, J. L., Garcia-Segura, L. M., and Anderson, R. G. W., 1984, Increase in membrane cholesterol: A possible trigger for degradation of HMG CoA reductase and crystalloid endoplasmic reticulum in UT-1-cells, *Cell* **36:**835–845.

Pacifici, M., and Iozzo, R. V., 1988, Remodeling of the rough endoplasmic reticulum during stimulation of procollagen secretion by ascorbic acid in cultured chondrocytes, *J. Biol. Chem.* **263:**2483–2492.

Palade, G. E., 1975, Intracellular aspects of the process of protein secretion, *Science* **189:**347–358.

Pelham, H. R. B., 1989, Control of protein exit from the endoplasmic reticulum, *Annu. Rev. Cell Biol.* **5:**1–23.

Pelham, H. R. B., Roberts, L. M., and Lord, J. M., 1992, Toxin entry: How reversible is the secretory pathway? *Trends Cell Biol.* **2:**183–185.

Pontremoli, S., and Melloni, E., 1986, Extralysosomal protein degradation, *Annu. Rev. Biochem.* **55:**455–481.

Randall, T. D., Brewer, J. W., and Corley, R. B., 1992, Direct evidence that J-chain regulates the polymeric structure of IgM in antibody-secreting B-cells, *J. Biol. Chem.* **267:**18002–18007.

Riordan, J. R., Rommens, J. M., Kerem, B., Alon, N., Rozmahel, R., Grzelczak, Z., Zielenski, J., Lok, S., Plavsic, N. Chou, J. L., Drumm, M. L., Iannuzzi, M. C., Collins, F. S., and Tsui, L. C., 1989, Identification of the cystic fibrosis gene: Cloning and characterization of complementary DNA, *Science* **245:**1066–1073.

Roitelman, J., Bar-Nun, S., Inoue, S., and Simoni, R. D., 1991, Involvement of calcium in the mevalonate-accelerated degradation of 3-hydroxy-3-methylglutaryl-CoA reductase, *J. Biol. Chem.* **266:**16085–16091.

Roitelman, J., Olender, E. H., Bar-Nun, S., and Dunn, W. A., Jr., 1992, Immunological evidence for eight spans in the membrane domains of 3-hydroxy-3-methylglutaryl coenzyme A reductase: Implications for enzyme degradation in the endoplasmic reticulum, *J. Cell Biol.* **117**:959–973.

Rose, J. K., and Doms, R. W., 1988, Regulation of protein export from the endoplasmic reticulum, *Annu. Rev. Cell Biol.* **4**:257–288.

Rotundo, R. I., 1988, Biogenesis of acetylcholinesterase molecular forms in muscle. Evidence for a rapidly turning over, catalytically inactive precursor pool, *J. Biol. Chem.* **263**;19398–19406.

Rotundo, R. I., Thomas, K., Porter-Jordan, K., Benson, R. J. J., Fernandez-Valle, C., and Fine, R. E., 1989, Intracellular transport, sorting, and turnover of acetylcholinesterase. Evidence for an endoglycosidase H-sensitive form in Golgi apparatus, sarcoplasmic reticulum, and clathrin-coated vesicles and its rapid degradation by a non-lysosomal mechanism, *J. Biol. Chem.* **264**:3146–3152.

Rötzschke, O., Falk, K., Deres, K., Schild, H., Norda, M., Metzger, J., Jung, G., and Rammensee, H. G., 1990, Isolation and analysis of naturally processed viral peptides as recognized by cytotoxic T cells, *Nature* **358**:510–512.

Sandvig, K., Garred, O., Prydz, K., Kozlov, J. V., Hansen, S. H., and van Deurs, B., 1992, Retrograde transport of endocytosed Shiga toxin to the endoplasmic reticulum, *Nature* **358**:510–512.

Sato, R., Imanaka, T., Takatsuki, A., and Takano, T., 1990, Degradation of newly synthesized apolipoprotein B-100 in a pre-Golgi compartment, *J. Biol. Chem.* **265**:11880–11884.

Semenza, J. C., Hardwick, K. G., Dean, N., and Pelham, H. R. B., 1990, ERD2, a yeast gene required for the receptor-mediated retrieval of luminal ER proteins from the secretory pathway, *Cell* **61**:1349–1357.

Shin, J., Lee, S., and Strominger, J. L., 1993, Translocation of TCR chains into the lumen of the endoplasmic reticulum and their degradation, *Science* **259**:1901–1904.

Sitia, R., and Meldolesi, J., 1992, Endoplasmic reticulum: A dynamic patchwork of specialized subregions, *Mol. Biol. Cell* **3**:1067–1072.

Sitia, R., Neuberger, M. S., and Milstein, C., 1987, Regulation of membrane IgM expression in secretory B cells: Translational and post-translational events, *EMBO J.* **6**:3969–3977.

Sitia, R., Alberini, C., Biassoni, R., Rubartelli, A., De Ambrosis, S., and Vismara, D., 1988, The control of membrane and secreted heavy chain biosynthesis varies in different immunoglobulin isotypes produced by a monoclonal B cell lymphoma, *Mol. Immunol.* **25**:189–197.

Sitia, R., Neuberger, M., Alberini, C., Bet, P., Fra, A., Valetti, C., Williams, G., and Milstein, C., 1990, Developmental regulation of IgM secretion: The role of the carboxy-terminal cysteine, *Cell* **60**:781–790.

Stafford, F. J., and Bonifacino, J. S., 1991, A permeabilized cell system identifies the endoplasmic reticulum as a site of protein degradation, *J. Cell Biol.* **115**:1225–1236.

Stoller, T. J., and Shields, D., 1989, The propeptide of preprosomatostatin mediates intracellular transport and secretion of α-globin from mammalian cells, *J. Cell Biol.* **108**:1647–1655.

Suzuki, C. K., Bonifacino, J. S., Lin, A. Y., Davis, M. M., and Klausner, R. D., 1991, Regulating the retention of T-cell receptor α chain variants within the endoplasmic reticulum: Ca^{2+}-dependent association with BiP, *J. Cell Biol.* **114**:189–205.

Tooze, J., Kern, H. F., Fuller, S. D., and Howell, K. E., 1989, Condensation-sorting events in the rough endoplasmic reticulum of exocrine pancreas cells, *J. Cell Biol.* **109**:35–50.

Tooze, J., Hollinshead, M., Ludwig, T., Howell, K., Hoflack, B., and Kern, H., 1990, In exocrine pancreas, the basolateral endocytic pathway converges with the autophagic pathway immediately after the early endosome, *J. Cell Biol.* **111**:329–345.

Tosetti, F., Ferrari, N., Pfeiffer, U., Brigati, C., and Vidali, G., 1992, Regulation of plasma retinol binding protein secretion in human HepG2 cells, *Exp. Cell Res.* **200:**467–472.

Tsao, Y. S., Ivessa, N. E., Adesnik, M., Sabatini, D. D., and Kreibich, G., 1992, Carboxy terminally truncated forms of ribophorin I are degraded in pre-Golgi compartments by a calcium dependent process, *J. Cell Biol.* **116:**57–67.

Valetti, C., Grossi, C. E., Milstein, C., and Sitia, R., 1991, Russell bodies: A general response of secretory cells to synthesis of a mutant immunoglobulin which can neither exit from, nor be degraded in, the endoplasmic reticulum, *J. Cell Biol.* **115:**983–994.

Vertel, B. M., Velasco, A., LaFrance, S., Walters, L., and Kaczman-Daniel, K., 1989, Precursor of chondroitin sulphate proteoglycan are segregated within a subcompartment of the chondrocyte endoplasmic reticulum, *J. Cell Biol.* **109:**1827–1836.

Wada, I., Rindress, D., Cameron, P. H., Ou, W. J., Doherty, H. H., Louvard, D., Bell, D., Thomas, D. Y., and Bergeron, J. J. M., 1991, The SSRα and associated calnexin are major calcium binding proteins of the endoplasmic reticulum membrane, *J. Biol. Chem.* **266:**19599–19610.

Wallace, J. C., Galili, G., Kawata, E. E., Cuellar, R. E., Shotwell, M. A., and Larkins, B. A., 1988, Aggregation of lysine-containing zeins into protein bodies in Xenopus oocytes, *Science* **240:**662–664.

Weiss, S., and Bogen, B., 1991, MHC class II-restricted presentation of intracellular antigen, *Cell* **64:**767–776.

Wieland, F. T., Gleason, M. L., Serafini, T. A., and Rothman, J. E., 1987, The rate of bulk flow from the endoplasmic reticulum to the cell surface, *Cell* **50:**289–300.

Wienands, J., Hombach, J., Radbruch, A., Riesterer, C., and Reth, M., 1990, Molecular components of the B cell antigen receptor complex of class IgD differ partly from those of IgM, *EMBO J.* **9:**449–455.

Wikström, L., and Lodish, H. F., 1991, Nonlysosomal, pre-Golgi degradation of unassembled asialoglycoprotein receptor subunits: A TLCK- and TCPK-sensitive cleavage within the ER, *J. Cell Biol.* **113:**997–1007.

Wikström, L., and Lodish, H. F., 1992, Endoplasmic reticulum degradation of a subunit of the asialoglycoprotein receptor in vitro. Vesicular transport from endoplasmic reticulum is unnecessary, *J. Biol. Chem.* **267:**5–8.

Wileman, T., Carson, G. R., Concino, M., Ahmed, A., and Terhorst, C., 1990, The γ and ϵ subunits of the CD3 complex inhibit pre-Golgi degradation of newly synthesized T cell antigen receptors, *J. Cell Biol.* **110:**973–986.

Wileman, T., Kane, L. P., Carson, G. R., and Terhorst, C., 1991, Depletion of cellular calcium accelerates protein degradation in the endoplasmic reticulum, *J. Biol. Chem.* **266:**4500–4507.

Williams, G. T., Venkitaraman, A. R., Gilmore, D. J., and Neuberger, M. S., 1990, The sequence of the μ transmembrane segment determines the tissue specificity of the transport of immunoglobulin M to the cell surface, *J. Exp. Med.* **171:**947–952.

Chapter 8

The Endoplasmic Reticulum Is the Site of Lipoprotein Assembly and Regulation of Secretion

Roger A. Davis

1. INTRODUCTION

The principal function of plasma lipoproteins is to transport water-insoluble lipid from its site of synthesis and absorption to peripheral cells. In modern industrialized societies where overconsumption of animal fat is common, plasma lipoproteins are dramatically increased leading to an increased incidence of atherosclerotic cardiovascular disease. In an attempt to design methods to prevent the occurrence of atherosclerosis, a great amount of effort has been applied to understanding the molecular mechanisms through which plasma lipoproteins are assembled and secreted. The major focus of this chapter is to describe these mechanisms, many of which are distinct from the paradigms describing the assembly, intracellular targeting, and metabolism of other lipid–protein macro-

Abbreviations used in this chapter: ALLN, *N*-acetyl-leucyl-leucyl-norleucinal; apo B, apolipoprotein B; CHO, Chinese hamster ovary; IDL, intermediate-density lipoprotein; LDL, low-density lipoprotein; VLDL, very-low-density lipoprotein.

Roger A. Davis   Department of Biology, San Diego State University, San Diego, California 92182.
Subcellular Biochemistry, Volume 21: Endoplasmic Reticulum, edited by N. Borgese and J. R. Harris. Plenum Press, New York, 1993.

molecular aggregates (e.g., membranes, vesicles, and viruses). These distinctions provide new insights into a regulatory secretion mechanism intimately linked to the physicochemical forces governing the translocation and integration of proteins in the endoplasmic reticulum (ER). Since very-low-density lipoprotein (VLDL) is the principal, if not sole, precursor of the other plasma lipoproteins, we will focus our attention on the molecular interactions involved in its assembly and secretion.

2. UNLIKE THE PHOSPHOLIPID BILAYER OF MEMBRANES, VLDL HAS A SURFACE MONOLAYER, IMPLYING A UNIQUE MECHANISM OF ASSEMBLY

The assembly of VLDL (Figure 1) requires the production of a thermodynamically stable macromolecular aggregate consisting of a hydrophobic core of triglyceride which is surrounded by a shell of amphipathic phospholipid and specific proteins [apolipoprotein B (apo B)]. The surface monolayer coat of amphipathic phospholipids is in contradistinction to the surface bilayer phospholipid coat of membranes, transport vesicles, and viruses. With regard to the function of VLDL (i.e., to carry hydrophobic triglycerides), the surface monolayer provides a structure able to maximize the content of the neutral lipid core. Moreover, since it is now well established that VLDL is derived from the bilayer of the ER, the mechanism involved in VLDL assembly is likely to involve a process whereby one leaflet of the membrane bilayer surrounds a neutral lipid core of triglyceride and the resulting sphere is "ejected" into the lumen for entrance into the secretory pathway. As discussed below, many of the processes are unique to the assembly of triglyceride-rich lipoproteins and are expressed only in tissues (i.e., liver and intestine) that actively assemble and secrete VLDL.

3. VLDL ASSEMBLY REQUIRES SEVERAL SPECIFIC GENE PRODUCTS

Nascent VLDL have been isolated from the Golgi of liver and have been found to resemble plasma VLDL with the exception that nascent VLDL contain less cholesterol ester and more phosphatidylethanolamine (Hamilton and Fielding, 1989). Thus, based on the known structure of VLDL (Figure 1), the assembly of VLDL requires the acquisition of three major lipids (phospholipid, free cholesterol, and triglyceride) and one major protein (apo B) into a liposomelike particle. The finding of uniform VLDL-size particles containing apo B in the lumen of the ER suggests that the particles are assembled at this intracellular site

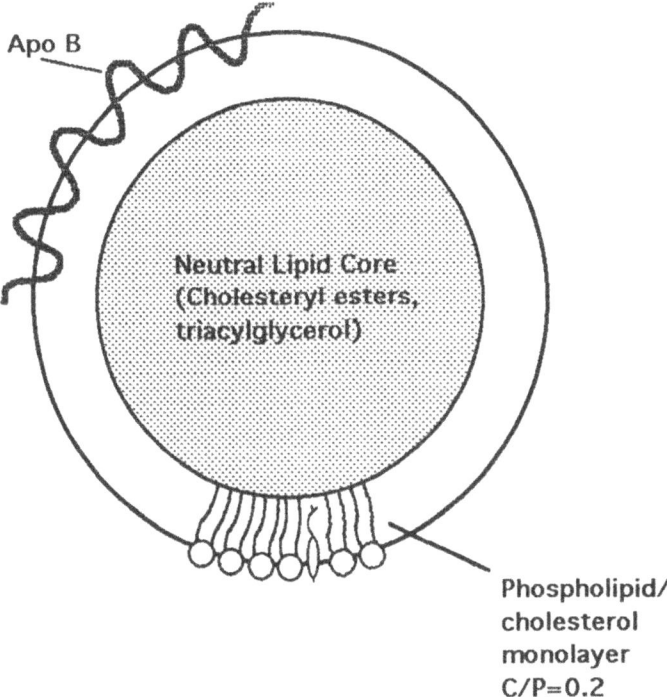

FIGURE 1. Schematic diagram of a VLDL particle. A VLDL particle has a polar lipid monolayer surface containing free cholesterol and phospholipid. This surface also contains apo B, which is thought to weave in and out of the monolayer surface. The monolayer surface surrounds a neutral lipid core containing mainly triglycerides and a small amount of cholesterol esters. The average diameter of a VLDL is 450 Å.

(Alexander *et al.*, 1976). Since the particle is derived directly from the ER membrane, there must be a process whereby triglyceride is concentrated into the core of the developing VLDL particle. This prediction is based on the finding that the amount of triglyceride in VLDL is approximately 60 mol%, whereas the maximal amount of triglyceride that can be accommodated in a phospholipid bilayer membrane is only 3 mol%. Thus, there must be a process that can concentrate triglyceride into the core of the nascent VLDL particle. Two mechanisms through which triglyceride might be assembled into a VLDL particle are: (1) the concentration of triglyceride into a pocket in the lumenal leaflet of the ER, which eventually buds off into the lumen forming a VLDL particle (Davis, 1991; Boren *et al.*, 1992), and (2) the "two-step" model whereby triglyceride is assembled into a triglyceride-rich particle in the smooth EP and this fat globule is combined with apo B in the junction of the smooth and rough ER (Alexander *et* al., 1976). The recent finding that a microsomal triglyceride transfer protein in the lumen of the ER is missing in the intestine of patients who cannot assemble VLDL (i.e., abetalipoproteinemia) supports the proposal that the association of triglyceride with the nascent VLDL particle is essential for VLDL secretion (Wetterau et al., 1992). The exact role of the microsomal triglyceride transfer protein in VLDL assembly has not been defined.

Several lines of evidence show that apo B is essential for VLDL assembly and secretion. There are several mutations in the human apo B gene that result in either the production of truncated forms of apo B or the deletion of apo B (i.e., hypobetalipoproteinemia) (Young, 1990). In many cases where apo B either is not produced or is shortened by about 65%, the assembly and secretion of VLDL is impaired (Young, 1990).

Apo B is one of the largest single-chain proteins known (Knott et al., 1986; Yang et al., 1986; Cladaras et al., 1986). It has structural motifs that afford unique amphipathic properties. Unlike other mammalian apolipoproteins, apo B does not exchange between lipoprotein particles; it behaves as an integral membrane protein. The structural motifs responsible for the association of apo B with lipids are complex and are likely to involve several domains having different secondary structures. Within the 4536 amino acids of mature apo B-100 are β sheets having hydrophobic and hydrophilic surfaces, short stretches of amphipathic α helices (thought to be too short to act as typical membrane-spanning domains), and β strands that have the potential to form amphipathic domains (Cladaras et al., 1986; Yang et al., 1989; Chen et al., 1989). Several independent studies show that throughout the human apo B100 molecule at least 13 separate lipid-binding domains exist allowing parts of the entire molecule to associate with lipid. The large number of dispersed lipid-binding domains in apo B may explain how it associates with the lumenal membrane bilayer of the ER (Bostrom et al., 1988; Davis et al., 1990).

The functional basis for the large size of apo B has been uncovered in elegant experiments examining the genetic basis for hypobetalipoproteinemia. A phenotype of this disorder has been produced by expressing truncated forms of apo B in rat hepatoma cells (Yao et al., 1991). Apo B molecules having a size smaller than 31% of the N-terminus of the maximal coding region transcribed by the apo B gene show an inability to assemble lipoproteins containing a neutral lipid core. Furthermore, the size of apo B varies inversely with the density of the lipoprotein produced, suggesting that the size of apo B determines the core volume of neutral lipids (Yao et al., 1991). Similar results have been reported using HepG2 cells (Boren et al., 1992; Graham et al., 1991; Spring et al., 1992). One interpretation of this finding is that apo B forms a lattice surrounding the neutral lipid core into which triglyceride is delivered. An intralumenal protein termed the microsomal triglyceride transfer protein has been implicated as being essential for the formation of lipoproteins in the intestine (Wetterau et al., 1992).

Data in support of this interpretation show that in human LDL, apo B100 "circumnavigates" the cholesterol ester core, as mapped by monoclonal antibodies (Chatterton et al., 1991).

The size of apo B cannot be the only determinant of VLDL size. Clearly, other factors (perhaps tissue specific) are important for producing larger triglyceride-rich lipoproteins in the intestine compared with the liver. In most species the intestine secretes apo B48 whereas the liver secretes apo B100 (human) or both apo B100 and apo B48 (rodents) (Krishnaiah et al., 1980; Kane, 1988). The triglyceride-rich lipoproteins secreted by the intestine (i.e., chylomicrons) contain apo B48, yet they are significantly larger than the triglyceride-rich lipoproteins secreted by the liver which contain apo B100 (Kane, 1983). Furthermore, the size of apo B100 containing VLDL secreted by rat hepatocytes is indistinguishable from that of apo B48 containing VLDL (Davis et al., 1989).

4. THERE ARE TWO MOLECULAR WEIGHT FORMS OF APO B DERIVED FROM A SINGLE GENE AND TRANSCRIPTION PRODUCT

The physiologic basis for producing two forms of apo B may involve the metabolism of plasma lipoproteins rather than the assembly of lipoproteins. The two distinct molecular weight forms of apo B differ by the carboxy-terminal half, which is thought to contain the LDL receptor binding domain (Powell et al., 1987; Chen et al., 1987; Hospattankar et al., 1988) (Figure 2). (The number corresponds to the percentage of the amino-terminal portion encoded by each form.) Both forms of apo B are derived from a single gene and transcription produce (i.e., hnRNA). At nucleotide 6666, a cytidine is deaminated forming a uridine (Garcia et al., 1992). The net result is the conversion of a Gln codon (CAA) to a termination codon (UAA). When edited, the mRNA encodes the truncated form of apo B (i.e., apo B48).

The gene product responsible for the enzymatic conversion of the CAA codon to the UAA codon has recently been identified and cloned (Teng et al., 1993). It is expressed in the intestine of mammals and in the liver of rodents. In rat liver the conversion of apo B100 mRNA to apo B48 mRNA is sensitive to thyroid hormone status (Davidson et al., 1990a) and nutritional state (Leighton et al., 1990; Davidson et al., 1990b). Several studies have identified an RNA sequence that appears to act as a signal for the deamination of the C to a U (Driscoll et al., 1989; Chen et al., 1990; Shah et al., 1991; Lau et al., 1990; Chan, 1993). Interestingly, in transgenic mice expressing human apo B100, the human apo B mRNA was not edited, whereas the endogenous (rat) apo B100 mRNA was edited (Xiong et al., 1992). In the rat, the level

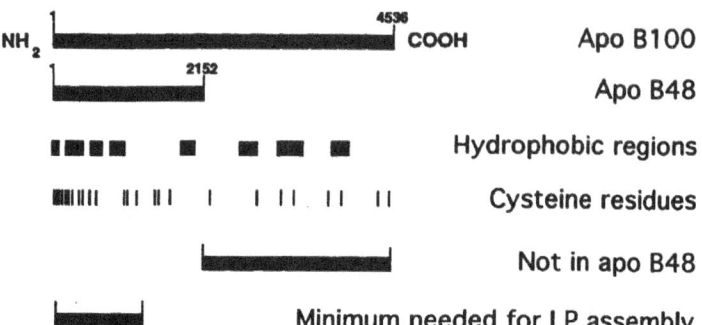

FIGURE 2. Domain regions of apo B, the structural protein of VLDL. Apo B has two major molecular weight forms, each designated by the percentage of the size of the largest form. Apo B100 is made principally in the liver of humans and rats. Apo B48 is a carboxy-terminally truncated form of apo B100. In humans, it is made exclusively in the intestine, whereas in the rat (and other rodents), it is made in both the liver and the intestine. There are important differences in structure between apo B48 and apo B100, as indicated. These differences dramatically affect the metabolism of the lipoprotein particles. They may also affect the mechanism of lipoprotein assembly.

of apo B mRNA editing in the liver appears to correlate with the rate of apo B secretion (Leighton *et al.*, 1990; Davidson *et al.*, 1990b).

5. THE ASSEMBLY AND SECRETION OF VLDL IS VARIABLE IN RESPONSE TO LIPID AVAILABILITY AND METABOLIC STATE

Considering that VLDL is composed of three major lipids (phospholipid, triglyceride, and cholesterol) and one specific protein (apo B), one might expect that the availability of each constituent could become rate-limiting for VLDL assembly. To a degree this proposal has been verified in several different experimental situations. Limiting the availability of each lipid can result in the impairment of VLDL assembly and secretion. For example, in cultured rat hepatocytes, blocking phosphatidylcholine synthesis dramatically decreases the assembly and secretion of VLDL, but not HDL (Yao and Vance, 1989). Blocking cholesterol synthesis also decreases VLDL secretion by cultured rat hepatocytes (Khan *et al.*, 1989, 1990). Decreasing triglyceride synthesis by fasting causes a parallel decrease in the secretion of VLDL (Davis *et al.*, 1985).

In contrast to the predicted impairment of VLDL assembly and secretion caused by restricting the availability of essential lipids, increasing the availability of a specific lipid does not always increase the assembly and secretion of VLDL. The ability of triglycerides to stimulate VLDL assembly and secretion appears to be determined by cell-type-specific processes. For example, in cultured rat hepatocytes, adding oleic acid to the medium to simulate the synthesis of glycerol lipids (triglycerides and phospholipid) did not affect the secretion of either apo B100 or apo B48 (Davis and Boogaerts, 1982; Patsch *et al.*, 1983). In marked contrast to the lack of an effect on apo B secretion in cultured rat hepatocytes, adding oleic acid to Hep G2 cells stimulates the secretion of both triglyceride and apo B100 (Bostrom *et al.*, 1988; Erickson and Fielding, 1986; Pullinger *et al.*, 1989; Dixon *et al.*, 1991). The distinctions between cultured rat hepatocytes and HepG2 cells are further reflected in their different ability to assemble VLDL-sized lipoproteins. HepG2 cells synthesize apo B100 on particles that are in the LDL size range (not the VLDL size range) (Thrift *et al.*, 1986). While stimulation of triglyceride synthesis in HepG2 cells increases particle size from a small LDL to an IDL-sized particle (Bostrom *et al.*, 1988; Ellsworth *et al.*, 1986), it does not result in the secretion of VLDL-sized particles. These data suggest that the inability of HepG2 cells to secrete apo B as a VLDL particle is not related to the size of apo B nor is it related to an inability to synthesize triglyceride. In contrast to HepG2 cells, cultured rat hepatocytes secrete both apo B100 and apo B48 on VLDL-sized particles (Leighton *et al.*, 1990; Davis *et al.*, 1979).

Because plasma concentrations of apo B usually are associated with similar changes in cholesterol, a great amount of effort has been paid to examining whether or not apo B secretion could be increased in response to increased availability of cholesterol. Feeding rats a fat-enriched diet (with and without cholesterol) did not change the hepatic secretion of apo B, but did increase its concentration in plasma (Davis *et al.*, 1982). Thus, while restricting the availability of cholesterol impairs VLDL assembly, an overabundance of cholesterol does not necessarily augment VLDL assembly. A possible explanation for this apparent discrepancy is that while limiting one essential component can make the availability of that component rate-limiting, increasing the availability of that component may enhance VLDL assembly only if the other essential components are also increased in a proportional manner necessary to satisfy the structural requirements for the additional VLDL particles formed.

There are, in fact, data supporting the view that physiologic induction of VLDL assembly is associated with a coordinate change in the synthesis of all essential VLDL components. Carbohydrate-rich diets are known to increase the hepatic lipogenesis and VLDL secretion (Boogaerts *et al.*, 1984). Examination of hepatocytes obtained from carbohydrate-fed rats showed that the secretion of all essential components of VLDL (i.e., triglycerides, cholesterol, phospholipids, and apo B) were increased (Boogaerts *et al.*, 1984). In contrast, fasting is associated with decreased hepatic lipogenesis and VLDL secretion (Davis *et al.*, 1985). Moreover, hepatocytes obtained from fasted rats showed a decrease in the secretion of all essential components of VLDL (Davis *et al.*, 1985). Finally, hepatic VLDL assembly and secretion is developmentally regulated (Coleman *et al.*, 1988). During the weaning period there is a coordinate induction of the secretion of all essential VLDL components (Coleman *et al.*, 1988). The combined data show that changes in the assembly and secretion of VLDL initiated by physiologic state (carbohydrate feeding, fasting, and development) are associated with a coordinate change in the secretion of all essential VLDL components. Additional data show that the changes in the secretion of essential lipids are caused by a coordinate change in their synthesis (Davis *et al.*, 1985; Boogaerts *et al.*, 1984; Coleman *et al.*, 1988). In contrast the changes in the secretion of apo B are posttranscriptional (Leighton *et al.*, 1990). An interesting and important question is: how can the secretion of apo B be linked coordinately with the synthesis of phosphatidylcholine, cholesterol, and triglyceride?

6. MOVEMENT OF APO B OUT OF THE ENDOPLASMIC RETICULUM IS RATE LIMITING

To gain an understanding of the processes regulating apo B secretion it is necessary to first determine the rate-limiting step. In rat hepatocytes, the majority of both apo B100 and apo B48 was confined to the ER suggesting that the rate-limiting step for apo B secretion is its movement out of this organelle (Davis *et al.*, 1989; Borchardt and Davis, 1987), which is the site where VLDL is assembled (Alexander *et al.*, 1976). The first-order rate constants for the intracellular movement of [^{35}S]methionine-labeled apoB were determined using pulse–chase protocols in cultured rat hepatocytes (Borchardt and Davis, 1987). The first-order rate constant describing the movement of both apo B48 and apo B100 out of the rough ER was the same as the rate constant describing the movement out of the hepatocyte. These data lead to the conclusion that movement out of the rough ER is the rate-limiting step in the VLDL assembly/secretion pathway. In HepG2 cells, movement of apo B out of the ER was similarly found to be the slowest step in the secretion pathway (Boren *et al.*, 1990). Interestingly, the movement of both apo B100 and apo B48 out of the ER was slower than that for albumin; apo B48 moved out of the rough ER slower than did apo B100 (Borchardt and Davis, 1987). The unusually slow movement of both molecular weight forms of apo B out of the rough ER is consistent with the idea that this organelle is the site of the posttranscriptional process regulating apo B secretion.

7. APO B IS DEGRADED INTRACELLULARLY BY A PROCESS THAT CAN ACCOUNT FOR POSTTRANSCRIPTIONAL REGULATION OF SECRETION

Surprisingly, only a fraction (30 to 60%) of the de novo synthesized apo B that was lost from the cell was recovered in the culture medium (Borchardt and Davis, 1987). Thus, a significant portion of apo B is degraded intracellularly.

Intracellular degradation has also been reported to occur in HepG2 cells (Dixon *et al.*, 1991; Boren *et al.*, 1990) and rat hepatoma cells (Yao *et al.*, 1991). In cultured rat hepatocytes, insulin decreases the secretion of apo B via a mechanism involving its degradation (Sparks and Sparks, 1990).

8. DECREASED INTRACELLULAR DEGRADATION CAN ACCOUNT FOR THE STIMULATION OF APO B SECRETION BY HepG2 CELLS

Pulse–chase studies were performed to examine the mechanism by which oleic acid stimulates apo B secretion in HepG2 cells (Dixon *et al.*, 1991). The results show that oleic acid decreases the intracellular degradation of apo B (Dixon *et al.*, 1991). Additional studies by the same group suggest that the state of lipidation of the lipoprotein particle in the ER determines the degree of apo B100 degradation (Furukawa *et al.*, 1992).

9. APO B DEGRADATION OCCURS IN A PRE-GOLGI COMPARTMENT

Proteolytic fragments of apo B were found in the rough and smooth microsomes, but were absent in the Golgi fraction (Davis *et al.*, 1989). These data suggest that apo B is degraded in a pre-Golgi compartment. Studies using HepG2 cells also lead to the conclusion that apo B100 is degraded in a pre-Golgi compartment (Sato *et al.*, 1990). More detailed studies using HepG2 cells conclusively showed that, indeed, the ER is the major site of apo B degradation (Furukawa *et al.*, 1992).

Degradation of apo B in the ER is reminiscent of the process responsible for the degradation of the α subunit of the T-cell receptor (Bonifacino *et al.*, 1990a,b, 1991) and of HMG-CoA reductase (Gil *et al.*, 1985; Chun *et al.*, 1990; Inoue *et al.*, 1991). The finding that the calpain I inhibitor blocks the degradation of HMG-CoA reductase, but not that of the T-cell receptor, suggests that their degradation pathways are different (Inoue and Simon, 1992). However, the findings that the ER degradation of both HMG-CoA reductase (Inoue *et al.*, 1991) and apo B (Thrift *et al.*, 1992) are blocked by the thiol protease inhibitor ALLN suggest the possibility that a common degradation pathway is involved. It is interesting to note that the ER degradation of both HMG-CoA reductase (Gil *et al.*, 1985) and apo B100 (Sato *et al.*, 1990) is stimulated by the uptake of LDL (Gil *et al.*, 1985).

The combined data suggest that intracellular degradation plays a role in regulating apo B secretion. This posttranscriptional mechanism can explain why the secretion of VLDL apo B is varied, while the relative abundance of apo B mRNA remains nearly constant (Leighton et al., 1990; Pullinger et al., 1989). Why would a cell degrade a newly synthesized secretory protein? More importantly, how is this apparently energy-wasteful degradative pathway regulated? What are the signals determining whether apo B is secreted or degraded? The answers to these questions are based on the unusual properties of apo B, which behaves as an integral membrane protein, but also requires translocation across the ER.

10. TRANSLOCATION OF APO B IS INEFFICIENT AND DETERMINES (IN PART) HOW MUCH IS DEGRADED OR SECRETED

Unlike other apolipoproteins, apo B does not exchange freely between lipoprotein particles, and thus behaves similarly to integral membrane proteins. The unique ability of apo B to integrate into the surface monolayer of the lipoprotein particle may play an essential role in lipoprotein assembly (Davis et al., 1990). However, the properties that allow proteins to integrate into membranes also act to impede translocation across the ER (Walter et al., 1984). Based on this rationale, we turned our attention to topographical distribution of apo B in membranes. Rough, smooth, and Golgi microsomal fractions were obtained from rat liver and the accessibility of apo B to the exogenous protease trypsin was determined (Davis et al., 1990). Two pools of apo B were found. One pool is fully translocated into the lumen of the ER and is protected from trypsin degradation. In contrast, the other pool remains untranslocated on the cytoplasmic surface and is completely degraded by trypsin. The untranslocated (trypsin accessible) pool of apo B is degraded, suggesting that translocation may be the first step governing how much apo B is secreted. Additional studies by others using chick hepatocytes (Dixon et al., 1992) and rat liver (Wilkinson et al., 1992) show that the majority of apo B in the ER is exposed on the cytoplasmic surface, findings consistent with the proposal that a significant portion of apo B remains untranslocated.

More detailed analysis of the translocation of apo B15 shows that its translocation is transiently arrested (i.e., "paused translocation") (Chuck et al., 1990). Recently, a sequence found in apo B15 containing charged amino acids has been identified as the sequence responsible for the "paused translocation" of apo B15 (Chuck and Lingappa, 1992). Not all studies using truncated forms of apo B have observed a "paused translocation" of apo B15 (Pease et al., 1991). The question remains: does incomplete translocation determine how much apo B is degraded or does degradation determine how much apo B is translocated?

11. TRANSLOCATION OF APO B REQUIRES A UNIQUE PROCESS WHICH MAY GOVERN ITS METABOLIC FATE: VLDL ASSEMBLY OR DEGRADATION

It is generally accepted that most secretory proteins utilize common intra-cellular processes for their synthesis, translocation, transport, and secretion. These processes display a remarkable degree of evolutionary conservation, al-lowing a protein indigenous to one organism to be expressed and secreted by tissues and cells of other phylogenetically distinct organisms (Garoff, 1985). For example, there are a plethora of proteins that are imported into the lumen of the ER, many of which contain distinct N-terminal signal sequences bearing little amino acid sequence homology. Yet, a wide variety of signal sequences can be correctly recognized and transported by the intracellular machinery of organisms as diverse as yeast and mammals. An exception to this paradigm is apo B. It requires a unique process for its secretion as demonstrated by the disease abet-alipoproteinemia. This disease is inherited recessively and is caused by a muta-tion in a gene other than the one coding for apo B. Livers of patients with abetalipoproteinemia have apo B mRNA of normal size (Lackner *et al.*, 1986; Bouma *et al.*, 1990). Moreover, altered relative abundance of apo B mRNA in liver and intestine cannot explain the phenotypic absence of plasma apo B (Lack-ner *et al.*, 1986; Bouma *et al.*, 1990). Since specific apo B epitopes recognized by monoclonal antibodies are present, apo B is synthesized (Bouma *et al.*, 1990; Dullaart *et al.*, 1986). However, almost no apo B100 (liver) and apo B48 (intes-tine) is secreted. The finding that the plasma concentrations of other proteins derived from the liver are not altered in abetalipoproteinemics indicates that the inability to secrete apo B is not caused by a general impairment of protein secretion, but rather by the loss of a process that may be uniquely required in order to secrete apo B. The genetic defects responsible for abetalipoproteinemia are not known. However, patients with abetalipoproteinemia lack detectable microsomal triglyceride transfer protein (Wetterau *et al.*, 1992). Until the muta-tion responsible for the loss of the microsomal triglyceride transfer protein is identified, it is not possible to conclude that the gene encoding this protein is the primary alteration responsible for the abetalipoproteinemic phenotype. More-over, considering the large differences of the abetalipoproteinemic phenotype (Lackner *et al.*, 1986; Bouma *et al.*, 1990; Talmud *et al.*, 1988; Huang *et al.*, 1990), it is likely that more than one genetic defect is involved. The combined data of the abetalipoproteinemic phenotype suggest that apo B requires specific processes for its secretion.

To gain an understanding of the possible cell-type-specific processes re-quired for apo B secretion, we examined the ability of nonhepatic cells (CHO cells) to translocate, secrete, and degrade apo B (Thrift *et al.*, 1992). When

Step 1. Translation of mRNA and Association with Translocation Complex via Signal Sequence

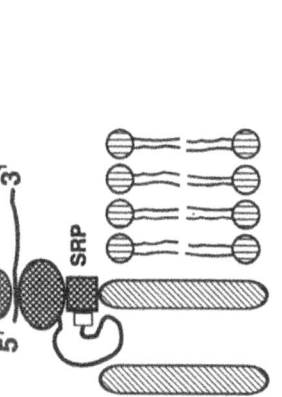

Step 2. Selection of destination determines fate of ApoB

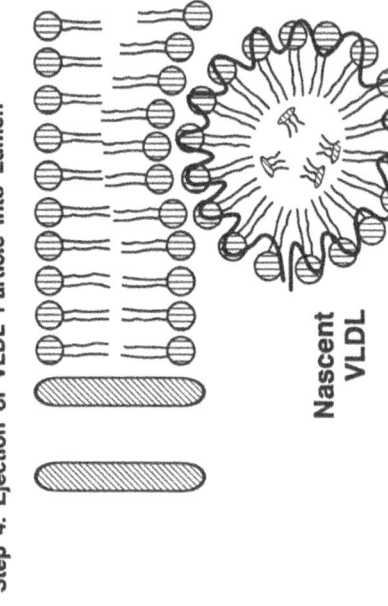

A) Translocation arrest, fate: degradation

B) Translocation requires unknown factor X fate: association with lumenal leaflet of membrane bilayer

Step 3. Accumulation of Triglycerides in "Pocket" of Lumenal Leaflet

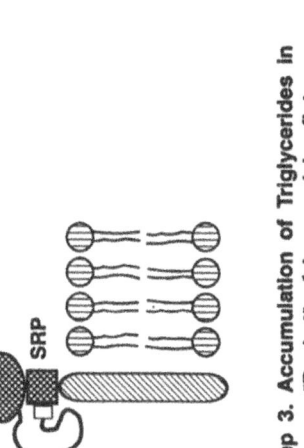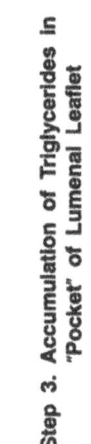

Triglycerides

Step 4. Ejection of VLDL Particle into Lumen

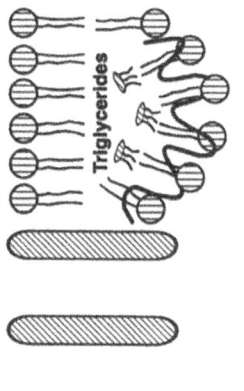

Nascent VLDL

expressed in hepatoma cells, apo B53 (the N-terminal 53% of apo B100, encompassing the sequence of the naturally occurring apo B48) assembles lipoproteins and is secreted as a lipoprotein complex (Yao *et al.*, 1991). In marked contrast, when apo B53 is expressed in nonhepatic COS cells, it is not secreted (Yao *et al.*, 1991). The inability of COS cells to secrete apo B53 could be an artifact of overexpression. Alternatively, this observation may reflect the absence of a tissue-specific process (such as the one lacking in abetalipoproteinemia) required for the secretion of apo B. This possibility was investigated by expressing two forms of apo B (apo B15 and apo B53) in nonhepatic CHO cells (Thrift *et al.*, 1992). Apo B15 was chosen for comparison with apo B53 because it has less than the minimum length required to assemble lipoproteins, and thus might not require the same unique processes for its secretion. Expression of apo B15 in CHO cells resulted in the accumulation of apo B15 protein in both medium and cells. In contrast, apo B was not detectable in medium or within CHO cells transfected with the plasmic encoding apo B53, despite the expression of apo B53 mRNA. Apo B53 did accumulate within transfected cells incubated with the thiol protease inhibitor ALLN, suggesting that it is synthesized but completely degraded in the absence of the inhibitor. Apo B53 was not secreted despite its presence within ALLN-treated cells. Essentially all the apo B53 that accumulated in microsomes from ALLN-treated cells was associated with the membrane and was susceptible to degradation by exogenous trypsin, indicating exposure on the cytoplasmic face of the membrane. Thus, translocation of B53 across the ER membrane is blocked. However, the apo B53 showed the capacity to bind to concanavalin A, suggesting that it is glycosylated and, therefore, partly exposed to the lumen (i.e., it traverses the membrane). Since both apo B15 and apo B53 are secreted when expressed in hepatoma cells, apo B requires a unique process, not expressed in CHO fibroblasts, for its complete translocation and entrance into the secretory pathway. This process might account for the inability of abetalipoproteinemic patients to secrete apo B.

FIGURE 3. Hypothesized steps in the assembly of VLDL. VLDL assembly begins early during the biogenesis of apo B. Clearly, apo B requires a unique process for its translocation across the endoplasmic reticulum. It is thought that at least a portion of the nascent apo B chain is translocated posttranslationally. At some stage in the translocation process, a decision is made either to fully translocate the protein into the lumen (the site of VLDL assembly) or to block further translocation. The pool of apo B that has undergone translocation arrest is thought to be degraded. Thus, regulated translocation can explain the posttranscriptional regulation of apo B secretion. Once in the lumen (step 3), apo B is thought to perturb the structure of the lumenal membrane leaflet, forming a pocket into which triglyceride can accumulate. At some stage (putative step 4) the nascent VLDL particle is ejected into the lumen of the endoplasmic reticulum. The mechanism through which the lumenal leaflet is disrupted forming a separate monolayer is unknown and unprecedented.

12. HYPOTHESIS FOR VLDL ASSEMBLY AND SECRETION

Based on the data of others and ours discussed above, we propose the following scheme regulating apo B translocation and degradation (Figure 3). We propose that there are four major steps in the VLDL assembly/secretion pathway: step 1, translation and association with the translocation complex; step 2, selection of destination (i.e., full translocation into the lumen leading to eventual secretion or translocation arrest and subsequent degradation); step 3, association with triglycerides in the lumen of the ER; step 4, ejection of the VLDL particle into the lumen and secretion.

We propose the following hypothesis: the amount of apo B that is secreted by the liver is determined by two interrelated posttranscriptional processes: translocation and intracellular degradation. Translocation of apo B across the ER requires a unique process not required by most (if not all) other secretory proteins. Moreover, this process is sensitive to the nutritional state of the liver. As a result, translocation is variable. *De novo* synthesized apo B which remains untranslocated or partially translocated is shunted into the degradation pathway, whereas the portion that is fully translocated into the lumen of the ER enters the VLDL assembly pathway.

13. CONCLUDING REMARKS

The molecular mechanisms responsible for the assembly of VLDL are complex and remain to a large degree unknown. In this review we have tried to emphasize what is established, what is controversial, and what is speculated. While the model for VLDL assembly is based on what has been observed in recent experiments, it represents our speculation (i.e., working hypotheses). Clearly, there remain questions concerning various aspects of this model that require further study. These questions and areas for further investigation include: identifying the topographical nature of apo B in the membrane of the ER, determining the process through which apo B is translocated across the ER (is it translocated through an aqueous pore or through the membrane bilayer?), identifying the processes responsible for the intracellular degradation of apo B, determining whether translocation efficiency determines how much apo B is degraded or the converse (i.e., does apo B degradation determine how much is translocated?), determining the mechanism through which triglycerides are added to the core of the nascent VLDL particle, determining how the nascent VLDL particle is "ejected" into the lumen of the ER, and determining what role the Golgi plays in VLDL assembly. Gaining answers to these questions will require the use of many different aspects of cell biology, biochemistry, and molecular genetics. The knowledge gained from these challenging experiments should provide new insights in macromolecular assembly. Furthermore, this information may ultimately lead to new methods to control VLDL assembly and prevent the pathology and disease that is associated with hyperlipidemia.

ACKNOWLEDGMENT. The author acknowledges support from the National Institutes of Health, HL41624.

14. REFERENCES

Alexander, C. A., Hamilton, R. L., and Havel, R. J. 1976, Subcellular localization of B apoprotein of plasma lipoproteins in rat liver, *J. Cell Biol.* **69**:241–263.

Bonifacino, J. S., Suzuki, S. K., and Klausner, R. D., 1990a. A peptide sequence confers retention and rapid degradation in the endoplasmic reticulum, *Science* **247**:79–82.

Bonifacino, J. S., Cosson, P., and Klausner, R. D., 1990b, Colocalized transmembrane determinants for ER degradation and subunit assembly explain the intracellular fate of TCR chains, *Cell* **63**:503–513.

Bonifacino, J. S., Cosson, P., Shah, N., and Klausner, R. D., 1991, Role of potentially charged transmembrane residues in targeting proteins for retention and degradation within the endoplasmic reticulum, *EMBO J.* **10**:2783–2793.

Boogaerts, J. R., Malone, M. M., Archambault, S. J., and Davis, R. A., 1984, Dietary carbohydrate induces lipogenesis and very-low-density lipoprotein synthesis, *Am. J. Physiol.* **246**:E77–E83.

Borchardt, R. A., and Davis, R. A., 1987, Intrahepatic assembly of very low density lipoproteins. Rate of transport out of the endoplasmic reticulum determines rate of secretion, *J. Biol. Chem.* **262**:16394–16402.

Boren, J., Wettesten, M., Sjoberg, A., Thorlin, T., Bondjers, G., Wiklund, O., and Olofsson, S.-O., 1990, The assembly and secretion of ApoB 100 containing lipoproteins in Hep G2 cells. Evidence for different sites for protein synthesis and lipoprotein assembly, *J. Biol. Chem.* **265**:10556–10564.

Boren, J., Graham, L., Wettesten, M., Scott, J., White, A., and Olofsson, S.-O., 1992, The assembly and secretion of apo B100-containing lipoproteins in Hep G2 cells. Apo B100 is cotranslationally integrated into lipoproteins, *J. Biol. Chem.* **267**:9858–9867.

Bostrom, K., Boren, J., Wettesten, M., Sjoberg, A., Bondjers, G., Wiklund, O., Carlsson, P., and Olofsson, S.-O., 1988, Studies on the assembly of apo B-100-containing lipoproteins in HepG2 cells, *J. Biol. Chem.* **263**:4434–4442.

Bouma, M. E., Beucler, I., Pessah, M., Heinzmann, C., Lusis, A. J., Naim, H. Y., Ducastelle, T., Leluyer, B., Schmitz, J., Infante, R., and Aggerbeck, L. P., 1990, Description of two different patients with abetalipoproteinemia: Synthesis of a normal-sized apolipoprotein B-48 in intestinal organ culture, *J. Lipid Res.* **31**:1–15.

Chan, L., 1993, RNA editing: Exploring one mode with apolipoprotein B mRNA, *BioEssays* **15**:33–41.

Chatterton, J. E., Phillips, M. L., Curtiss, L. K., Milne, R. W., Marcel, Y. L., and Schumaker, V. N., 1991, Mapping apolipoprotein B on the low density lipoprotein surface by immunoelectron microscopy, *J. Biol. Chem.* **266**:5955–5962.

Chen, G. C., Hardman, D. A., Hamilton, R. L., Mendel, C. M., Schilling, J. W., Zhu, S., Lau, K., Wong, J. S., and Kane, J. P., 1989, Distribution of lipid-binding regions in human apolipoprotein B-100, *Biochemistry* **28**:2477–2484.

Chen, S.-H., Habib, G., Yang, C.-Y., Gu, G.-W., Lee, B. R., Wang, S.A., Silberman, S. R., Cai, S.-J., Deslypere, J. P., Rosseneu, M., Jr., Gotto, A. M., Li, W.-H. and Chan, L., 1987, Apolipoprotein B-48 is the product of a messenger RNA with an organ-specific in-frame stop codon, *Science* **238**:363–366.

Chen, S.-H., Li, X., Liao, W., Wu, J. H., and Chan, L., 1990, RNA editing of apolipoprotein B mRNA. Sequence specificity determined by in vitro coupled transcription editing, *J. Biol. Chem.* **265**:6811–6816.

Chuck, S. L., and Lingappa, V. R., 1992, Pause transfer: A topogenic sequence in apolipoprotein B mediates stopping and restarting of translocation, *Cell* **68**:9–21.

Chuck, S. L., Yao, Z., Blackhart, B. D., McCarthy, B. J., and Lingappa, V. R., 1990, New variation on the translocation of proteins during early biogenesis of apolipoprotein B, *Nature* **346**:382–385.

Chun, K. T., Bar, N. S., and Simoni, R. D., 1990, The regulated degradation of 3-hydroxy-3-methylglutaryl-CoA reductase requires a short-lived protein and occurs in the endoplasmic reticulum, *J. Biol. Chem.* **265**:22004–22010.

Cladaras, C., Hadzopoulou-Cladaras, M., Nolte, R. T., Atkinson, D., and Zannis, V. I., 1986, The complete sequence and structural analysis of human apolipoprotein B-100: Relationship between apoB-100 and apoB-48 forms, *EMBO J.* **13**:3495–3507.

Coleman, R. A., Haynes, E. B., Sand, T. M., and Davis, R. A., 1988, Developmental coordinate expression of triacylglycerol and small molecular weight apo B synthesis and secretion by rat hepatocytes, *J. Lipid Res.* **29**:33–42.

Davidson, N. O., Carlos, R. C., and Lukaszewicz, A. M., 1990a, Apolipoprotein B mRNA editing is modulated by thyroid hormone analogs but not growth hormone administration in the rat, *Mol. Endocrinol.* **4**:779–785.

Davidson, N. O., Carlos, R. C., Sherman, H. L., and Hay, R. V., 1990b, Modulation of apolipoprotein B-100 mRNA editing: Effects on hepatic very low density lipoprotein assembly and intracellular apoB distribution in the rat, *J. Lipid Res.* **31**:899–908.

Davies, M. S., Wallis, S. C., Driscoll, D. M., Wynne, J. K., Williams, G. W., Powell, L. M., and Scott, J., 1989, Sequence requirements for apolipoprotein B RNA editing in transfected rat hepatoma cells, *J. Biol. Chem.* **264**:13395–13398.

Davis, R. A., 1991, Lipoprotein structure and secretion in *Biochemistry of Lipids and Membranes*, Elsevier, Amsterdam, pp. 403–424.

Davis, R. A., and Boogaerts, J. R., 1982, Intrahepatic assembly of very low density lipoproteins. Effect of fatty acids on triacylglycerol and apolipoprotein synthesis, *J. Biol. Chem.* **257**:10908–10913.

Davis, R. A., Engelhorn, S. C., Pangburn, S. H., Weinstein, D. B., and Steinberg, D., 1979, Very low density lipoprotein synthesis and secretion by cultured rat hepatocytes, *J. Biol. Chem.* **254**:2010–2016.

Davis, R. A., Malone, M. M., and Moses, R. L., 1982, Intrahepatic assembly of very low density lipoprotein. Competition by cholesterol esters for the hydrophobic core, *J. Biol. Chem.* **257**:2634–2640.

Davis, R. A., Boogaerts, J. R., Borchardt, R. A., Malone, M. M., and Archambault-Schexnayder, J., 1985, Intrahepatic assembly of very low density lipoproteins: Varied synthetic response of individual apolipoproteins to fasting, *J. Biol. Chem.* **260**:14137–14144.

Davis, R. A., Prewett, A. B., Chan, D., Thompson, J. J., Borchardt, R. A., and Gallaher, W. R., 1989, Intrahepatic assembly of very low density lipoproteins: Immunologic characterization of apolipoprotein B in lipoproteins and hepatic membrane fractions and its intracellular distribution, *J. Lipid Res.* **30**:1185–1196.

Davis, R. A., Thrift, R. N., Wu, C. C., and Howell, K. E., 1990, Apolipoprotein B is both integrated into and translocated across the endoplasmic reticulum membrane. Evidence for two functionally distinct pools, *J. Biol. Chem.* **265**:10005–10011.

Dixon, J. L., Furukawa, S., and Ginsberg, H. N., 1991, Oleate stimulates secretion of apolipoprotein B-containing lipoproteins from Hep G2 cells by inhibiting early intracellular degradation of apolipoprotein B, *J. Biol. Chem.* **266**:5080–5086.

Dixon, J. L., Chattapadhyay, R., Huima, T., Redman, C. M., and Banerjee, D., 1992, Biosynthesis of lipoprotein: Location of nascent apoAI and apoB in the rough endoplasmic reticulum of chicken hepatocytes, *J. Cell Biol.* **117**:1161–1169.

Driscoll, D. M., Wynne, J. K., Wallis, S. C., and Scott, J., 1989, An in vitro system for the editing of apolipoprotein B mRNA, *Cell* **58**:519–525.

Dullaart, R. P. F., Speelberg, B., Schuurman, H. J., Milne, R. W., Havekes, L. M., Marcel, Y., Geuze, H. J., Hulshof, M. M., and Erkelens, D. W., 1986, Epitopes of apolipoprotein B-100 and B-48 in both liver and intestine. Expression and evidence for local synthesis in recessive abetalipoproteinemia, *J. Clin. Invest.* **78**:1397–1404.

Ellsworth, J. L., Erickson, S. K., and Cooper, A. D., 1986, Very low and low density lipoprotein synthesis and secretion by the human hepatoma cell line Hep-G2: Effects of free fatty acid, *J. Lipid Res.* **27**:858–874.

Erickson, S. K., and Fielding, P. E., 1986, Parameters of cholesterol metabolism in the human hepatoma cell line, Hep-G2, *J. Lipid Res.* **27**:875–883.

Furukawa, S., Sakata, N., Ginsberg, H. N., and Dixon, J. L., 1992, Studies of the sites of intracellular degradation of apolipoprotein B in Hep G2 cells, *J. Biol. Chem.* **267**:22630–22638.

Garcia, Z. C., Poksay, K. S., Bostrom, K., Johnson, D. F., Balestra, M. E., Shechter, I., and Innerarity, T. L., 1992, Characterization of apolipoprotein B mRNA editing from rabbit intestine, *Arterioscler. Thromb.* **12**:172–179.

Garoff, H., 1985, Using recombinant DNA techniques to study protein targeting in the eukaryotic cell, *Annu. Rev. Cell Biol.* **1**:403–445.

Gil, G., Faust, J. R., Chin, D. J., Goldstein, J. L., and Brown, M. S., 1985, Membrane-bound domain of HMG CoA reductase is required for sterol-enhanced degradation of the enzyme, *Cell* **41**:249–258.

Graham, D. L., Knott, T. J., Jones, T. C., Pease, R. J., Pullinger, C. R., and Scott, J., 1991, Carboxyl-terminal truncation of apolipoprotein B results in gradual loss of the ability to form buoyant lipoproteins in cultured human and rat liver cell lines, *Biochemistry* **30**:5616–5621.

Hamilton, R. L., and Fielding, P. E., 1989, Nascent very low density lipoproteins from rat hepatocytic Golgi fractions are enriched in phosphatidylethanolamine, *Biochem. Biophys. Res. Commun.* **160**:162–167.

Hospattankar, A. V., Higuchi, K., Law, S. W., Meglin, N., Cartwright, J., and Brewer, H. B., Jr., 1988, Identification of a novel in-frame translational stop codon in human intestine apoB mRNA, *Biochem. Biophys. Res. Commun.* **148**:279–285.

Huang, L. S. Jnne, P. A., De, G. J., Cooper, M., Decklebaum, R. J., Kayden, H., and Breslow, J. L., 1990, Exclusion of linkage between the human apolipoprotein B gene and abetalipoproteinemia, *Am. J. Hum. Genet.* **46**:1141–1148.

Inoue, S., and Simoni, R. D., 1992, 3-Hydroxy-3-methylglutaryl-coenzyme A reductase and T cell receptor alpha subunit are differentially degraded in the endoplasmic reticulum, *J. Biol. Chem.* **267**:9080–9086.

Inoue, S., Bar, N. S., Roitelman, J., and Simoni, R. D., 1991, Inhibition of degradation of 3-hydroxy-3-methylglutaryl-coenzyme A reductase in vivo by cysteine protease inhibitors, *J. Biol. Chem.* **266**:13311–13317.

Kane, J. P., 1983, Apolipoprotein B: Structural and metabolic heterogeneity, *Annu. Rev. Physiol.* **45:**637–650.

Khan, B., Wilcox, H. G., and Heimberg, M., 1989, Cholesterol is required for secretion of very-low-density lipoprotein by rat liver, *Biochem. J.* **258:**807–816.

Khan, B. V., Fungwe, T. V., Wilcox, H. G., and Heimberg, M., 1990, Cholesterol is required for the secretion of the very-low-density lipoprotein: In vivo studies, *Biochim. Biophys. Acta* **1044:**297–304.

Knott, T. J., Pease, R. J., Powell, L. M., Wallis, S. C., Rall, S. C., Innerarity, T. L., Blackhart, B., Taylor, W. H., Marcel, Y., Milne, R., Johnson, D., Fuller, M., Lusis, A. J., McCarthy, B. J., Mahley, R. W., Levy-Wilson, B., and Scott, J., 1986, Complete protein sequence and identification of structural domains of human apolipoprotein B, *Nature* **323:**734–738.

Krisnaiah, K. L., Walker, L. R., Borensztain, J., Schonfeld, G., and Getz, G. S., 1980, Apolipoprotein B variant derived from rat intestine, *Proc. Natl. Acad. Sci. USA* **77:**3806–3810.

Lackner, K. J., Monge, J. C., Gregg, R. E., Hoeg, J. M., Triche, T. J., Law, S. W., and Brewer, H. B., Jr., 1986, Analysis of the apolipoprotein B gene and messenger ribonucleic acid in abetalipoproteinemia, *J. Clin. Invest.* **78:**1707–1712.

Lau, P. P., Chen, S. H., Wang, J. C., and Chan, L., 1990, A 40 kilodalton rat liver nuclear protein binds specifically to apolipoprotein B mRNA around the RNA editing site, *Nucleic Acids Res.* **18:**5817–5821.

Leighton, J. K., Joyner, J., Zamarripa, J., Deines, M., and Davis, R. A., 1990, Fasting decreases apolipoprotein B mRNA editing and the secretion of small molecular weight apoB by rat hepatocytes: Evidence that the total amount of apoB secreted is regulated post-transcriptionally, *J. Lipid Res.* **31:**1663–1668.

Patsch, W., Tamai, T., and Schonfeld, G., 1983, Effect of fatty acids on lipid and apoprotein secretion and association in hepatocyte cultures, *J. Clin. Invest.* **72:**371–378.

Pease, R. J., Harrison, G. B., and Scott, J., 1991, Cotranslocational insertion of apolipoprotein B into the inner leaflet of the endoplasmic reticulum, *Nature* **353:**448–450.

Powell, L. M., Wallis, S. C., Pease, R. J., Edwards, Y. H., Knott, T. J., and Scott, J., 1987, A novel form of tissue-specific RNA processing produces apolipoprotein-B48 in intestine, *Cell* **50:**831–840.

Pullinger, C. R., North, J. D., Powell, L. M., Wallis, S. C., and Scott, J., 1989, The apolipoprotein B gene is constitutively expressed in HepG2 cells: Regulation of secretion by oleic acid, albumin, and insulin, and measurement of the mRNA half-life, *J. Lipid Res.* **30:**1065–1077.

Sato, R., Imanaka, T., Takatsuki, A., and Takano, T., 1990, Degradation of newly synthesized apolipoprotein B-100 in a pre-Golgi compartment, *J. Biol. Chem.* **265:**11880–11884.

Shah, R. R., Knott, T. J., Legros, J. E., Navaratnam, N., Greeve, J. C., and Scott, J., 1991, Sequence requirements for the editing of apolipoprotein B mRNA, *J. Biol. Chem.* **266:**16301–16304.

Sparks, J. D., and Sparks, C. E., 1990, Insulin modulation of hepatic synthesis and secretion of apolipoprotein B by rat hepatocytes, *J. Biol. Chem.* **265:**8854–8862.

Spring, D. J., Chen-Liu, L. W., Chatterton, J. E., Elovson, J., and Schumaker, V. N., 1992, Lipoprotein assembly: Apolipoprotein B determines lipoprotein core circumference, *J. Biol. Chem.* **267:**14839–14845.

Talmud, P. J., Lloyd, J. K., Muller, D. P., Collins, D. R., Scott, J., and Humphries, S., 1988, Genetic evidence from two families that the apolipoprotein B gene is not involved in abetalipoproteinemia, *J. Clin. Invest.* **82:**1803–1806.

Teng, B., Burant, C. F., and Davidson, N. O., 1993, Molecular cloning of an apolipoprotein B messenger RNA editing protein. *Science* **260;**1816–1819.

Thrift, R. N., Forte, T. M., Cahoon, B. E., and Shore, V. G., 1986, Characterization of lipoproteins produced by the human liver cell line, Hep G2, under defined conditions, *J. Lipid Res.* **27**:236–250.

Thrift, R. N., Drisko, J., Dueland, S., Trawick, J. D., and Davis, R. A., 1992, Translocation of apolipoprotein B across the endoplasmic reticulum is blocked in a nonhepatic cell line, *Proc. Natl. Acad. Sci. USA* **89**:9161–9165.

Walter, P., Gilmore, R., and Blobel, G., 1984, Protein translocation across the endoplasmic reticulum, *Cell* **38**:5–8.

Wetterau, J. R., Aggerbeck, L. P., Bouma, M.-E., Eisenberg, C., Munck, A., Hermier, M., Schmitz, J., Gay, G., Rader, D., Jr., and Gregg, R. E., 1992, Absence of microsomal triglyceride transfer protein in individuals with abetalipoproteinemia, *Science* **258**:999–1101.

Wilkinson, J., Higgins, J. A., Groot, P., Gherardi, E., and Bowyer, D. E., 1992, Membrane-bound apolipoprotein B is exposed at the cytosolic surface of liver microsomes, *FEBS Lett.* **304**:24–26.

Xiong, W., Zsigmond, E., Gotto, A. M., Jr., Reneker, L. W., and Chan, L., 1992, Transgenic mice expressing full-length human apolipoprotein B100. Full-length human apolipoprotein B mRNA is essentially not edited in mouse intestine or liver, *J. Biol. Chem.* **267**:21412–21420.

Yang, C.-Y., Chen, S.-H., Gianturco, S. H., Bradley, W. A., Sparrow, J. T., Tanimura, M., Li, W.-H., Sparrow, D. A., DeLoof, H., Rosseneu, M., Lee, F., Gu, Z.-W., Gotto, A. M., and Chan, L., 1986, Sequence, structure, receptor-binding domains and internal repeats of human apolipoprotein B-100, *Nature* **323**:738–742.

Yang, C., Gu, Z., Weng, S., Kim, T. W., Chen, S., Pownall, H. J., Sharp, P. M., Liu, S., Li, W., Gotto, A. M., Jr., and Chan, L., 1989, Structure of apolipoprotein B-100 of human low density lipoproteins, *Arteriosclerosis* **9**:96–108.

Yao, Z., and Vance, D. E., 1989, Head group specificity in the requirement of phosphatidylcholine biosynthesis for very low density lipoprotein secretion from cultured hepatocytes, *J. Biol. Chem.* **264**:11373–11380.

Yao, Z., Blackhart, B. D., Linton, M. F., Taylor, S. M., Young, S. G., and McCarthy, B. J., 1991, Expression of carboxyl-terminally truncated forms of human apolipoprotein B in rat hepatoma cells. Evidence that the length of apolipoprotein B has a major effect on the buoyant density of the secreted lipoproteins, *J. Biol. Chem.* **266**:3300–3308.

Young, S. G., 1990, Recent progress in understanding apolipoprotein B, *Circulation* **82**:1574–1594.

Chapter 9

Endoplasmic Reticulum and the Control of Ca^{2+} Homeostasis

Jacopo Meldolesi and Antonello Villa

1. INTRODUCTION

Control of Ca^{2+} homeostasis is one of the fundamental activities of eukaryotic cells. In the extracellular medium the concentration of the free ion is maintained in the millimolar range whereas in the cytosol the resting value is approximately four orders of magnitude lower, yet the total calcium content per cell amounts to several millimoles per liter. This means that most of this calcium is not free, but bound to appropriate molecules and/or segregated within intracellular organelles. At each given time, the distribution of Ca^{2+} among the segregated, the cytosolic, and the extracellular compartments is the result of a complex, dynamic equilibrium that relies on the coordinated activation of specific pumps, transporters, and channels.

Abbreviations used in this chapter: BiP, heavy-chain binding protein; $[Ca^{2+}]_i$, cytosolic Ca^{2+} concentration; CICR, Ca^{2+}-induced Ca^{2+} release; CR, calreticulin; CSQ, calsequestrin; ER, endoplasmic reticulum; Ins-P_3, inositol 1,4,5-trisphosphate; PDI, protein disulfide isomerase; Ry, ryanodine; SERCA, sarcoplasmic–endoplasmic reticulum Ca^{2+}-ATPase; SR, sacroplasmic reticulum.

Jacopo Meldolesi and Antonello Villa   Department of Pharmacology, CNR Cytopharmacology and B. Ceccarelli Centers, and San Raffaele Institute, University of Milan, Milan, Italy.
Subcellular Biochemistry, Volume 21: Endoplasmic Reticulum, edited by N. Borgese and J. R. Harris. Plenum Press, New York, 1993.

The existence of this multicompartment equilibrium offers the cells remarkable function privileges. Ca^{2+} not only participates (together with Na^+, K^+, H^+, and Cl^-) in the regulation of membrane potential, but in addition it is the only ion that plays the role of general intracellular messenger, acting in the cytosol as well as within various intracellular structures. This role depends on the existence of appropriate proteins that modify their conformation—in the case of enzymes or enzyme regulators, also the activity—when Ca^{2+} binds to their specific sites. Interestingly, the properties of these proteins differ considerably also in relation to the Ca^{2+} concentration existing in their compartments of residence. Thus, at least two classes of high-affinity ($K_d \simeq 0.1$ μM) Ca^{2+} binding proteins—those including the so-called E–F hand motif, such as calmodulin; and the annexins—are contained in the cytosol, and some of them appear also in the nucleus. Moreover, high-affinity binding has been observed with some dehydrogenases of mitochondria, a compartment where the resting Ca^{2+} concentration appears not far from that in the cytosol (Rizzuto et al., 1992). In contrast, within other organelles (the endoplasmic and sarcoplasmic reticula, the Golgi complex, and secretory granules) where the concentration of the ion is much higher than in the cytosol, the affinity of Ca^{2+} binding by resident proteins is most often low ($K_d \simeq$ 1 μM).

The role of Ca^{2+} in the cytosol is of fundamental importance because $[Ca^{2+}]_i$ can be quickly (msec) and considerably (severalfold) increased by the opening of even a few specific channels and the ensuing flux of the ion along its steep electrochemical gradient. Because of the presence of the (numerous) high-affinity Ca^{2+} binding proteins mentioned above, the rates of Ca^{2+} diffusion measured in the cytosol (10–100 $\mu m^2/s$) are in fact much lower (1–10%) than those in free solution, and therefore these $[Ca^{2+}]_i$ changes do not extend immediately to the entire cytosol. Interestingly, diffusion of the second messenger generated at the receptor level and competent for intracellular Ca^{2+} release—inositol 1,4,5-trisphosphate (Ins-P_3)—is not expected to be much slower in the cytoplasm than in solution. The generation of Ins-P_3, and the involvement of intracellular Ca^{2+} stores, are therefore more advantageous for the cell than simple activation of surface Ca^{2+} channels when a quick spreading of the $[Ca^{2+}]_i$ signal is required throughout the cytosol (for further information along these lines, see Carafoli, 1987; Pietrobon et al., 1990; Tsien and Tsien, 1990).

The points briefly discussed so far already give an explanation for the existence within all cells of intracellular Ca^{2+} stores that respond to activation with the rapid release of the cation, followed by its reaccumulation within a few seconds (rapidly exchanging Ca^{2+} stores). Simple increases of $[Ca^{2+}]_i$ could in fact be easily sustained by the activation of surface channels, drawing the cation out from the practically inexhaustible extracellular Ca^{2+} pool. In this case, however, the $[Ca^{2+}]_i$ increases would invariably begin at the cytosolic rim immediately beneath the plasma membrane, and involvement of the deep areas would take place only after an appreciable delay because of the slow diffusion of the cation in the whole cytosolic environment. A similar dissociation, but in the reverse direction, would occur at the end of the stimulation. In contrast, the existence of rapidly exchanging Ca^{2+} stores extends to the entire cytoplasm the ability of Ca^{2+} uptake and release. This results in a much more stringent coordination throughout the cell both of the $[Ca^{2+}]_i$ changes and of the corresponding regulated activities. Moreover, since these stores are not homogeneous (see below), their physiological role varies, i.e., they can be involved primarily in the generation, positive modulation, and possibly also in the blunting of $[Ca^{2+}]_i$ transients. This can result in heterogeneities of $[Ca^{2+}]_i$ caused not by the distance from the surface but by the specific distribution of the stores within the cell. An important function of the stores, which will not be discussed in detail here, is the generation of rhythmic $[Ca^{2+}]_i$ oscillations, an activity observed in a variety of cells after moderate stimulation or even at rest. Oscillations can give rise to $[Ca^{2+}]_i$ waves and gradients, and this has led to the recent provocative definition of the cytoplasm as an excitable structure (see Petersen and Wakui, 1990; Tsien and Tsien, 1990; Berridge, 1990, 1993; Berridge and Moreton, 1991).

For quite a long time, the cytological nature of the rapidly exchanging Ca^{2+} stores failed to attract much interest. Only in the case of the SR of striated muscle fibers, extensive evidence, accumulated primarily during the 1970s, led to a definite identification (see Fleischer and Inui, 1989). ER was first considered when Ins-P$_3$ was shown to induce release of accumulated ^{45}Ca from a total microsomal fraction isolated from the pancreatic tissue (Streb *et al.*, 1983; reviewed by Berridge and Irvine, 1984). Subsequent cell fractionation and immunocytochemical studies (reviewed by Meldolesi *et al.*, 1990) cast doubts on the simple identification of the entire ER with the rapidly exchanging store, and considered the possibility of this function being located either in distinct organelles, recovered together with the ER-derived elements in the microsomal fraction, or in specialized areas of the ER itself. Before we proceed, however, we would like to discuss the criteria to be employed for the identification of ER structures within the cells. Up until now, such an identification has been based primarily on simple morphological criteria, i.e., the appearance of individual membrane-bound structures in conventional electron micrographs. Thus, any ribosome-studded cisterna was attributed to the rough-surfaced ER (or the nuclear envelope, when this was obviously the case); most anastomosed tubules and vesicle profiles, different from both the Golgi complex and the endosome–lysosome system, to the smooth-surfaced ER. As pointed out recently elsewhere, we propose that the morphological criteria—which remain largely valid—be integrated with the distribution of widely expressed (possibly ubiquitous) ER markers, i.e., heavy-chain binding protein (BiP) and protein disulfide isomerase (PDI), two lumenal proteins that can be revealed not only at the biochemical, but also at the immunocytochemical level (Sitia and Meldolesi, 1992). In this review we will extensively rely on this criterion. The overall picture of the rapidly exchanging Ca^{2+} stores that emerges from the results obtained during the last few years, using a variety of experimental techniques (not only high-resolution immunocytochemistry, but also $[Ca^{2+}]_i$ measurement at both the cell suspension and single cell level; subcellular fractionation; protein biochemistry and molecular biology), is that of specialized areas, where the ER general markers mentioned above coexist with specialization markers present at concentrations much higher than in the rest of the endomembrane system. These ER areas correspond to the definition of calciosomes, a nomenclature proposed by our group based on initial biochemical and immunocytochemical studies in a variety of cell types (Volpe et al., 1988; Hashimoto et al., 1988). Specialized ER areas probably exist not only in relation to Ca^{2+} storage, but also in relation to other functions, including protein degradation (see Fra and Sit . this volume). Thus, the ER can now be envisaged as a patchwork of specialized subcompartments (Sitia and Meldolesi, 1992).

2. MOLECULAR COMPOSITION OF RAPIDLY EXCHANGING Ca^{2+} STORES

In order to carry out the functions outlined in the preceding pages, a Ca^{2+} store needs to express at least three components: (1) a high-affinity Ca^{2+} transport system operative across its limiting membrane, (2) one (or more) Ca^{2+} binding protein(s) of appropriate affinity (low) and capacity (high), to reversibly trap the transported ion within their membrane-bound lumen, and therefore keep their lumenal Ca^{2+} concentration at appropriate levels, and (3) one (or more) type(s) of membrane channel, gated at the cytosolic surface by its (their) interaction with second messengers, and thus able to rapidly discharge the segregated Ca^{2+} pool. In the Ca^{2+} storage structures, these specific components coexist with additional ER proteins, including the lumenal markers BiP and PDI (Villa et al., 1991, 1992, 1993; Takei et al., 1992), as discussed in detail below.

Knowledge about the Ca^{2+} store-specific components has rapidly increased during the last few years, and therefore the picture now emerging is comprehensive, although possibly not yet complete. Uptake of Ca^{2+} has been found in all cases to be caused by a group of Ca^{2+}-ATPases that, because of their occurrence in both the SR and the ER, are now indicated under the acronym SERCA. These enzymes, of molecular weight varying in the 110–100,000 range, require high concentrations of vanadate to be inhibited and are in contrast highly sensitive to various organic blockers, in particular thapsigargin and cyclopiazonic acid (Thastrup et al., 1990; Lytton et al., 1991; Zacchetti et al., 1991). Suggestions for the involvement of additional, thapsigargin-insensitive Ca^{2+}-ATPases have emerged from results in a few cell types (Foskett *et al.*, 1991; Papp *et al.*, 1991; Robinson and Burgoyne, 1991), but have not been confirmed in many others.

As far as the Ca^{2+} binding proteins are concerned, the best known example is still calsequestrin (CSQ). The latter is expressed in two molecularly distinct isoforms, one typical of the fast-twitch skeletal muscle, the other of the heart (Fliegel *et al.*, 1987; Scott *et al.*, 1988). Both of those proteins possess a long ($\sim$ one-third of the molecule), highly acidic C-terminal extension where as many as 40–50 Ca^{2+} ions can be bound with a K_d of $\sim$ 1 mM. In addition to striated muscle fibers, CSQ (cardiac isoform) has been shown to be expressed by the cerebellar Purkinje neurons of birds (Volpe *et al.*, 1990; Villa *et al.*, 1991; Takei *et al.*, 1992) and by smooth muscle fibers of mammals (Wuytack *et al.*, 1987; Villa *et al.*, 1993). Another Ca^{2+} binding protein that resembles CSQ in both functional properties and structural organization, but is almost entirely distinct in sequence, is calreticulin (CR) (Fliegel *et al.*, 1989a; Smith and Koch, 1989; Treves *et al.*, 1990; Baksh and Michalak, 1991; Michalak *et al.*, 1992), expressed by many cell types, although at markedly variable levels (Fliegel *et al.*, 1989b). Calnexin, a widely expressed ER transmembrane protein that protrudes largely in the lumen, has been shown to be partially homologous to CR and to bind Ca^{2+} (Wada *et al.*, 1991). Other ER lumenal proteins, such as a group of glycoproteins referred to as endoplasmin (or grp 94); two members of the thioredoxin family (apparent M_r, 59 and 80 kDa) as well as BiP and PDI themselves, also bind Ca^{2+}, but to an extent and with affinities that have not been precisely established (see Macer and Koch, 1988; Nguyen Van *et al.*, 1989; Peter *et al.*, 1992). This list does not exhaust the complement of the ER lumen Ca^{2+} binding proteins, because preliminary evidence for the existence of additional members of the family has been obtained by us and others working with microsomal fractions from various tissues and cell types. Whether these new proteins operate in Ca^{2+} storage within the lumen of the entire ER or of its areas specialized in Ca^{2+} storage remains to be established.

Intracellular Ca^{2+} channels belong to two different families. The first includes the receptors for Ins-P₃, the second messenger generated at the receptor level by the hydrolysis of the membrane phospholipid, phosphatidylinositol 4,5-bisphosphate. As the latter reaction has been revealed in all cells so far investigated, Ins-P₃ receptors are believed to be ubiquitous (but see Moreno *et al.*, 1992). The first isoform (type I) of this receptor was initially purified (Supattapone *et al.*, 1988) and then cloned (Furuichi *et al.*, 1989) from the cerebellum, where it is expressed at high levels by Purkinje neurons (Ross *et al.*, 1989; Mignery *et al.*, 1989; Maeda *et al.*, 1989). Channels are composed of four identical 313-kDa subunits (apparent M_r in SDS-PAGE, 260,000), each including an Ins-P₃ receptor binding site at its voluminous N-terminal domain protruding into the cytosol, and a hydrophobic C-terminal domain inserted across the membrane (Mignery and Südhof, 1990; Miyawaki *et al.*, 1991). Recent studies have documented the heterogeneity of the Ins-P$_3$ receptor based on both alternative splicing and the existence of two additional, homologous but distinct genes, at least one of which (type III) expressed preferentially outside the brain. Whether the various receptor isoforms are addressed to different Ca^{2+} stores or, in a few cell types, also to the plasma membrane, remains to be established (for a recent review see Meldolesi, 1992).

The channels of the second class take their name from a plant alkaloid, ryanodine (Ry), that they bind with high affinity. Their general organization resembles that of Ins-P$_3$ receptors (four identical subunits, $M_r \sim 550,000$). The sequence is, however, largely different. For quite some time, only two isoforms of the Ry receptors were believed to exist, expressed by skeletal (Takeshima *et al.*, 1989; Zorzato *et al.*, 1990) and cardiac (K. Otsu *et al.*, 1990) muscle fibers, respectively. This heterogeneity appeared reasonable, because the skeletal muscle Ry receptors are known to be physically coupled (in the muscle triads) to L-type Ca^{2+} channels clustered in the T channels of the plasma membrane, and to be rapidly activated following the depolarization of the latter membrane, via the voltage-sensing ability of their partners (Rios *et al.*, 1991). In contrast, cardiac Ry receptors work independently from other channels and are activated by the increase of [Ca^{2+}]$_i$ (Ca^{2+}-induced Ca^{2+} release, CICR), a mechanism active also on the skeletal muscle receptor. Recently, cardiac-type Ry receptors have been revealed in various other cell types, including eggs (McPherson *et al.*, 1992) and neurons (Lipscombe *et al.*, 1988; McPherson and Campbell, 1990; McPherson *et al.*, 1991). Moreover, a completely new class of CICR-activated Ry receptors, different from both the cardiac and skeletal muscle channels based on both molecular and pharmacological criteria, has been identified in a number of cell types, including various epithelial cell lines (Giannini *et al.*, 1992). The duality of the intracellular Ca^{2+} release systems—Ins-P$_3$ and Ry receptors—seems therefore to be a property not only of specialized muscle fibers, but of many, possibly all cell types.

3. THE ER SUBCOMPARTMENTS SPECIALIZED IN RAPID Ca^{2+} EXCHANGE

Based on the criteria discussed so far, we will include in this group the intracellular structures that, on the one hand, express bona fide ER markers, such as BiP and PDI, and, on the other hand, possess the molecular properties of the rapidly exchanging Ca^{2+} stores discussed in this section. The best known example of this type of structure is the SR of skeletal muscle. This elaborate endomembrane system consists of two distinct, yet continuous portions, the longitudinal SR and the terminal cisternae. The first is composed of cisternae and tubules arranged as a highly anastomosed water jacket all around the contractile apparatus, continuous on both sides with the terminal cisternae. These latter structures are voluminous and laterally anastomosed expansions, attached with their junctional membranes to the T tubules, the plasmalemma transverse infoldings that run perpendicularly to the major axis of the fiber. This arrangement is ideally suited to sustain the contraction–relaxation activity of the fibers. Ry receptors are in fact clustered at high density in the junctional membrane, and their direct coupling with the L-type channels of the T tubules makes their activation possible without delay as soon as the plasmalemma is depolarized. CSQ, on the other hand, is also highly concentrated within the terminal cisternae, arranged to yield dense masses that are resistant to permeabilization of the limiting membrane with low concentrations of detergent. These masses appear to be directly anchored to the membrane by filaments composed of peculiar proteins. Because of this arrangement of CSQ, the Ca^{2+} translocated by the ATPase across the entire surface (except the junctional membrane) of the SR finds its way to accumulate within the terminal cisternae, immediately available for release when the Ry receptors are activated.

Until recently, the understanding of the SR function that we have briefly summarized was not matched by adequate information on the cytological nature of this endomembrane system. In fact, the ER nature of the SR, initially proposed based on the observation of lumenal continuities in developing muscle fibers (Porter and Palade, 1957; Ezerman and Ishikawa, 1967), had been questioned based on the marked biochemical differences concerning in particular the protein complement (complex in the ER, highly simplified in the SR; see Campbell, 1986). The recent demonstration that skeletal muscle SR fractions contain PDI (Fliegel et al., 1990), and that BiP is distributed not only within the lumen of the conventional ER cisternae (concentrated in mature fibers at a pole of each nucleus), but also throughout the entire SR lumen, including the terminal cisternal space surrounding the CSQ masses (Volpe et al., 1992; Figure 1), has provided new and strong evidence in favor of the classical hypothesis, which should therefore now be fully accepted.

In the other types of muscle fibers, the information about the ER is not as detailed as in skeletal fibers. In the heart, the sites of distribution of both CSQ and high calcium, possibly together with the Ry receptors, include vacuoles located at some distance from the T tubules (corbular SR vacuoles) in addition to the terminal cisternae (Jorgensen et al., 1985, 1988). In smooth muscle fibers the entire ER has often been referred to as SR. Recent immunogold labeling studies of the rat vas deferens fibers have, however, provided evidence for a molecular distinction between conventional ER cisternae and a population of small membrane-bound structures, some flat and apparently attached to the plasmalemma, others characterized by a moderately dense content, located both at the surface and within the fibers. Only these structures were found to express high

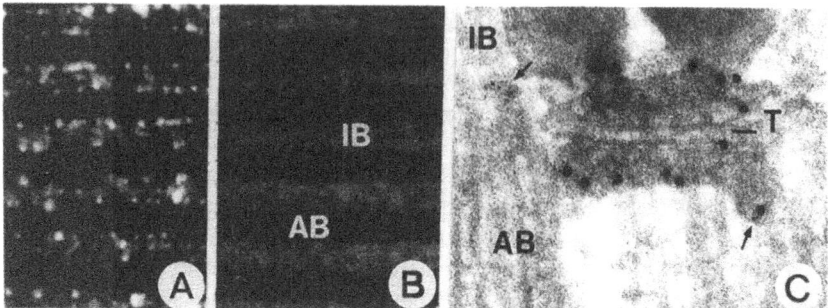

FIGURE 1. Rabbit skeletal muscle cryosections immunolabeled for CSQ (A, C) and BiP (B, C). The immunofluorescent images of 1-μm-thick cryosections show that CSQ (panel A) is clustered in rows of bright spots coinciding with the boundary between isotropic and anisotropic bands (IB and AB in panel B) whereas BiP (panel B) is more evenly distributed in the isotropic band and present also in the anisotropic band. Panel C shows a high-resolution dual immunogold labeling for the two proteins carried out on an ultrathin cryosection. CSQ (large gold particles) labels the dense masses within the two terminal cisternae coupled to the plasmalemma T tubule, whereas BiP (small gold, arrows) is restricted to the space between the dense masses and the limiting membrane of the cisternae. Magnification: A, B, 1500×; C, 90,000×. Reproduced with permission from Volpe *et al.* (1992).

concentrations of both CSQ and the Ins-P_3 receptor, and to contain also the general marker, PDI (Villa *et al.*, 1993). These organelles correspond therefore to the definition of rapidly exchanging Ca^{2+} stores given above, i.e., specialized areas of the ER. The distribution of the Ry receptor, which is also expressed by smooth muscle fibers, has not yet been investigated.

Among nonmuscle cells, the only one where the cytological nature of Ca^{2+} stores has been investigated in detail is the Purkinje neuron of the chicken. The reason these neurons were chosen is the high level of expression of Ins-P_3 receptors ($\sim$ 50-fold with respect to average cells) and the good complement of both Ry receptors and Ca^{2+}-ATPase, together with CSQ as the Ca^{2+} storage protein. The results (illustrated by cryosection dual labeling images for Ca^{2+}-ATPase, Ins-P_3 receptor, and CSQ in Figures 2 and 3) identified various structures at least potentially capable of Ca^{2+} handling. Of these, conventional ER cisternae (rough and smooth surfaced) were found to express all of the various Ca^{2+} store components, although at relatively low levels; peculiar stacked ER cisternae were rich in Ins-P_3 receptor but poor in CSQ and, especially, Ca^{2+}-ATPase; finally, small vacuoles (calciosomes: Volpe *et al.*, 1988) of moderately dense content rich in CSQ and Ca^{2+}-ATPase were of two types, expressing either the Ins-P_3 receptor or (probably) the Ry receptor. All those structures exhibited BiP in their segregated content, although the concentration was low in the small vacuoles. Finally the spines, peculiar dendritic expansions of considerable physiological importance, failed to express CSQ and Ry recep-

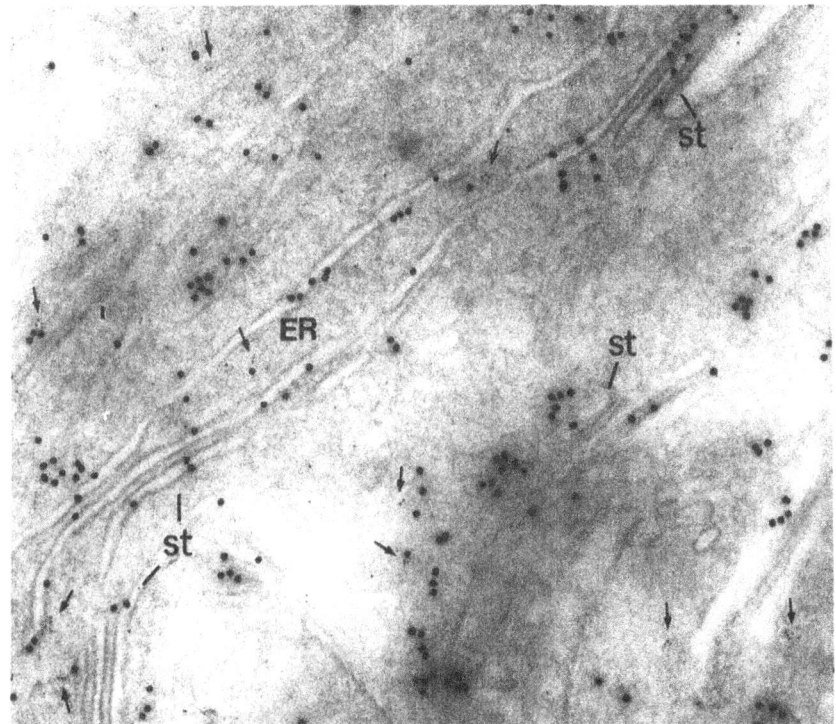

FIGURE 2. Chicken Purkinje neuron ultrathin cryosection doubly immunolabeled for CSQ and Ca²⁺-ATPase (large and small gold particles, respectively). The field shows ER cisternae running obliquely and exhibiting a moderate CSQ labeling. Where indicated (st), adjacent cisternae become parallel to each other and give rise to stacks that also exhibit some CSQ labeling. The structures more heavily positive for the Ca²⁺ binding protein are the vacuoles of moderately dense content, distributed in the proximity of, but not visibly continuous with, the ER cisternae. The Ca²⁺-ATPase immunolabeling (arrows) occurs on both cisternae and vacuoles, but not on the stacks. Most often the small gold particles appear arranged in small clusters revealed by grazing sections of the limiting members. 65,000×.

tors. However, their ER cisternae (spine apparatus) were rich in Ca²⁺-ATPase and Ins-P₃ receptors as well as BiP (Villa *et al.*, 1991; H. Otsu *et al.*, 1990; Ellisman *et al.*, 1990; Volpe *et al.*, 1991; Walton *et al.*, 1991; Takei *et al.*, 1992). Taken together, these observations document that the complexity of the Ca²⁺ stores in Purkinje neurons is greater than previously envisaged, with all of the structures mentioned above probably participating in the control of Ca²⁺ homeostasis, but apparently with different properties and therefore with different functional roles. Of importance is the separation in distinct sets of Ca²⁺ stores of the Ins-P₃ and Ry receptors, a result recently obtained also in sea urchin eggs

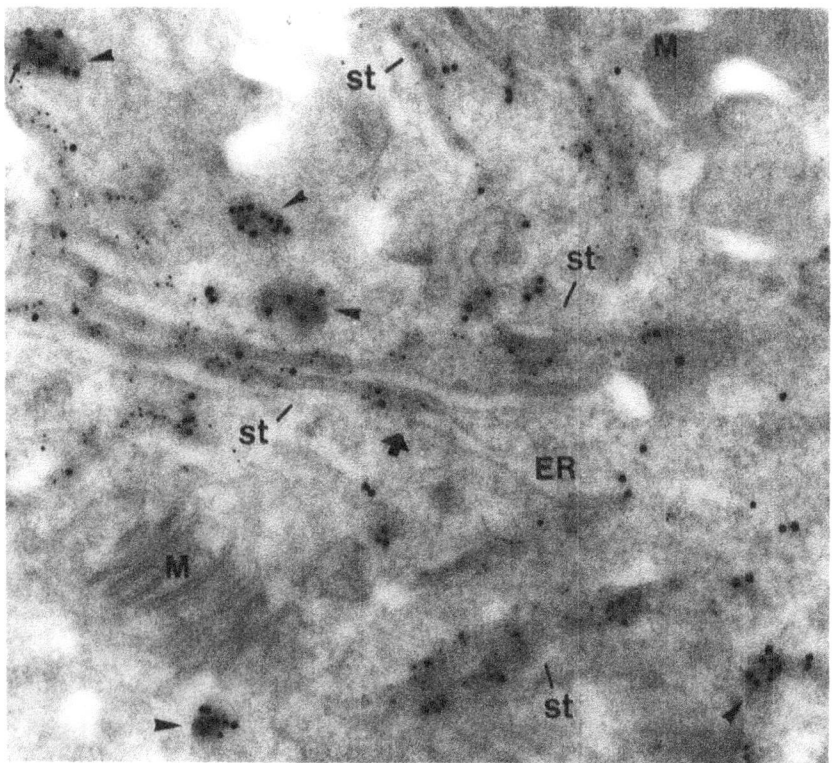

FIGURE 3. Chicken Purkinje neuron ultrathin cryosection doubly immunolabeled for CSQ and Ins-P₃ receptor (large and small gold particles, respectively). The Ins-P₃ receptor appears heavily concentrated over the cisterna stacks (st), and sparse or absent over the ER cisternae (note the abrupt change of labeling density at the transition between the two structures, large arrow). CSQ labeling is low on both of these structures. The dense vacuoles (arrowheads) are in contrast very rich in CSQ. Ins-P₃ receptor labeling is visible only on the vacuole at the top left (small arrow), the others appear negative. Mitochondria (M) and the cytosol appear unlabeled. 67,000×.

(McPherson *et al.*, 1992), which will be discussed from the functional point of view in a subsequent section.

In other tissues CSQ is not expressed, while pumps and channels are present at low levels which, although permitting functional competence, are hard to reveal by biochemical and, especially, immunocytochemical techniques. Thus, the information about Ca^{2+} stores is still largely incomplete in these cells. Concentration of CR within specialized structures, probably part of the ER, is supported, on the one hand, by subcellular fractionation studies in HL60 cells, with isolation of a microsomal subfraction highly enriched in the protein (Krause *et al.*, 1990; Van Delden *et al.*, 1992); and, on the other hand, by immunocytochemical results in hepatocytes, L6 myogenic and other cell types (Treves *et al.*, 1990; Arber *et al.*, 1992). The latter studies are of particular interest because CR was found to colocalize with a protein, cyclophilin S, endowed with a specific C-terminal target sequence, discussed in the following section. CR is contained also within conventional ER cisternae (Treves *et al.*, 1990; Peter *et al.*, 1992; Villa *et al.*, 1993).

4. BIOGENESIS OF Ca^{2+} STORES

In this section we will discuss not only the mechanisms by which the specific components are synthesized and retained in the ER, but also those by which these components assemble within the ER to give rise to the Ca^{2+} stores. Information in this field is still very limited. In skeletal muscle, Ca^{2+}-ATPase is synthesized by bound polysomes and rapidly transferred to the SR membranes, most likely by simple diffusion along the ER–SR continuities (Chyn *et al.*, 1979). In contrast, the two intracellular channels, Ins-P$_3$ and Ry receptors, have not yet been investigated for their site of synthesis.

Some information exists on the assembly of the SR during development. Yuan *et al.* (1991) have in fact recently shown that, in rabbit skeletal muscle fibers, Ry receptors appear toward the end of gestation, concentrated in discrete cytoplasmic vesicles. Shortly thereafter, when these vesicles fuse with the growing SR, the receptors remain clustered in discrete membrane patches that adhere to other vesicles rich in the L-type Ca^{2+} channels. Fusion of the latter with the growing T tubules of the plasmalemma ultimately yields the definite organization of the system. Whatever the precise mechanisms, it is clear that both the Ry and the L-type Ca^{2+} channels have a tendency to aggregate in the plane of the SR and T tubule membranes, respectively. The coupling of Ry receptors with the L-type channels is probably not necessary for their aggregation because in sea urchin eggs (McPherson *et al.*, 1992) the phenomenon occurs also in structures located at some distance from the plasma membrane, i.e., where coupling with L channels is not possible. In the case of Ins-P$_3$ receptors, aggregation has been observed in peculiar structures of Purkinje neurons, the stacks of smooth-surfaced ER cisternae (Figure 2; Satoh *et al.*, 1990; H. Otsu *et al.*, 1990; Villa *et al.*, 1991; Takei *et al.*, 1992; Rusakov *et al.*, 1993). In each stack, cisternae are attached to each other by multiple, perpendicular bridges, probably composed of two Ins-P$_3$ receptor molecules arranged in register. Whether the lateral aggregation of Ry and Ins-P$_3$ receptors is the result of the mutual, direct interaction of the channel molecules themselves, or is mediated by other membrane components, is not established.

In the case of lumenal proteins, accumulation within the Ca^{2+} storage lumen is probably the result of multiple mechanisms. As discussed in detail by Rowling and Freedman and by Lippincott-Schwartz (this volume; see also Pelham, 1989), many lumenal ER proteins express at their C-terminus a peculiar motif, KDEL and analogues, that is recognized by a specific receptor. The latter is believed to shuttle empty from the ER to a distal compartment, loaded with its ligands on its way back. By this mechanism the escaped lumenal proteins are retrieved and can therefore be considered as resident of the ER lumen. Of the Ca^{2+} binding proteins, not only BiP and PDI, but also endoplasmin and CR express the motif. Surprisingly, however, these various proteins seem to follow different pathways within the cell, as documented by their oligosaccharide chains. In the case of BiP, PDI, and endoplasmin, these chains show no processing, as expected for proteins recycled from a pre-Golgi compartment, which therefore do not reach bona fide Golgi cisternae. In contrast, the rat liver CR was found to exhibit the complex hybrid configuration, with terminal galactoses, typical of proteins that have reached the distal (*trans*) Golgi cisterna (Peter *et al.*, 1992). Taken together, these data suggest the KDEL receptor distal compartment to correspond to both the pre-Golgi and the Golgi compartments. Individual proteins endowed with the appropriate C-terminal motif could interact with the receptor in either one of these distal locations, possibly depending on the different environmental conditions existing there. In this respect it should be noted that the evidence for recycling from compartments beyond the pre-Golgi is recent, and that the possibility of long trips along the Golgi complex for a resident protein of the ER lumen had not even been considered previously.

In contrast, the rat liver CR was the uneven distribution of CR, with concentration in discrete areas of the ER lumen (Treves *et al.*, 1990; Van Delden *et al.*, 1992; Arber *et al.*, 1992). A possible explanation for the latter observation could be suggested by the colocalization, observed in various cell types, of CR with cyclophilin S. The latter, widely expressed lumenal protein includes at its C-terminus not the KDEL, but another motif, i.e., VEPFAIAKE. Deletion of this motif results in the secretion of cyclophilin S, while attachment of it to a typical secretory protein results in the retention of the latter in the same ER areas where cyclophilin S and CR are concentrated. The VEPFAIAKE motif might therefore be responsible for both the retention and subcompartmentalization of cyclophilin S, and the colocalization of CR could be the result of a physical aggregation between the two proteins (see Arber *et al.*, 1992).

In contrast to the lumenal proteins discussed so far, CSQ and a few minor SR proteins express neither the KDEL nor the VEPFAIAKE motifs at their C-terminus (Fliegel *et al.*, 1987). Also, in the case of CSQ the intracellular path to its final destination includes the Golgi complex, but only the *cis–*medial cisternae (Thomas *et al.*, 1989). For this protein the mechanism of retention might depend on its insolubilization to yield the dense masses visible within the terminal cisternae. These masses are apparently not free but attached to specific membrane proteins that account for the discrete filaments protruding into the lumen (Franzini-Armstrong *et al.*, 1987; Mitchell *et al.*, 1988). This type of organization might exist not only within the terminal cisterne, but also else-where. In fact, a cross talk of some kind, responsible for the co-accumulation of membrane and lumenal proteins, is necessary in all ER areas specialized for Ca^{2+} handling. So far, however, no clear experimental evidence exists outside the SR.

5. ABOUT THE FUNCTIONING OF Ca^{2+} STORES

As already mentioned, the role of the various rapidly exchanging Ca^{2+} stores is not the same, although some general properties of these structures have emerged. Experiments in a variety of cells have shown that, when the SERCA Ca^{2+}-ATPase is blocked, for example by thapsigargin, Ca^{2+} leaks out from the stores yielding measurable $[Ca^{2+}]_i$ increases that, although distinctly slower, resemble those induced via the activation of the Ca^{2+} release receptors. These results demonstrate that the limiting membrane of Ca^{2+} stores is not imperme-able to the cation even under resting conditions. The Ca^{2+} content of the stores depends therefore on the ongoing activity of their Ca^{2+}-ATPase.

As far as Ca^{2+} release is concerned, initial ideas about the differential activation of the two channels (one by Ins-P_3, freely diffusing throughout the cytoplasm; the other by any increase of $[Ca^{2+}]_i$, working by CICR) have to be reconsidered. Recent developments document in fact that the action of Ins-P_3 is modulated by $[Ca^{2+}]_i$ and possibly also by the lumenal $[Ca^{2+}]$ existing within the store (Missiaen et al., 1992); moreover, Ry receptors are potently stimulated by an endogeneous molecule, cyclic ADP ribose, that at the moment should be considered as a putative second messenger (see Galione, 1992). The possibility that the activation mechanisms of both receptors are similar (regulation by a diffusible second messenger, but under the modulatory control of the $[Ca^{2+}]_i$ in the immediately surrounding cytoplasm) needs therefore to be further investi-gated. These considerations do not apply to the skeletal muscle Ry receptors that are controlled by the plasma membrane potential (see Rios et al., 1991) as extensively discussed in previous sections.

Finally, attention should be given to the heterogeneity of Ca^{2+} stores, reflecting the different distribution of the three components: Ca^{2+} pump, Ca^{2+} binding proteins and release channels. This heterogeneity has been identified so far in chicken Purkinje neurons (Villa et al., 1991; Volpe et al., 1991; Takei et al., 1992) but may be present also in other cell types. In this respect it should be emphasized that ion transport by the pumps and, especially, the channels is considerable (of the order of 10^2 and 10^5 ions/s, respectively) while the capacity of even the best Ca^{2+} binding proteins is limited (50 mol/mol for both CSQ and CR; less for the others). In order to sustain large Ca^{2+} release responses the stores need therefore to contain large amounts of these proteins. Indeed, the CSQ concentration in the apparently insolubilized dense masses of terminal cisternae is so high that values of calcium of up to 50–100 mmol/liter have been measured (Somlyo et al., 1986). In chicken Purkinje neurons and rat vas deferens smooth muscle fibers, the ER structures that contain large amounts of CSQ are the dense vacuoles (calciosomes) that are thus expected to play a major role in Ca^{2+} homeostasis (Villa et al., 1991, 1993; Takei et al., 1992). The complement of both Ca^{2+}-ATPase and release channels of these structures, although moderate (Villa et al., 1991; Volpe et al., 1991), might in fact be sufficient for uptake and release purposes. In contrast, structures like the Ins-P_3 receptor-rich ER stacks, which are poor in Ca^{2+} binding protein, could only yield prompt, but very short-lasting, release responses. Alternatively, these structures could serve primarily to blunt Ins-P_3 transients, an activity apparently consistent with recent data of Purkinje neuron physiology (see Llano et al., 1991). Lastly, the ability to accumulate some Ca^{2+} appears to be localized not only to the specialized areas discussed so far, but shared also by the "conventional" ER cisternae. Here Ca^{2+} appears to be involved in the interactions of lumenal proteins, among themselves and with appropriate membrane proteins, such as calnexin and possibly also the protein known as the signal sequence receptor (Wada et al., 1991). In these cisternae the concentration of free Ca^{2+} is expected to be relatively stable because marked reductions have been shown to yield drastic alterations, with accelerated turnover, secretion, and appearance at the cell surface of ER lumenal and membrane proteins, respectively (Booth and Koch, 1989; Wileman et al., 1991; Suzuki et al., 1991).

We conclude that, in Purkinje neurons, smooth muscle fibers, and possibly in other cells, ER areas that differ in their molecular organization play different, and in a few cases even antagonistic, roles in Ca^{2+} control. Within the cells, the distribution of these areas appears not random, thus their contribution to the local homeostasis could be variable, with different roles in the general Ca^{2+} homeostasis and the regulation of cell activity. The physical state of these areas, whether discrete or in direct lumenal continuity with the rest of the ER, is still an open question. Physiological experiments in various cell types (mostly neurons; Thayer et al., 1988a,b; Brorson et al., 1991) have shown that the Ca^{2+} stores responsive to Ins-P_3 and Ry can be discharged independently from each other, whereas in other cells they cannot (Zacchetti et al., 1991; Giannini et al., 1992). While these last results suggest intermixing of the two types of receptors or at least close lumenal continuities of the two stores, the first require the stores, if continuous, to be at least far away from each other. In view of the present understanding of the ER as a highly motile and pleomorphic system (Dabora and Sheetz, 1988; Lee and Chen, 1988; Terasaki and Jaffe, 1991), we imagine that distant structures do not remain permanently continuous but can separate and rejoin, via frequent pinching off and fusion events. Since, however, a direct experimental approach appears at the moment difficult to even envisage, our interpretation appears destined to remain only hypothetical for quite some time ahead.

6. REFERENCES

Arber, S., Krause, K.-H., and Caroni, P., 1992, s-Cyclophilin is retained intracellularly via a unique COOH-terminal sequence and colocalizes with the calcium storage protein calreticulin, *J. Cell Biol.* **116:**113–125.

Baksh, S., and Michalak, M., 1991, Expression of calreticulin in Escherichia coli and identification of its Ca²⁺ binding domains, *J. Biol. Chem.* **266:**21458–21465.

Berridge, M. J., 1990, Calcium oscillations, *J. Biol. Chem.* **265:**9583–9586.

Berridge, M. J., 1993, Inositol trisphosphate and calcium signalling, *Nature* **361:**315–325.

Berridge, M. J., and Irvine, R. F., 1984, Inositol trisphosphate, a novel second messenger in cellular signal transduction, *Nature* **312:**315–321.

Berridge, M., and Moreton, R. B., 1991, Calcium waves and spirals, *Curr. Biol.* **1:**296–297.

Booth, C., and Koch, G. L. E., 1989, Perturbation of cellular calcium induces secretion of lumenal ER proteins, *Cell* **59:**729–737.

Brorson, J. R., Bleakman, D., Gibbons, S. J., and Miller, R. J., 1991, The properties of intracellular Ca²⁺ stores in cultured rat cerebellar neurons, *J. Neurosci.* **12:**127–138.

Campbell, K. P., 1986, Protein components and their roles in sarcoplasmic reticulum function, in: *Sarcoplasmic Reticulum in Muscle Physiology* (M. L. Entman and W. B. Van Winkle, eds.), pp. 65–99, CRC Press, Boca Raton, Fla.

Carafoli, E., 1987, Intracellular calcium homeostasis, *Annu. Rev. Biochem.* **56:**395–433.

Chyn, T. L., Martonosi, A. N., Morimoto, T., and Sabatini, D. D., 1979, In vitro synthesis of the Ca²⁺ transport ATPase by ribosomes bound to sarcoplasmic reticulum membranes, *Proc. Natl. Acad. Sci. USA* **76:**1241–1245.

Dabora, S. L., and Sheetz, M. P., 1988, The microtubule-dependent formation of a tubulovesicular network with characteristics of the ER from cultured cell extracts, *Cell* **54:**27–35.

Ellisman, M. H., Deerinck, T. J., Ouyang, Y., Beck, C. F., Tanksley, S. J., Walton, P. D., Airey, J. A., and Sutko, J. L., 1990, Identification and localization of ryanodine binding proteins in the avian central nervous system, *Neuron* **5:**135–146.

Ezerman, E. B., and Ishikawa, H., 1967, Differentiation of the sarcoplasmic reticulum and T-system in developing chick skeletal muscle in vitro, *J. Cell Biol.* **35:**405–420.

Fleischer, S., and Inui, M., 1989, Biochemistry and biophysics of excitation–contraction coupling, *Annu. Rev. Biophys. Chem.* **18:**333–364.

Fliegel, L., Ohnishi, M., Carpenter, M. R., Khanna, V. H., Reinhart, A. F., Reithmeier, R. A. F., and MacLennan, D. H., 1987, Amino acid sequence of rabbit fast-twitch skeletal muscle calsequestrin deduced from cDNA and peptide sequencing, *Proc. Natl. Acad. Sci. USA* **84:**1167–1171.

Fliegel, L., Burns, K., MacLennan, D. H., Reithmeier, R. A. F., and Michalak, M., 1989a, Molecular cloning of the high affinity calcium binding protein (calreticulin) of skeletal muscle sarcoplasmic reticulum, *J. Biol. Chem.* **264:**21522–21528.

Fliegel, L., Burns, K., Opas, M., and Michalak, M., 1989b, The high-affinity calcium binding protein of sarcoplasmic reticulum. Tissue distribution and homology with calregulin, *Biochim. Biophys. Acta* **982:**1–8.

Fliegel, L., Newton, E., Burns, K., and Michalak, M., 1990, Molecular cloning of cDNA encoding a 55 KDa multifunctional thyroid hormone binding protein of skeletal muscle sarcoplasmic reticulum, *J. Biol. Chem.* **265:**15496–15502.

Foskett, J. K., Roifmans, C. M., and Wong, D., 1991, Activation of calcium oscillation by thapsigargin in parotid acinar cells, *J. Biol. Chem.* **266**:2778–2782.

Franzini-Armstrong, C., Kenney, L. J., and Varriano-Marston, E., 1987, The structure of calsequestrin in triads of vertebrate skeletal muscle: A deep-etch study, *J. Cell Biol.* **105**:49–56.

Furuichi, T., Yoshikawa, S., Myawaki, A., Wada, K., Maeda, N., and Mikoshiba, K., 1989, Primary structure and functional expression of the inositol 1,4,5-trisphosphate-binding protein P400, *Nature* **342**:32–38.

Galione, A., 1992, Ca^{2+}-induced Ca^{2+} release and its modulation by cyclic ADP-ribose, *Trends Pharmacol. Sci.* **13**:356–360.

Giannini, G., Clementi, E., Ceci, R., Marziali, G., and Sorrentino, V., 1992, Identification of a broadly expressed ryanodine receptor-Ca^{2+} channel regulated by TGFβ, *Science* **257**:91–94.

Hashimoto, S., Bruno, B., Lew, D. P., Pozzan, T., Volpe, P., and Meldolesi, J., 1988, Immunocytochemistry of calciosomes in liver and pancreas, *J. Cell Biol.* **107**:2524–2531.

Jorgensen, A. O., Shen, A.-C. I., and Campbell, K. P., 1985, Ultrastructural localization of calsequestrin in adult rat atrial and ventricular muscle cells, *J. Cell Biol.* **101**:257–268.

Jorgensen, A. O., Broderick, R., Somlyo, A. P., and Somlyo, A. V., 1988, Two structurally distinct calcium storage sites in rat cardiac sarcoplasmic reticulum: An electron microprobe analysis study, *Circ. Res.* **63**:1060–1069.

Krause, K. H., Simmerman, H. K. B., Jones, L. R., and Campbell, K. P., 1990, Sequence similarity of calreticulin with a Ca^{2+} binding protein that co-purifies with an $Ins(1,4,5)P_3$-sensitive Ca^{2+} store in HL-60 cells, *Biochem. J.* **270**:545–548.

Lee, C., and Chen, L. B., 1988, Dynamic behavior of endoplasmic reticulum in living cells, *Cell* **54**:37–46.

Lipscombe, D., Madison, D. V., Poenie, M., Reuter, H., Tsien, R. Y., and Tsien, R. W., 1988, Spatial distribution of calcium channels and cytosolic calcium transients in growth cones and cell bodies of sympathetic neurons, *Proc. Natl. Acad. Sci. USA* **85**:2398–2402.

Llano, I., Dreessen, J., Kano, M., and Konnerth, A., 1991, Intradendritic release of calcium induced by glutamate in cerebellar Purkinje cells, *Neuron* **7**:577–583.

Lytton, J., Westlin, M., and Hanley, M. R., 1991, Thapsigargin inhibits the sarcoplasmic or endoplasmic reticulum Ca-ATPase family of calcium pumps, *J. Biol. Chem.* **266**:17067–17071.

Macer, D. R. J., and Koch, G. L. E., 1988, Identification of a set of calcium-binding proteins in reticuloplasm, the luminal content of the endoplasmic reticulum, *J. Cell Sci.* **91**:61–70.

McPherson, P. S., and Campbell, K. P., 1990, Solubilization and biochemical characterization of the high affinity [³H] ryanodine receptor from rabbit brain membranes, *J. Biol. Chem.* **265**:18454–18460.

McPherson, P. S., Kim, Y.-K., Valdivia, H., Knudson, C. M., Takekura, H., Franzini-Armstrong, C., Coronado, R., and Campbell, K. P., 1991, The brain ryanodine receptor: A caffeine-sensitive calcium release channel, *Neuron* **7**:17–25.

McPherson, S. M., McPherson, P. S., Matheus, L., Campbell, K. P., and Longo, F. J., 1992, Cortical localization of a calcium release channel in sea urchin eggs, *J. Cell Biol.* **116**:1111–1121.

Maeda, N., Niinobe, M., Inoue, I., and Mikoshiba, K., 1989, Developmental expression and intracellular location of P400 protein characteristic of Purkinje cells in the mouse cerebellum, *Dev. Biol.* **133**:67–76.

Meldolesi, J., 1992, Multivarious IP₃ receptors, *Curr. Biol.* **2**:393–394.

Meldolesi, J., Madeddu, L., and Pozzan, T., 1990, Intracellular Ca^{2+} storage organelles in nonmuscle cells: Heterogeneity and functional assignment, *Biochim. Biophys. Acta* **1055**:130–140.

Michalak, M., Milner, R. E., Burns, K., and Opas, M., 1992, Calreticulin, *Biochem. J.* **285**:681–692.

Mignery, G. A., and Südhof, T. C., 1990, The ligand binding site and transduction mechanism in the inositol 1,4,5-trisphosphate receptor, *EMBO J.* **9**:3893–3898.

Mignery, G. A., Südhof, T. C., Takei, K., and De Camilli, P., 1989, Putative inositol 1,4,5-trisphosphate receptor similar to ryanodine receptor, *Nature* **342**:192–195.

Missiaen, L., De Smedt, H., Droogmans, G., and Casteels, R., 1992, Ca²⁺ release induced by inositol 1,4,5-trisphosphate is a steady-state phenomenon controlled by luminal Ca²⁺ in permeabilized cells, *Nature* **357**:599–601.

Mitchell, R. D., Simmerman, H. K. B., and Jones, L. R., 1988, Calcium binding effects on protein conformation and protein interactions of canine cardiac calsequestrin, *J. Biol. Chem.* **263**:1376–1381.

Miyawaki, A., Furuichi, T., Ryou, S., Nakagawa, T., Saitoh, T., and Mikoshiba, K., 1991, Structure–function relationships of the mouse inositol 1,4,5-trisphosphate receptor, *Proc. Natl. Acad. Sci. USA* **88**:4911–4915.

Moreno, S. N. J., Docampo, R., and Vercesi, A. E., 1992, Calcium homeostasis in procyclic and bloodstream forms of Trypanosoma brucei, *J. Biol. Chem.* **267**:6020–6026.

Nguyen, Van, P. N., Peter, F., and Söling, H.-D., 1989, Four intracisternal calcium-binding glycoproteins from rat liver microsomes with high affinity for calcium, *J. Biol. Chem.* **264**:17494–17501.

Otsu, H., Yamamoto, A., Maeda, N., Mikoshiba, K., and Tashiro, Y., 1990, Immunogold localization of inositol 1,4,5-trisphosphate (InsP₃) receptor in mouse cerebellar Purkinje cells using three monoclonal antibodies, *Cell Struct. Funct.* **15**:163–173.

Otsu, K., Willard, H. F., Khanna, V. K., Zorzato, F., Green, N. M., and MacLennan, D. H., 1990, Molecular cloning of cDNA encoding the Ca²⁺ release channel (ryanodine receptor) of rabbit cardiac muscle sarcoplasmic reticulum, *J. Biol. Chem.* **265**:13472–13483.

Papp, B., Enyedi, A., Kovacs, T., Sarkadi, B., Wuytack, F., Thastrup, O., Gardos, G., Bredoux, R., Levy-Toledano, S., and Enouf, J., 1991, Demonstration of two forms of calcium pumps by thapsigargin inhibition and radioimmunoblotting in platelet membrane vesicles, *J. Biol. Chem.* **266**:14593–14596.

Pelham, H. R. B., 1989, Control of protein exit from the endoplasmic reticulum, *Annu. Rev. Cell Biol.* **5**:1–23.

Peter, F., Nguyen Van, P., and Söling, H. D., 1992, Different sorting of KDEL proteins in rat liver, *J. Biol. Chem.* **267**:10631–10637.

Petersen, O. H., and Wakui, M., 1990, Oscillating intracellular Ca²⁺ signals evoked by activation of receptors linked to inositol lipid hydrolysis: Mechanism of generation, *J. Membr. Biol.* **118**:93–105.

Pietrobon, D., Di Virgilio, F., and Pozzan, T., 1990, Structural and functional aspects of calcium homeostasis in eukaryotic cells, *Eur. J. Biochem.* **193**:599–622.

Porter, K. R., and Palade, G. E., 1957, Studies on the sarcoplasmic reticulum. III. Its form and distribution in striated muscle cells, *J. Biophys. Biochem. Cytol.* **3**:269–300.

Rios, E., Ma, J., and Gonzales, A., 1991, The mechanical hypothesis of excitation–contraction (EC) coupling in skeletal muscle, *J. Muscle Res. Cell Motil.* **12**:127–135.

Rizzuto, R., Simpson, A. W. M., Brini, M., and Pozzan, T., 1992, Rapid changes of mitochondrial Ca²⁺ revealed by specifically targeted recombinant aequorin, *Nature* **358**:325–327.

Rusakov, D. A., Podini, P., Villa, A., and Meldolesi, J., 1993, Tridimensional organization of Purkinje neuron cisternal stacks, a specialized endoplasmic reticulum subcompartment rich in inositol 1,4,5-trisphosphate receptors, *J. Neurocytol.* **22**:273–282.

Robinson, I. M., and Burgoyne, R. D., 1991, Characterization of distinct inositol 1,4,5-trisphosphate-sensitive and caffeine-sensitive calcium stores in digitonin-permeabilised adrenal chromaffin cells, *J. Neurochem.* **56**:1587–1593.

Ross, C. A., Meldolesi, J., Milner, T. A., Satoh, T., Supattapone, S., and Snyder, S. H., 1989, Inositol 1,4,5-trisphosphate receptor localized to endoplasmic reticulum in cerebellar Purkinje neurons, *Nature* **339:**468–470.

Satoh, T., Ross, C. A., Villa, A., Supattapone, S., Pozzan, T., Snyder, S. H., and Meldolesi, J., 1990, The inositol 1,4,5-trisphosphate receptor in cerebellar Purkinje cells: Quantitative immunogold labeling reveals concentration in an ER subcompartment, *J. Cell Biol.* **111:**615–624.

Scott, B. T., Simmerman, H. K. B., Collins, J. H., Nadal-Ginard, B., and Jones, L. R., 1988, Complete amino acid sequence of canine cardiac calsequestrin deduced by cDNA cloning, *J. Biol. Chem.* **263:**8958–8964.

Sitia, R., and Meldolesi, J., 1992, The endoplasmic reticulum: A dynamic patchwork of specialized subregions, *Mol. Biol. Cell* **3:**1067–1072.

Smith, M. J., and Koch, G. L. E., 1989, Multiple zones in the sequence of calreticulin (CRP 55, calregulin, HACBP), a major calcium binding ER/SR protein, *EMBO J.* **8:**3581–3586.

Somlyo, A. V., Bond, M., Shuman, H., and Somlyo, A. P., 1986, Electron-probe X ray microanalysis of in situ calcium and other ion movements in muscle and liver, *Ann. N.Y. Acad. Sci.* **483:**229–240.

Streb, H., Irvine, R. F., Berridge, M. J., and Schultz, I., 1983, Release of Ca^{2+} from a nonmitochondrial intracellular store in pancreatic acinar cells by inositol-1,4,5-trisphosphate, *Nature* **306:**67–69.

Supattapone, S., Worley, P. F., Baraban, J. M., and Snyder, S. H., 1988, Solubilization, purification and characterization of an inositol trisphosphate receptor, *J. Biol. Chem.* **263:**1530–1534.

Suzuki, C. K., Bonifacino, J. S., Lin, A. Y., Davis, M. M., and Klausner, R. D., 1991, Regulating the retention of T cell receptor chain variants within the endoplasmic reticulum: Ca^{2+}-dependent association with BiP, *J. Cell Biol.* **114:**189–205.

Takei, K., Stukenbrok, H., Metcalf, A., Mignery, G., Südhof, T., Volpe, P., and De Camilli, P., 1992, Ca^{2+} stores in Purkinje neurons: Endoplasmic reticulum subcompartments demonstrated by the heterogeneous distribution of the Ins-P_3 receptor, Ca^{2+}-ATPase and calsequestrin, *J. Neurosci.* **12:**489–505.

Takeshima, H., Nishimura, S., Matsumoto, T., Ishida, H., Kangawa, K., Minamino, N., Matsu, H., Ueda, M., Hanaoka, M., Hirose, T., and Numa, S., 1989, Primary structure and expression from complementary DNA of skeletal muscle ryanodine receptor, *Nature* **339:**439–445.

Terasaki, M., and Jaffe, L. A., 1991, Organization of the sea urchin egg endoplasmic reticulum and its reorganization at fertilization, *J. Cell Biol.* **114:**929–940.

Thastrup, O., Cullen, P. J., Drobak, B. K., Hanley, M. R., and Dawson, A. P., 1990, Thapsigargin, a tumor promoter, discharges intracellular Ca^{2+} stores by specific inhibition of the endoplasmic reticulum Ca^{2+}-ATPase, *Proc. Natl. Acad. Sci. USA* **87:**2466–2470.

Thayer, S. A., Hirning, L. D., and Miller, R. J., 1988a, The role of caffeine-sensitive calcium stores in the regulation of the intracellular free calcium concentration in rat sympathetic neurons in vitro, *Mol. Pharmacol.* **34:**664–673.

Thayer, S. A., Perney, T. M., and Miller, R. J., 1988b, Regulation of calcium homeostasis in sensory neurons by bradykinin, *J. Neurosci.* **8:**4089–4097.

Thomas, K., Navarro, J., Benson, R. J. J., Campbell, K. P., Rotundo, R. L., and Fine, R. E., 1989, Newly synthesized calsequestrin, destined for the sarcoplasmic reticulum, is contained in early/intermediate Golgi-derived clathrin-coated vesicles, *J. Biol. Chem.* **264:**3140–3145.

Treves, S., DeMattei, M., Lanfredi, M., Villa, A., Green, N. M., MacLennan, D., Meldolesi, J., and Pozzan, T., 1990, Calreticulin is a candidate for a calsequestrin-like function in Ca^{2+}-storage compartments (calciosomes) of liver and brain, *Biochem. J.* **271:**473–480.

Tsien, R. W., and Tsien, R. Y., 1990, Calcium channels, stores and oscillations, *Annu. Rev. Cell Biol.* **6:**715–760.

Van Delden, C., Favre, C., Spat, A., Cerny, E., Krause, K.-H., and Lew, D. P., 1992, Purification of an inositol 1,4,5-trisphosphate-binding, calreticulin-containing intracellular compartment of HL-60 cells, *Biochem. J.* **281**:651–656.

Villa, A., Podini, P., Clegg, D. O., Pozzan, T., and Meldolesi, J., 1991, Intracellular Ca²⁺ stores in chicken Purkinje neurons: Differential distribution of the low affinity–high capacity calcium binding protein, calsequestrin, of Ca²⁺ ATPase and of the ER lumenal protein, BiP, *J. Cell Biol.* **113**:779–791.

Villa, A., Sharp, A. H., Racchetti, G., Podini, P., Bole, D. G., Dunn, W. A., Pozzan, T., Snyder, S. H., and Meldolesi, J., 1992, The endoplasmic reticulum of Purkinje neuron body and dendrites: Molecular identity and specialization for Ca²⁺ transport, *Neuroscience* **49**:467–477.

Villa, A., Podini, P., Panzeri, M. C., Söling, H. D., Volpe P., and Meldolesi, J., 1993, The endoplasmic-sarcoplasmic reticulum of smooth muscle fibers, *J. Cell Biol.* **121**:1041–1051.

Volpe, P., Krause, K.-H., Hashimoto, S., Zorzato, F., Pozzan, T., Meldolesi, J., and Lew, D. P., 1988, "Calciosome", a cytoplasmic organelle: The inositol 1,4,5-trisphosphate-sensitive Ca²⁺ store of non muscle cells? *Proc. Natl. Acad. Sci. USA* **85**:1091–1095.

Volpe, P., Alderson-Lang, B. H., Madeddu, L., Damiani, E., Collins, J. H., and Margreth, A., 1990, Calsequestrin, a component of the inositol 1,4,5-trisphosphate-sensitive Ca²⁺ store of chicken cerebellum, *Neuron* **5**:713–721.

Volpe, P., Villa, A., Damiani, E., Sharp, A. H., Podini, P., Snyder, S. H., and Meldolesi, J., 1991, Heterogeneity of microsomal Ca²⁺ stores in chicken Purkinje neurons, *EMBO J.* **10**:3183–3189.

Volpe, P., Villa, A., Podini, P., Martini, A., Nori, A., Panzeri, M. C., and Meldolesi, J., 1992, The ER–SR connection. Distribution of the endoplasmic reticulum markers in the sarcoplasmic reticulum of skeletal muscle fibers, *Proc. Natl. Acad. Sci. USA* **89**:6142–6146.

Wada, I., Rindress, D., Cameron, P. H., Ou, W. J., Doherty, H. H., Louvard, D., Bell, D., Thomas, D. Y., and Bergeron, J. J. M., 1991, The SSR and associated calnexin are major calcium binding proteins of the endoplasmic reticulum membrane, *J. Biol. Chem.* **266**:19599–19610.

Walton, P. D., Airey, J. A., Sutko, J. L., Beck, C. F., Mignery, G. A., Südhof, T. C., Deerinck, T. J., and Ellisman, M. H., 1991, Ryanodine and inositol trisphosphate receptors coexist in avian cerebellar Purkinje neurons, *J. Cell Biol.* **113**:1145–1157.

Wileman, T., Kane, L. P., Carson, G. R., and Terhorst, C., 1991, Depletion of cellular calcium accelerates protein degradation in the endoplasmic reticulum, *J. Biol. Chem.* **266**:4500–4507.

Wuytack, F., Raeymaekers, L., Verbist, J., Jones, L. R., and Casteels, R., 1987, Smooth-muscle endoplasmic reticulum contains a cardiac-like form of calsequestrin, *Biochem. Biophys. Acta* **899**:151–158.

Yuan, S., Arnold, W., and Jorgensen, A. O., 1991, Biogenesis of transverse tubules and triads: Immunolocalization of the 1,4-dihydropyridine receptor, TS28, and the ryanodine receptor in rabbit skeletal muscle developing in situ, *J. Cell Biol.* **112**:289–301.

Zacchetti, D., Clementi, E., Fasolato, C., Lorenzon, P., Zottini, M., Grohovaz, F., Fumagalli, G., Pozzan, T., and Meldolesi, J., 1991, Intracellular Ca²⁺ pools in PC12 cells. A unique, rapidly exchanging pool is sensitive to both inositol 1,4,5 trisphosphate and caffeine-ryanodine, *J. Biol. Chem.* **266**:20152–20158.

Zorzato, F., Fujii, J., Otsu, K., Phillips, M., Green, N. M., Lai, F. A., Meissner, G., and MacLennan, D. H., 1990, Molecular cloning of cDNA encoding human and rabbit forms of the Ca²⁺ release channel (ryanodine receptor) of skeletal muscle, *Biol. Chem.* **265**:2244–2256.

Chapter 10

Antigen Processing and Presentation
The Role of the Endoplasmic Reticulum

Vincenzo Cerundolo

1. INTRODUCTION

The immune system has evolved to recognize and destroy pathogens. Invasion by extracellular microorganisms is mainly controlled by B lymphocytes which, using surface immunoglobulins (Ig) as receptors, recognize antigens in their native conformation either free in solution or on the surface of cells. Invasion by intracellular parasites is mainly controlled by cytotoxic T lymphocytes (CTL), which recognize foreign protein antigens on the surface of infected cells by a highly specific receptor and lyse the cells. Both CTL and B lymphocyte responses are specifically enhanced by lymphokines released by antigen-specific T helper cells. Unlike Ig, the T-cell receptor of both CTL and T helper cells does not recognize native proteins but it recognizes a binary complex formed by the

---

Abbreviations used in this chapter: ABC, ATP binding cassette; BFA, brefeldin A; β_2m, β_2-microglobulin; *cim,* class I modifier; CTL, cytotoxic T lymphocytes; EMS, ethyl methane sulfonate; ER, endoplasmic reticulum; HA, hemagglutinin; HAM, histocompatibility antigen modifier; Ig, immunoglobulin; MHC, major histocompatibility complex; *mtp,* MHC-linked transporter protein; NP, nucleoprotein; PSF, peptide supplying factor; TAP, transporter associated with antigen processing.

Vincenzo Cerundolo   Institute of Molecular Medicine, John Radcliffe Hospital, Headington, Oxford OX3-9DU, United Kingdom.

Subcellular Biochemistry, Volume 21: Endoplasmic Reticulum, edited by N. Borgese and J. R. Harris. Plenum Press, New York, 1993.

association of peptides, derived from degradation of target proteins, with class I and class II molecules encoded in the major histocompatibility complex (MHC). Processing of target proteins occurs through two distinct pathways. Exogenous proteins are degraded in an endolysosome compartment, which intersects the biosynthetic pathway of MHC class II molecules (Neefjes *et al.*, 1990). Thus, peptides derived from degradation of endocytosed proteins bind to MHC class II molecules. Newly synthesized proteins are degraded in the cytosol, and peptides generated from them are transported into the endoplasmic reticulum (ER) where they associate with MHC class I molecules. This review focuses on the latter pathway dealing with processing and presentation of intracellular proteins.

Mutant cells have been described that have a phenotype consistent with loss of a mechanism required for transport of peptides from the cytosol to the ER (Townsend *et al.*, 1989a; Cerundolo *et al.*, 1990; Kelly *et al.*, 1992). Over the last couple of years, a cluster of genes, encoded in the human (Trowsdale *et al.*, 1990; Spies *et al.*, 1990), mouse (Monaco *et al.*, 1990), and rat (Deverson *et al.*, 1990) MHC, was cloned. It was shown that the mutant cells have deletion (DeMars *et al.*, 1985; Cerundolo *et al.*, 1990) or mutation (Kelly *et al.*, 1992) of these MHC-encoded genes. The aim of this chapter is to review these recent findings.

2. MHC CLASS I MOLECULES BIND PEPTIDES IN THE ER

Class I molecules are highly polymorphic transmembrane glycoproteins of 45 kDa encoded in the MHC and expressed at the surface of the majority of cells. During biosynthesis, class I molecules associate with β_2-microglobulin (β_2m) in the ER and the heterodimer is transported through the Golgi to the cell surface (Ploegh *et al.*, 1979; Owen *et al.*, 1980).

In 1974 Zinkernagel and Doherty demonstrated that CTL-mediated lysis is restricted by MHC class I molecules. Their experiments showed that virus-specific CTL kill virus-infected target cells only if they share the same class I allele with target cells. The nature of the antigen recognized by CTL, however, remained obscure and for several years it was thought to be transmembrane glycoproteins (Klein, 1986). The isolation of CTL clones and the use of recombinant influenza A viruses showed that some influenza A-specific CTL clones do not recognize viral glycoproteins (Townsend and Skehel, 1982; Bennink *et al.*, 1982). In 1985 Townsend *et al.* demonstrated that intact proteins are not required for CTL recognition and showed that virus-specific CTL recognize a binary complex formed by the association of peptides, derived from degradation of viral proteins, with MHC class I molecules. Transfection of deletion mutants of the influenza nucleoprotein (NP) gene into L cells showed that cytosolic NP fragments, lacking known signal sequences to cross the ER membrane, sensitize target cells for lysis by influenza NP-specific CTL. These results led to the hypothesis that CTL recognize peptides derived by degradation of viral antigens in the cytosol, in association with MHC class I molecules. This prediction was confirmed in 1986 (Townsend *et al.*, 1986a) by experiments showing that epitopes recognized by NP-specific CTL can be replaced *in vitro* by synthetic peptides corresponding to the influenza NP sequence. In addition, infection with a recombinant vaccinia virus, encoding in the cytosol a 15-amino-acid fragment of influenza NP, sensitizes cells for lysis by peptide-specific CTL (Gould *et al.*, 1989). Similar results were obtained with glycoproteins lacking the N-terminal hydrophobic signal sequence, normally required for cotranslational transport into the ER lumen. Cells infected with a recombinant vaccinia virus coding for a signal sequence-deleted influenza hemagglutinin (HA) were very efficiently recognized as targets by HA-specific CTL (Townsend *et al.*, 1986b). In these experiments, the signal-deleted HA was rapidly degraded in the cytosol, with a half-life of less than 30 min, as compared with greater than 4 h for wild-type HA.

Killed virus particles or purified viral proteins do not sensitize target cells for class I-restricted mediated lysis although they do sensitize target cells for class II restricted recognition (Morrison *et al.*, 1986). These findings suggest that two pathways for presenting antigens to T cells exist: one dealing with newly synthesized protein of virus-infected cells, the other concerned with endocytosed soluble proteins. Although there are some possible exceptions (Stearz *et al.*, 1988; Tevethia *et al.*, 1980; Yide *et al.*, 1988; Wraith and Vessey, 1986), the majority of the results obtained are consistent with this model. *De novo* synthesis of antigenic proteins, however, is not an absolute requirement for class I-restricted recognition. Introduction of an exogenous protein into the cytosol by pinosome lysis (Moore *et al.*, 1988) or infection of target cells with heat-inactivated influenza virus, which retains the ability to fuse and reach the cytosol (Yewdell *et al.*, 1988), result in presentation of epitopes to class I-restricted CTL. Thus, the requirement appears to be for exposure of a protein antigen in the cytosol.

Analysis of the effects of the antibiotic brefeldin A (BFA) and of the adenovirus E3/19K glycoprotein on antigen processing and presentation, shows that class I molecules bind peptides in the ER. BFA blocks transport of newly synthesized proteins from the ER to the Golgi apparatus, but it also has an effect on the medial and *trans* Golgi, endosomes, and lysosomes (Lippincott-Schwartz *et al.*, 1991; see Lippincott-Schwartz, this volume). After BFA treatment, proteins and lipids from the Golgi complex redistribute into the ER (Lippincott-Schwartz *et al.*, 1989, 1990; Pelham, 1989, 1991). Although BFA has no effect on viral protein synthesis, BFA-treated cells are not able to present viral antigens to class I-restricted CTL, whereas the capacity to present exogenous peptide epitopes is retained (Nutchern *et al.*, 1989; Yewdell and Bennick, 1989). These results indicate that class I molecules associate with peptides prior to transport out of the ER.

Similar results were obtained after infection of target cells with recombinant vaccinia virus coding for the adenovirus protein E3/19K. This protein associates in the ER with some class I alleles (H-$2D^b$, H-$2K^b$, H-$2K^d$, and H-$2L^d$ but not H-$2D^d$ or H-$2K^k$) and retains them in the ER (Cox *et al.*, 1990, 1991). After infection, target cells are not able to present viral antigens in association with K^d molecules but there is no effect on D^d-restricted killing. Thus, class I molecules bind peptides in the compartment in which they are retained by the E3/19K protein. These results are consistent with the finding that in normal cells class I heavy chains associate with β_2m within minutes of synthesis. After a few minutes of [^{35}S]-Met pulse, newly synthesized class I molecules form a stable complex with β_2m in a compartment in which glycoproteins are sensitive to endoglycosidase H digestion (ER and *cis*-Golgi). Since peptide binding to class I molecules is required to stabilize class I/β_2m complex (see below), the early formation of a stable class I/β_2m complex is consistent with the model that class I molecules bind peptides in the ER.

Since class I molecules are cotranslationally synthesized in the ER, peptide epitopes generated in the cytosol must cross the ER membrane, in order to associate with class I molecules and to be recognized by CTL. The evidence derived from studies on influenza NP and HA suggested the existence of a signal-independent transport mechanism that passed peptides from the cytosol to the ER (Townsend *et al.*, 1985). Recent data supporting these ideas have come predominantly from the study of mutant cells that are unable to present viral antigens to antigen-specific CTL.

3. HUMAN AND MOUSE PROCESSING CELL MUTANTS

The description of the events leading to the presentation of intracellular antigens to MHC class I molecules suggested that any defect preventing either the intracellular generation of peptides, or the transport of peptides to the ER or the assembly of class I molecules with β_2m or the trafficking to the cell surface would impair the recognition of cells by class I-restricted CTL. Indeed, β_2m-deficient cells that lack surface expression of class I molecules (Parnes and Seidman, 1982) are unable to present intracellular antigens to class I-restricted CTL (Vitiello *et al.*, 1990). Free class I molecules are retained in the ER (Ploegh *et al.*, 1979) and are not detected by conformation-specific mAb (Elliott *et al.*, 1991).

Until recently, association of class I molecules with β_2m was thought to be independent of peptide binding to class I molecules. The generally held view was that processing of intracellular proteins and generation of peptides were not linked to folding of class I molecules, assembly with β_2m, and transport of the class I/β_2m complex to the cell surface. In 1989 Townsend *et al.* (1989a) demonstrated that class I assembly and peptide binding to class I molecules are coupled phenomena. These findings were made possible by the use of mutant cells, which have a defect in the formation of stable class I/β_2m complexes.

The murine T lymphoma RMA-S (Townsend *et al.*, 1989a; Ljunggren *et al.*, 1989) was the first described processing mutant cell. RMA-S was selected for low expression of class I molecules (H-2D^b and H-2K^b) at the cell surface after treatment with the mutagen ethyl methane sulfonate (EMS) followed by exposure to class I-specific antibodies and complement. The cells were originally derived by Karre and colleagues as a tool for their work on recognition by NK cells, which can detect target cells that express low levels of class I molecules at the cell surface (Karre *et al.*, 1986). The RMA-S phenotype is characterized by two features: (1) a selective defect in presentation of intracellular viral antigens to class I-restricted CTL, with retention of the capacity to present defined epitopes as extracellular peptides; (2) low levels of class I molecules at the cell surface (5% of the surface expression in the parental cell line RMA) and formation of unstable class I/β_2m complexes. When the mutant cells were lysed and class I molecules (D^b and K^b) precipitated after overnight preclearing in dilute solution, they were not associated with and could not be precipitated by β_2m conformation-specific monoclonal antibodies (Ljunggren *et al.*, 1989).

Townsend postulated that the stability of the class I/β_2m complex might depend on occupation of the binding site by a peptide ligand. If so, the two phenotypic traits of RMA-S could be the result of a single genetic defect that prevented access of cytosolic peptides to the class I binding site, possibly through loss of the signal-independent transport mechanism postulated earlier (Townsend *et al.*, 1985, 1989a,b).

Indeed, the addition of a peptide ligand to both live RMA-S (Townsend *et al.*, 1989b) and RMA-S lysate (Townsend *et al.*, 1990) increased the stability of class I/β_2m complexes and stabilized a conformation of the $\alpha1$ and $\alpha2$ domains of the heavy chain detected by conformation-specific monoclonal antibodies. These data are consistent with the three-dimensional crystal structure of HLA A2 molecules (Bjorkman *et al.*, 1987a,b; Saper *et al.*, 1991). HLA Aw68 (Garrett *et al.*, 1989), and HLA B27 (Madden *et al.*, 1991) showing that peptide ligands are deeply embedded in a groove between the $\alpha1$ and $\alpha2$ domains of class I molecules, and that $\alpha1$ and $\alpha2$ domains make contacts with several residues of the β_2m (see below).

Three other processing human cell mutants have recently been described with a phenotype similar to that of RMA-S: LBL721.174 (.174) (DeMars *et al.*, 1985; Cerundolo *et al.*, 1990) [and its derivative T2 (Salter *et al.*, 1985; Salter and Cresswell, 1986)], LBL721.134 (.134) (DeMars *et al.*, 1985; Spies and DeMars, 1991), and BM36.1 (Kelly *et al.*, 1992). Both .174 and .134 cells were selected by irradiating the B cell line LBL721 followed by repeated treatment with Ab anti-class I and class II and complement. BM36.1 was selected by irradiating the parental B cell line (28.7) followed by Ab anti class I and complement.

The analysis of the phenotype of the human mutant cell lines confirmed the role of peptide in stabilizing class I assembly and gave the opportunity to demonstrate that genes encoded in the MHC control the presentation of viral antigens (DeMars et al., 1985; Cerundolo et al., 1990). .174 cells synthesize normal amounts of HLA-A2 (A2) and HLA-B5 (B5) molecules. However, they do not express B5 at the cell surface and express 20–30% A2 compared with the level expressed by the parental line 721. Newly synthesized A2 and B5 molecules dissociate readily in vitro to a form that is not detected by conformation-specific mAbs, and is not associated with β_2m (Salter and Cresswell, 1986; Cerundolo et al., 1990; Townsend et al., 1990). In addition, .174 cells are not able to present endogenous antigens to A2-restricted CTL whereas they present exogenous peptide epitopes 100-fold more efficiently than the parental cell line 721 (Cerundolo et al., 1990). Somatic hybrids of .174 with the T lymphoma CEM showed that the defect of .174 is recessive and that it can be restored by genes encoded by the fusion partner (T1 cells). Loss of both copies of human chromosome 6, provided by CEM cells, results in a cell line that has the same phenotype as .174 (T2 cells). This indicated that a trans-acting element encoded on chromosome 6 is responsible for restoring the normal phenotype in .174. Since .174 cells bear a large homozygous deletion in the class II region of the MHC, which goes from the cluster of complement genes to the DP α gene, this proved that the genetic defect of .174 can be mapped within this deleted region of the MHC. Similarly, the RMA-S phenotype is recessive, being restored by fusion with mouse L cells (Ohlen et al., 1990).

The nature of this "trans-acting factor" able to restore .174 and RMA-S normal phenotype was unknown. However, the finding that the unstable class I phenotype could be reversed by the occupation of the antigen binding site by a peptide ligand suggested that these cells have a defect impairing either generation of peptides in the cytosol or transport of peptides from the cytosol to the ER. Alternatively, they might have a defect in loading peptides to class I molecules in the ER.

Class I molecules synthesized by .174/T2 (data not shown), and also by RMA-S (Elliott et al., 1991) and BM36.1 (Kelly et al., 1992), are associated in vivo with β_2m. Precipitation of class I molecules, immediately after lysis, shows that they are folded and associated with β_2m. However, if the complex is left for several hours in the lysate at 4°C, the heavy chains dissociate from β_2m and lose the folded conformation. This observation indicates that class I assembly is not impaired in the mutant cells but the stability of the class I/β_2m complex is decreased. Peptide ligands, added immediately after cell lysis, stabilize the complexes by increasing the binding affinity of heavy chain for β_2m. In addition, rapid degradation of an unstable ubiquitin–NP fusion protein (Townsend *et al.*, 1988) is not impaired either in RMA-S (Townsend *et al.*, 1989b) or in .174 cells (data not shown). These data lead to two alternative hypotheses: either a defect in the transport across the ER membrane or loss of an ER mechanism required to load peptides into the antigen binding site of class I molecules. The latter case is unlikely because (1) class I molecules purified from T2 cells bind exogenous peptides both in lysate and in buffer (Cerundolo *et al.*, 1991), indicating that no special mechanism is required for peptide loading; (2) analysis of peptides eluted from class I molecules synthesized in T2 cells showed that A2 molecules (but not other class I molecules) contain peptides derived from ER degradation of signal sequences (Henderson *et al.*, 1992; Wei and Cresswell, 1992); (3) finally, peptide epitopes, which are cotranslationally synthesized in the ER of the mutant cells T2, are efficiently presented to class I-restricted CTL, whereas peptide epitopes synthesized in the cytosol are not presented (Anderson *et al.*, 1991). These results show that T2 have not lost the ability to load peptides *in vivo*, and are consistent with the hypothesis that T2 have a defect in the transport of peptides from the cytosol to the ER. The recent findings than the mutant phenotype of .134 (Spies and DeMars, 1991; Spies, 1992), BM36.1 (Kelly *et al.*, 1992), and RMA-S (Powis *et al.*, 1991b; Attaya *et al.*, 1992) are reversed by transfecting MHC genes encoding proteins with structural similarities with other transporter proteins support this hypothesis (see below).

Further evidence that genes encoded in the MHC control presentation of intracellular antigens has come from studies with rat class I molecules. Livingstone *et al.* (1989) described a rat MHC-linked locus (named *cim* for class I modifier), mapped in the rat MHC class II region, which controls the alloantigenic structure and rate of intracellular transport of the rat class I molecule RT1.A^a. Different alleles of the *cim* locus (*cim^a* and *cim^b*) regulate the specificity of class I antigen presentation (Livingstone *et al.*, 1989) and the time of residence of newly synthesized class I molecules in the ER (Powis *et al.*, 1991a): (1) RT1.A^a-specific CTL can distinguish between RT1.A^a synthesized in a strain coding for the *cim^a* or *cim^b* allele. (2) RT1.A^a shows fast transport kinetics out of the ER in a *cim^a* strain, whereas the same class I molecule shows a much slower rate of transport in a *cim^b* strain. F$_1$ hybrids (*cim^a* × *cim^b* strain) show that *cim^a* is dominant over *cim^b*.

The findings that genes in the rat and human MHC control antigen processing and presentation led 1 year later to the identification and cloning of a cluster of genes in the MHC of human (Trowsdale *et al.*, 1990; Spies *et al.*, 1990), rat (Deverson *et al.*, 1990), and mouse (Monaco *et al.*, 1990) that are structurally related to the ATP binding cassette (ABC) superfamily of transporters, and which after transfection were able to restore the normal phenotype in the mutant cells.

Table I
Antigen Processing Genes in the MHC

	Revised Name:	TAP 1	TAP 2
Human (Trowsdale *et al.*, 1990)		RING 4	RING 11
Human (Spies *et al.*, 1990)		PSF 1	PSF 2
Mouse (Monaco *et al.*, 1990)		HAM 1	HAM 2
Rat (Deverson *et al.*, 1990)		mtp 1	mtp 2

4. ABC TRANSPORTERS ENCODED IN THE MHC

The analysis of the series of human MHC deletion mutant cells, obtained from the immunoselection of 721 (DeMars *et al.*, 1985), narrowed down the region in the MHC containing the genes responsible for class I mutant phenotype from a DNA segment of ~ 1 Mb to one of 50 kb (Spies *et al.*, 1990). In this region, between HLA-DNA and DOB, two genes were cloned (Spies *et al.*, 1990; Trowsdale *et al.*, 1990). Both genes share sequence homology with members of the ABC transporter superfamily and have been named TAP 1 and TAP 2 for *t*ransporter *a*ssociated with *a*ntigen *p*rocessing (formerly RING 4/PSF1 and RING 11/PSF2). The rat [*mtp 1* and *mtp 2* (Deverson *et al.*, 1990), MHC-linked *t*ransporter *p*rotein] and the mouse genes [HAM 1 and HAM 2 (Monaco *et al.*, 1990), *h*istocompatibility *a*ntigen *m*odifier] have also been cloned (see Table I).

The ABC transporter superfamily includes over 30 members, found in both prokaryotic and eukaryotic cells (Higgins *et al.*, 1988; Hyde *et al.*, 1990). All of them are closely related in terms of domain organization, structure, ATP hydrolysis, and probably evolutionary origin. Each member of this family requires four membrane-bound domains: two highly hydrophobic domains, spanning the membrane 12 times (6 times per domain) and assumed to be responsible for translocating the substrate, and two hydrophilic domains, localized on the cytoplasmic face of the membrane. Each domain can be either fused together [e.g., multidrug resistance gene product (mdr) (Chen *et al.*, 1986; Gros *et al.*, 1987) and cystic fibrosis gene CFTR (Riordan *et al.*, 1989)] or encoded by a separate polypeptide chain [e.g., oligopeptide permease (Opp B-F) (Hiles *et al.*, 1987)], or consisting of one hydrophobic domain fused to an ATP binding domain (*Drosophila* white and brown loci and HlyB hemolysin export system of *E. coli*). The TAP genes would fall into the last of these structural organizations.

The hydrophilic domains of different ABC transporters share considerable sequence homology (ca. 50% identity) over 200 amino acids, including two short ATP binding motifs (Walker A and B motifs) (Higgins *et al.*, 1990). Despite these similarities, each of the ABC transporters is specific for a different substrate. They can either import or export a wide range of substrates ranging from ions to proteins (Higgins *et al.*, 1990). There are members of the family that, in prokaryotic cells, import ribose (*E. coli* RbsA) (Bell *et al.*, 1986) or histidine (*Salmonella typhimurium* HisP) (Higgins *et al.*, 1982) and others that export hemolysin A (*E. coli* HlyB). The Opp B-F of the bacterium *S. typhimurium* (Hiles *et al.*, 1987) transports short peptides (two to five amino acids) across the bacterial membrane. Similarly to the bacterial Opp transporter, the STE6 protein, in yeast, transports the alpha factor pheromone, a short farnesylated peptide. In mammalian cells the Mdr-1 transporter has a very broad specificity exporting lipophilic drugs out of cells. However, its normal physiological substrate remains unknown.

The MHC-encoded ABC transporters, then, could consist of one-half of a typical member of the ABC transporter proteins: one hydrophobic domain and one ATP binding domain. The TAP 1 gene encodes a protein of 808 amino acids. Three allelic forms of the TAP 1 gene have been described with substitutions at positions 333 and 637 (Colonna *et al.*, 1992). Three allelic forms of the TAP 2 gene have been identified with substitutions at positions 379, 665, and 687 (Powis *et al.*, 1992b; Colonna *et al.*, 1992). The TAP 2A and TAP 2C genes encode a protein of 686 amino acids and the TAP 2B gene encodes a protein of 703 amino acids (see Figure 1). TAP 2A has a frequency of 79% among the Caucasoid population. The ATP binding domains of TAP 1 and TAP 2 share 61% homology and the hydrophobic domains share 30% homology (Figure 2).

The genetic analysis of the processing mutant cells gave the opportunity to demonstrate that deletion or mutation of either TAP 1 and/or TAP 2 proteins can determine loss of presentation of viral antigens and unstable class I/β_2m complexes. .174 cells lack both TAP 1 and TAP 2 genes. Transfection of the TAP 1 gene into .174 failed to restore the normal phenotype (Spies and DeMars, 1991), implying that the TAP 2 gene or other gene(s) deleted in .174 cells are required

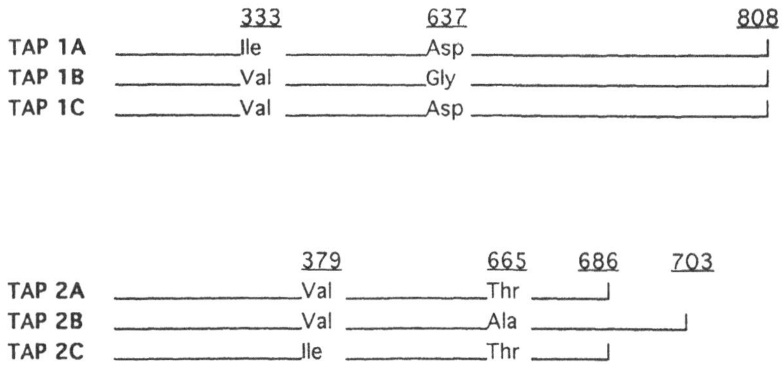

FIGURE 1. Human TAP allelic variants.

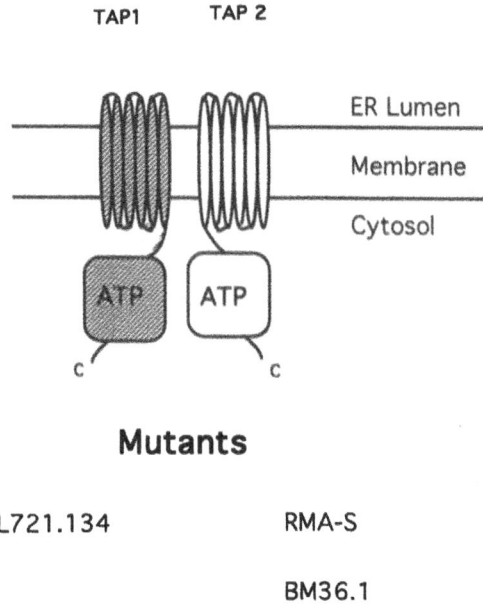

FIGURE 2. Transporter associated with presentation of intracellular antigens. TAP 1 and TAP 2 proteins are inserted in the ER membrane. Processing mutant cells are shown directly below their defective TAP proteins.

for the normal phenotype. The mutant cell .134 lacks only the product of the TAP 1 gene (probably because of a mutation in the promoter) although it synthesizes the TAP 2B protein (Spies *et al.*, 1992). Transfection of .134 with the TAP 1 cDNA clone restores normal expression of class I molecules at the cell surface (Spies and DeMars, 1991), the stability of class I/β_2m complexes, and the ability to present viral antigens to class I-restricted CTL (Spies *et al.*, 1992). These data show that the TAP 1 gene product controls presentation of endogenous antigens.

Evidence for a role of the TAP 2 protein in the processing of intracellular antigens was provided by the analysis of the mutant lymphoblastoid cell line BM36.1 (Kelly *et al.*, 1992) and RMA-S (Powis *et al.*, 1991b; Attaya *et al.*, 1992). Immunoprecipitation of the TAP 1 protein, using an antibody raised to a peptide from the C-terminus of the TAP 1 ATP binding domain, demonstrated that TAP 1 associates with TAP 2 protein forming a complex (Kelly *et al.*, 1992; Spies *et al.*, 1992). These results strongly support the notion that, like other members of the ABC superfamily, the functional TAP protein is a heterodimer with two transmembrane and two ATP binding domains. In heterozygous cells, both TAP 2 alleles are coprecipitated with TAP 1 indicating the absence of allelic exclusion of these gene products. The TAP 2A protein in BM36.1 was shown to be larger than the parental TAP 2A by $\sim$ 6 kDa, suggesting that it may be a mutant form (Kelly *et al.*, 1992). Sequence analysis of TAP 2 cDNA from BM36.1 revealed a deletion of 2 bp 3' to the Walker B ATP binding domain, causing a frameshift with replacement of 52 amino acids in the ATP binding domain, and a 51-amino-acid extension of the protein. Transfection of the TAP 2A gene into BM36.1 reversed the mutant phenotype, and wild-type protein competed with the elongated mutant form for binding to the TAP 1 protein (Kelly *et al.*, 1992). Presentation of viral antigens and the proportion of stable class I molecules were restored to levels comparable to the parental cell line. These results showed that the mutation in the ATP binding domain of the TAP 2 protein is responsible for the mutant phenotype of BM36.1 cells which might result from loss of ATP hydrolysis required for its function *in vivo*.

Analysis of the cellular distribution of TAP 1, by light microscopy and immunoelectron microscopy, showed that the TAP 1/TAP 2 complex is localized in the ER and *cis*-Golgi and is oriented with its ATP binding domain in the cytosol (Kleijmeer *et al.*, 1992). These findings provide strong circumstantial evidence that the products of the TAP genes are pumps transporting peptides from the cytosol into the ER.

Recent results, obtained with microsomes from T2 cells, argue against a role of the TAP complex in the transport of peptides. Levy *et al.* (1991) showed that translocation of peptides through the ER membrane of T2 cells is not impaired whereas assembly of *in vitro* translated B27 molecules is tenfold lower than in the parental cell line (T1). Thus, these results suggest that the inability to assemble class I molecules in T2 is independent of peptide transport and is probably the result of a defect in the lumen of the ER. Although evidence obtained using class I molecules from whole cells is not consistent with this hypothesis (see above), formal demonstration of TAP gene function will require purification of the transporter complex.

Like the defect in BM36.1, the RMA-S phenotype can be restored by transfection of either the mouse (Attaya *et al.*, 1992), rat (Powis *et al.*, 1991b), or human (J. Trowsdale, personal communication) TAP 2 homologues. As RMA-S was mutagenized with EMS, which makes point mutations, it is possible that a point mutation in the TAP 2 sequence is responsible for its lack of activity. Indeed, sequence analysis of the TAP 2 gene revealed a mutation at position 97 (C $\rightarrow$ T), causing a premature stop codon (Yang *et al.*, 1992).

Recent experiments have shown that *mtp 2* alleles control HPLC profiles of peptides eluted from the rat class I molecules RT1.A^a (Powis *et al.*, 1992a). This was shown to be the case only for RT1.A^a class I molecules since other rat class I molecules do not discriminate between different *mtp 2* alleles. This finding, which may be directly relevant for the *cim* phenomenon (see above), is consistent with a functional polymorphism at the level of peptide transport in the ER. Alternatively, *mtp 2* alleles, expressed by the target cell line, do not transport a different set of peptides into the ER lumen, but regulate the capacity of RT1.A[a] class I molecules to bind peptides.

5. PROTEASES ENCODED IN THE MHC

Closely linked to the two putative peptide transporter genes (TAP 1 and TAP 2), two other genes LMP 2 and LMP 7 (formerly RING 10 and RING 12) have been recently cloned. LMP 2 and LMP 7 have sequence homology with subunits of a large intracellular protease complex, called proteasome (Martinez and Monaco, 1991; Glynne *et al.*, 1991; Kelly *et al.*, 1991; Brown *et al.*, 1991). The proteasome consists of ~ 16 subunits whose molecular weights range from 21,000 to 35,000. They form a complex of 580,000 organized in four rings with a cylindrical shape, which is localized in the nucleus and in the cytosol. MHC-encoded subunits are polymorphic in mice (Monaco and McDevitt, 1984; Martinez and Monaco, 1991) but as yet there is no described polymorphism of the human subunits. Proteasome degrades protein and peptide substrates and can cleave bonds on the carboxyl side of basic, hydrophobic, and acidic amino residues (Rivett, 1989). Unlike other proteases, this broad specificity is achieved by catalysis at more than one type of catalytic site within the same complex (Dick *et al.*, 1991). Since proteasome is found in all organisms, including bacteria, it is likely that it plays a central role in the degradation of cytosolic proteins. This is supported by experiments showing that mutations of proteasome subunits in *Saccharomyces cerevisiae* are lethal (Heinemeyer *et al.*, 1991). By contrast, deletion of LMP 2 and LMP 7 genes in .174 cells did not impair their viability. This suggests that the two MHC-encoded proteasome subunits are not directly involved in "housekeeping" intracellular protein turnover. As yet there is no evidence that the proteasome has a role in the processing antigens, and the identity of the proteolytic system involved in the presentation of cytoplasmic protein remains unknown. However, the close association of LMP 2 and LMP 7 with TAP 1 and TAP 2 genes and the finding that interferon-γ upregulates both TAP 1 and TAP 2 and LMP 2 and LMP 7 (Trowsdale *et al.*, 1990; Ortiz-Navarrete *et al.*, 1991) suggest that the products of these genes may be playing a role in the antigen processing and presentation pathway.

6. THE ROLE OF PEPTIDES IN THE ASSEMBLY OF MHC CLASS I MOLECULES

We have discussed up to this point the mechanisms by which peptides might be generated in the cytosol and transported into the ER. In the ER peptides bind to class I molecules. It is not known, however, whether peptides interact with the $\alpha 1$ and $\alpha 2$ domain of class I molecules first, thus enabling $\beta_2 m$ to bind and stabilize the complex, or whether $\beta_2 m$ and heavy chain may associate without peptide and are then stabilized by the binding of those peptides that will fit into the groove. The three-dimensional structures of HLA-A2 (Bjorkman et al., 1987a), -Aw68 (Garrett et al., 1989), and -B27 (Madden et al., 1991) show a binding cleft between the $\alpha 1$ and $\alpha 2$ domains which could accommodate short peptides in an extended conformation. A more detailed examination of the A2 binding groove to 2.6-Å resolution reveals six pockets (A–F), which appear suited for binding side chains from antigenic peptides (Saper et al., 1991). Indeed, in A2, Aw68, and B27, peptidelike material can be seen in the binding site with side chains embedded into deep pockets of the binding groove. The structural analysis of A2 crystals at a resolution of 2.6 Å shows that $\beta_2 m$ makes contacts with several residues on the underside of the $\alpha 1$ and $\alpha 2$ domains.

Over the last year, direct access to class I-bound peptides was made possible by the development of acid-extraction methods from both immunoprecipitated class I molecules (Van Bleek and Nathenson, 1990) and whole cells (Falk et al., 1991a). The application of this technique has resulted in the identification of the natural fragments of several viral (Van Bleek and Nathenson, 1990; Falk et al., 1991a,b) and host (Rötzschke et al., 1990a,b; Falk et al., 1990; Wallny and Rammensee, 1990) antigens presented to CTL. Direct sequencing of the eluted peptides demonstrated that naturally produced class I-bound peptides share similar properties They have a homogeneous length of eight or nine amino acids and have conserved residues which determine the class I allele binding specificity (Falk et al., 1991a; Jardetzky et al., 1991).

It is not known how and where nine-amino-acid-long peptides are generated. One possibility is that peptides longer than nine amino acids bind to class $I/\beta_2 m$ complexes and are then trimmed in the ER. Alternatively, peptide trimming might occur in the cytosol or during the transport. Two recent studies showed that A2 molecules synthesized in T2 are loaded with peptides derived from ER signal sequences (Henderson et al., 1992; Wei and Cresswell, 1992). Two out of three A2-bound peptides are longer than nine amino acids. This suggests that ER trimming is not very efficient and it is consistent with the hypothesis either that cytosolic proteases generate nine-amino-acid-long peptides or that trimming occurs during the transport into the ER (Townsend, 1992).

The homogeneous length of naturally produced peptides can be explained by measuring the kinetics of peptide binding in vitro. These studies demonstrated that peptides of different lengths can bind to class $I/\beta_2 m$ complexes but only peptides of nine amino acids form stable complexes. Addition or deletion of a single residue at the C- and N-terminus of a peptide of optimum length greatly reduces its capacity to form a stable complex with class I molecules (Cerundolo et al., 1991). For instance, the rate of dissociation of peptides from the H-2D[b] class I molecule increases by 100-fold if the 9-mer is extended by one residue on the C-terminus (Cerundolo et al., 1991). These results suggest that in order to form a high-affinity interaction between peptides and class I molecules, both amino acid side chains and the length of the peptides are important. This is consistent with the peptide N- and C-termini forming stabilizing contacts with the class I binding site. Indeed, analysis of the B27 crystal, with finer resolution of bound peptide, shows the presence of two pockets at the ends of the groove (A and F), where the cleft narrows, interacting with the charged N- and C-termini of the bound peptide (Madden et al., 1991). In addition, we have demonstrated that the α amino and carbonyl groups at the N- and C-termini play a major role in both inducing the conformational change in free heavy chain (see below) and formation of a stable class I–peptide complex (Elliott et al., 1992). The longer peptides can be anchored in the specificity pockets without forming stabilizing contacts at the ends of the cleft. Thus, both long and short peptides may bind in the ER, but long peptides are likely to have dissociated by the time the class I molecules arrive at the cell surface. The rapid off rate of longer peptides would therefore lead to enrichment of class I molecules containing peptides of eight or nine residues.

Analysis of class I assembly in mutant cells (Townsend et al., 1989a, 1990; Cerundolo et al., 1990; Elliott et al., 1991) and in cells lacking β_2m (Elliott et al., 1991) showed that peptide and β_2m bind cooperatively to class I heavy chains.

These studies demonstrated that both peptide and β_2m can bind to free class I molecules and that the stability of each intermediate complex (peptide–class I and β_2m–class I) is increased by the presence of the other ligand. Both long and short peptides can bind to class $1/\beta_2$m complex. However, only peptides of optimum length can bind to free heavy chains inducing a conformational change detectable by conformation-specific mAbs (Elliott et al., 1991). Thus, peptides of optimum length influence the conformation of free heavy chains (Elliot et al., 1991) and form extremely stable complexes in vitro with class I molecules (Cerundolo et al., 1991). However, the basis for the formation of long-lived peptide class I complexes in vivo is not known. The findings that peptides are deeply embedded in the class I groove (Bjorkman et al., 1987a,b; Garrett et al., 1989; Madden et al., 1991), and that optimum length peptides induce a conformational change of free class I heavy chains (Elliott et al., 1991) suggest that the formation of long-lived peptide–class I molecule complexes is based on a "peptide trapping" mechanism analogous to that proposed for class II–peptide interaction (Sadegh-Nasseri and Germain, 1991). Indeed, the comparison of the association binding kinetics of long and short peptides to class I molecules demonstrated that long peptides bind with a simple kinetics and rapid dissociation rates, whereas short peptide binding is consistent with two sets of kinetics (Cerundolo et al., manuscript in preparation). This dependence of the binding kinetics on the length of the peptide correlates with the length-dependent conformational change induced in the free class I heavy chain (Elliott *et al.*, 1991). Class I and class II molecules may employ analogous conformational transitions for "trapping" peptides with inherently low binding affinities.

7. CONCLUSIONS

The description of the events leading to the presentation of intracellular antigens to CTL indicates three theoretical levels at which selection of the peptide repertoire presented to the immune system could occur: (1) degradation of intracellular proteins; (2) peptide transport specificity; (3) polymorphisms of class I molecules.

The very presence of class I allele-specific motifs implies a selection operated by class I molecules in the peptide repertoire. Falk *et al.* (1990) have shown that the peptide, defining any particular antigen, can be isolated only from cells that express the class I molecule known to function as a restriction element for that antigen. The class I molecules in a cell, therefore, define the set of peptides that can be isolated by acid extraction. This finding implies that class I molecules themselves are responsible for this selective effect and that unbound peptides might have an extremely short half-life.

So far no functional polymorphisms in the degradation of intracellular proteins or in the peptide transport into the ER have been reported in mouse and human cells. There are several examples of mouse CTL that can kill human target cells transfected with the appropriate mouse class I molecule and vice versa of human CTL that kill mouse target cells transfected with human class I molecules. In addition, the human TAP 2A gene can restore the normal phenotype in the murine RMA-S cells. Thus, murine and human processing pathways appear to be interchangeable without any effect on the presentation of individual epitopes. The rat *cim* phenomenon is possibly the only reported example of functional polymorphism of the processing of intracellular antigens. It will be of interest to look for a similar phenomenon in human cells. Functional polymorphism at the level of peptide transport in the ER might play a central role in the development of the T cell repertoire and in the susceptibility to infections and autoimmune diseases.

8. REFERENCES

Anderson, K., Cresswell, P., Gammon, M., Hermes, J., Williamson, A., and Zweerink, H., 1991, Endogenously synthesized peptide with an endoplasmic reticulum signal sequence sensitizes antigen processing mutant cells to class I-restricted cell-mediated lysis, *J. Exp. Med.* **174**:489–492.

Attaya, M., Jameson, S., Martinez, C., Hermel, E., Aldrich, C., Forman, J., Fischer Lindahl, K., Bevan, M., and Monaco, J., 1992, HAM-2 corrects the class I antigen-processing defect in RMA-S cells, *Nature* **355**:647–649.

Bell, A., Buckel, J., Groarke, J., Hope, J., Kingsley, D., and Hermodson, M., 1986, The nucleotide sequences of the rbsD, rbsA and rbsC genes of Escherichia coli K12, *J. Biol. Chem.* **261**:7652–7658.

Bennink, J., Yewdell, J., and Gerard, W., 1982, A viral polymerase involved in recognition of influenza-infected cells by a cytotoxic T cell clone, *Nature* **296**:75–76.

Bjorkman, P., Saper, M., Samraoui, B., Bennett, W., Strominger, J., and Wiley, D., 1987a, Structure of the human class I histocompatibility antigen HLA-A2, *Nature* **329**:506–512.

Bjorkman, P., Saper, M., Samraoui, B., Bennett, W., Strominger, J., and Wiley, D., 1987b, The foreign antigen binding site and T cell recognition regions of class I histocompatibility antigens, *Nature* **329**:512–518.

Braciale, T., Braciale, V., Winkler, M., Strynowski, I., Hood, L., Sabrook, J., and Gething, M., 1987, On the role of the transmembrane anchor sequence of influenza hemagglutinin in target cell recognition by class I MHC restricted, hemagglutinin-specific cytotoxic T lymphocytes, *J. Exp. Med.* **166**:678–692.

Brown, M., Driscoll, J., and Monaco, J., 1991, Structural and serological similarity of MHC-linked LMP and proteasome (multicatalytic proteinase) complexes, *Nature* **353**:355–357.

Cerundolo, V., Alexander, J., Anderson, K., Lamb, C., Cresswell, P., McMichael, A., Gotch, F., and Townsend, A., 1990, Presentation of viral antigens controlled by a gene in the MHC, *Nature* **345**:449–456.

Cerundolo, V., Elliott, T., Elvin, J., Bastin, J., Rammensee, H.-G., and Townsend, A., 1991, The binding affinity and dissociation rates of peptides for class I MHC molecules, *Eur. J. Immunol.* **21**:2069–2075.

Chen, C., Chin, J., Ueda, K., Clark, D., Pastan, I., Gottesman, M., and Roninson, I., 1986, Internal duplication and homology with bacterial transport proteins in the mdr1 (P glycoprotein) gene from multidrug resistance human gene, *Cell* **47**:381–386.

Colonna, M., Bresnahan, M., Bahram, S., and Strominger, J., 1992, Allelic variants of the human putative transporter involved in antigen processing, *Proc. Natl. Acad. Sci. USA* **89**:3932–3936.

Cox, J., Yewdell, J., Eisenlohr, L., Johnson, P., and Bennink, J., 1990, Antigen presentation requires transport of MHC class I molecules from the endoplasmic reticulum, *Science* **247**:715–718.

Cox, J., Bennink, J., and Yewdell, J., 1991, Retention of adenovirus E19 glycoprotein in the endoplasmic reticulum is essential to its ability to block antigen presentation, *J. Exp. Med.* **174**:1629–1637.

DeMars, R., Rudersdorf, R., Chang, C., Peterson, J., Strandtmann, J., Korn, N., Sidwell, B., and Orr, H., 1985, Mutations that impair a post transcriptional step in expression of HLA-A and -B antigens, *Proc. Natl. Acad. Sci. USA* **82**:8183–8187.

Deverson, E., Gow, I., Coadwell, J., Monaco, J., Butcher, G., and Howard, J., 1990, MHC class II region encoding proteins related to the multidrug resistance family of transmembrane transporters, *Nature* **348**:738–741.

Dick, L., Moomaw, C., DeMartino, G., and Slaughter, C., 1991, Degradation of oxidised insulin B chain by the multiproteinase complex macropain (proteasome), *Biochemistry* **30**:2725–2734.

Elliott, T., Cerundolo, V., Elvin, J., and Townsend, A., 1991, Peptide-induced conformational change of the class I heavy chain, *Nature* **351**:402–405.

Elliott, T., Elvin, J., Cerundolo, V., Allen, H., and Townsend, A., 1992, Structural requirements for the peptide induced conformational change of free MHC class I heavy chains, *Eur. J. Immunol.* **22**:2085–2091.

Falk, K., Rötzschke, O., and Rammensee, H.-G., 1990, Cellular composition governed by major histocompatibility complex class I molecules, *Nature* **348**:248–251.

Falk, K., Rötzschke, O., Stevanovic, S., Jung, G., and Rammensee, H.-G., 1991a, Allele-specific motifs revealed by sequencing of self-peptides eluted from MHC molecules, *Nature* **351**:290–296.

Falk, K., Rötzschke, O., Deres, K., Metzger, J., Jung, G., and Rammensee, H.-G., 1991b, Identification of naturally processed viral nonapeptides allows their quantification in infected cells and suggests an allele-specific T cell epitope forecast, *J. Exp. Med.* **174**:425–434.

Garrett, T., Saper, M., Bjorkman, P., Strominger, J., and Wiley, D., 1989, Specificity pockets for the side chains of peptide antigens in HLA-Aw68, *Nature* **342**:692–696.

Glynne, R., Powis, S., Beck, S., Kelly, A., Kerr, L., and Trowsdale, J., 1991, A proteasome-related gene between the two ABC transporter loci in the class II region of the human MHC, *Nature* **353**:357–360.

Gould, K., Cossins, J., Bastin, J., Brownlee, G., and Townsend, A., 1989, A 15 amino acid fragment of influenza nucleoprotein synthesized in the cytoplasm is presented to class I-restricted cytotoxic T lymphocytes, *J. Exp. Med.* **170**:1051–1056.

Gros, P., Croop, J., and Housman, D., 1986, Mammalian multi drug resistance gene: Complete cDNA sequence indicates strong homology to bacterial transport proteins, *Cell* **47**:371–380.

Heinemeyer, W., Kleinschmidt, J., Sasidiwsky, J., Escher, C., and Wolf, D., 1991, Proteinase yscE, the yeast proteasome/multicatalytic-multifunctional proteinase: Mutants unravel its function in stress induced proteolysis and uncover its necessity for cell survival, *EMBO J.* **10**:555–567.

Henderson, R., Michel, H., Sakaguchi, K., Shabanowitz, J., Appella, E., Hunt, D., and Engelhard, V., 1992, HLA-A2.1-associated peptides from a mutant cell line: A second pathway of antigen presentation, *Science* **255**:1264–1266.

Higgins, C., Haag, P., Nikaido, K., Ardeshir, F., Garcia, G., and Ames, G., 1982, Complete nucleotide sequence and identification of the histidine transport operon of S. typhimurium, *Nature* **298**:723–727.

Higgins, C., Gallagher, M., Mimmack, M., and Pearce, S., 1988, A family of closely related ATP-binding subunits from prokaryotic and eukaryotic cells, *BioEssay* **8**:111–116.

Higgins, C., Hyde, S., Mimmack, M., Gileadi, U., Gill, D., and Gallagher, M., 1990, Binding protein dependent transport system, *J. Bioenerg. Biomembr.* **22**:571–592.

Hiles, I., Gallagher, M., Jamieson, D., and Higgins, C., 1987, Molecular characterization of the oligo-peptide permease of Salmonella typhimurium, *J. Mol. Biol.* **195**:125–129.

Hyde, S., Emsley, P., Hartshorn, M., Mimmack, M., Gileadi, U., Pearce, R., Gallagher, P., Gill, D., Hubbard, R., and Higgins, C., 1990, Structural model of ATP-binding associated with cystic fibrosis, multidrug resistance and bacterial transport, *Nature* **346**:362–365.

Jardetzky, T., Lan, W., Robinson, R., Madden, D., and Wiley, D., 1991, Identification of self-peptides bound to purified HLA-B27, *Nature* **353**:326–329.

Karre, K., Ljunggren, H. G., Ointek, G., and Kiessling, R., 1986, Selective rejection of H-2-deficient lymphoma variants suggests alternative immune defence strategy, *Nature* **319**:675.

Kelly, A., Powis, S., Glynne, R., Radley, E., Beck, S., and Trowsdale, J., 1991, Second proteasome-related gene in the human MHC class II region, *Nature* **353**:667–668.

Kelly, A., Powis, S., Kerr, L., Mockridge, I., Elliott, T., Bastin, J., Uchanska-Ziegler, B., Ziegler, A., Trowsdale, J., and Townsend, A., 1992, Assembly and function of the two ABC transporter proteins encoded in the human major histocompatibility complex, *Nature* **355**:641–644.

Kleijmeer, M., Kelly, A., Geuze, H., Slot, J., Townsend, A., and Trowsdale, J., 1992, MHC-encoded transporters are located in the ER and cis-Golgi, *Nature* **357**:342–344.

Klein, J., 1986, *Natural History of the Major Histocompatibility Complex*, Wiley, New York.

Levy, F., Gabathuler, R., Larsoon, R., and Kvist, S., 1991, ATP is required for in vitro assembly of

MHC class I antigens but not for transfer of peptides across the ER membrane, *Cell* **67**:265–274.

Lippincott-Schwartz, J., Yuan, L., Bonifacino, J., and Klausner, R., 1989, Rapid redistribution of Golgi proteins in the ER in cell treated with brefeldin A: Evidence for membrane cycling from Golgi to ER, *Cell* **56**:801–813.

Lippincott-Schwartz, J., Donaldson, J., Schweizer, A., Berger, E., Hauri, H., Yuan, L., and Klausner, R., 1990, Microtubule-dependent retrograde transport of proteins into the ER in the presence of brefeldin A suggests a ER recycling pathway, *Cell* **60**:821–836.

Lippincott-Schwartz, J., Yuan, L., Tipper, C., Amherdt, M., Orci, L., and Klausner, R., 1991, Brefeldin A effects on endosomes, lysosomes, and the TGN suggest a general mechanism for regulating organelle structure and membrane traffic, *Cell* **67**:601–616.

Livingstone, A., Powis, S., Diamond, A., Butcher, G., and Howard, J., 1989, A trans acting major histocompatibility complex-linked gene whose alleles determine gain and loss changes in the antigenic structure of a classical class I molecule, *J. Exp. Med.* **170**:777–795.

Ljunggren, H., Paabo, S., Cochet, M., Kling, G., Kourilsky, P., and Karre, K., 1989, Molecular analysis of H-2 deficit lymphoma lines. Distinct defects in biosynthesis and dissociation of MHC class I heavy chains and β-2m observed in cells with increased sensitivity to NK cell lysis, *J. Immunol.* **142**:2911–2915.

Madden, D., Gorga, J., Strominger, J., and Wiley, D., 1991, The structure of HLA-B27 reveals nonamer self-peptides bound in an extended conformation, *Nature* **353**:321–325.

Martinez, C., and Monaco, J., 1991, Homology of the proteasome subunits to a major histocompatibility complex-linked LMP gene, *Nature* **353**:664–667.

Monaco, J., and McDevitt, H., 1984, H-2-linked low-molecular weight polypeptide antigens assemble into an unusual macromolecular complex, *Nature* **309**:797–799.

Monaco, J., Cho, J., and Attaya, M., 1990, Transport protein genes in the murine MHC: Possible implications for antigen processing, *Science* **250**:1723–1726.

Moore, M., Carbone, F., and Bevan, M., 1988, Introduction of soluble protein into the class I pathway of antigen presentation, *Cell* **54**:777–785.

Morrison, L., Lukacker, A., Braciale, V., Fan, D., and Braciale, T., 1986, Differences in antigen presentation to MHC class I and class II restricted influenza virus-specific cytotoxic T lymphocyte clones, *J. Exp. Med.* **163**:903–910.

Neefjes, J., Stollorz, V., Peters, P., Geuze, H., and Ploegh, H., 1990, The biosynthetic pathway of MHC class II but not class I molecules intersects the endocytic route, *Cell* **61**:171–183.

Nutchern, J., Bonifacino, J., Biddison, W., and Klausner, R., 1989, Brefeldin A implicates egress from endoplasmic reticulum in class I restricted antigen presentation, *Nature* **339**:223–226.

Ohlen, C., Bastin, J., Ljunggren, H.-G., Imreh, S., Klein, G., Townsend, A., and Karre, K., 1990, Restoration of H-2^b expression and processing of endogenous antigens in the MHC class I pathway by fusion of a lymphoma mutant to L cells of the H-2^k haplotype, *Eur. J. Immunol.* **20**:1873–1876.

Ortiz-Navarrete, V., Seeling, A., Gernold, M., Frentzel, S., Kloetzel, P., and Hammerling, G., 1991, Subunits of the "20S" proteasome (multicatalytic proteinase) encoded by the major histocompatibility complex, *Nature* **353**:662–664.

Owen, M., Kissonerghis, A., and Lodish, H., 1980, Biosynthesis of HLA-A and HLA-B antigens in vivo, *J. Biol. Chem.* **255**:9678–9684.

Parnes, J., and Seidman, J., 1982, Structure of wild-type and mutant mouse β-2 microglobulin genes, *Cell* **29**:661–669.

Pelham, H. R. B., 1989, Control of protein exit from the endoplasmic reticulum, *Annu. Rev. Cell Biol.* **5**:1–23.

Pelham, H. R. B., 1991, Recycling of proteins between the endoplasmic reticulum and Golgi complex, *Curr. Biol.* **3**:585–591.

Ploegh, H., Cannon, E., and Strominger, J., 1979, Cell-free translation of the mRNAs for the heavy and light chains of HLA-A and HLA-B antigens, *Proc. Natl. Acad. Sci. USA* **76**:2273–2277.

Powis, S., Howard, J., and Butcher, G., 1991a, The major histocompatibility complex class II linked cim locus controls the kinetics of intracellular transport of a classical class I molecule, *J. Exp. Med.* **173**:913–921.

Powis, S., Townsend, A., Deverson, E., Bastin, J., Butcher, G., and Howard, J., 1991b, Restoration of antigen presentation to the mutant cell line RMA-S by an MHC-linked transporter, *Nature* **354**:356.

Powis, S., Deverson, E., Coadwell, J., Ciruela, A., Huskisson, N., Smith, H., Butcher, G., and Howard, J., 1992a, Effect of polymorphism of an MHC-linked transporter on the peptides assembled in a class I molecule, *Nature* **357**:211–215.

Powis, S., Mockridge, I., Kelly, A., Kerr, L., Glynne, R., Gileadi, U., Beck, S., and Trowsdale, J., 1992b, Polymorphism in a second ABC transporter gene located within the class II region of the human MHC, *Proc. Natl. Acad. Sci. USA* **89**:1463–1467.

Riordan, J. R., Rommens, J. M., Kerem, B., Alon, N., Rozmahel, R., Grzelczak, Z., Zielenski, J., Lok, S., Plavsic, N., Chou, J. L., Drumm, M., Iannuzzi, M., Collins, F. S., and Tsui, L. C., 1989, Identification of the cystic fibrosis gene: Cloning and characterization of complementary DNA, *Science* **245**:1066–1073.

Rivett, J., 1989, The multicatalytic proteinase, *J. Biol. Chem.* **264**:12215–12219.

Rötzschke, O., Falk, K., Wallny, H., Faath, S., and Rammensee, H.-G., 1990a, Characterization of naturally occurring minor histocompatibility peptides including H-4 and H-Y, **Science 249**:283–287.

Rötzschke, O., Falk, K., Deres, K., Schild, H., Norda, M., Metzger, J., Jung, G., and Rammensee, H.-G., 1990b, Isolation and analysis of naturally processed viral peptides as recognized by cytotoxic T cells, *Nature* **348**:252–254.

Sadegh-Nasseri, S., and Germain, R., 1991, A role for peptide in determining MHC class II structure, *Nature* **353**:167–169.

Saper, M., Bjorkman, P., and Wiley, D., 1991, Refined structure of the human histocompatibility antigen HLA-A2 at 2.6 Å resolution, *J. Mol. Biol.* **219**:277–319.

Salter, R., and Cresswell, P., 1986, Impaired assembly and transport of HLA-A and -B antigens in a mutant T × B cell hybrid, *EMBO J.* **5**:943–949.

Spies, T., and DeMars, R., 1991, Restored expression of major histocompatibility class I molecules by gene transfer of a putative peptide transporter, *Nature* **351**:323–324.

Spies, T., Bresnahan, M., Bahram, S., Arnold, D., Blank, G., Mellins, E., Pious, D., and DeMars, R., 1990, A gene in the human major histocompatibility complex class II region controlling the class I antigen presentation pathway, *Nature* **348**:744–747.

Spies, T., Cerundolo, V., Colonna, M., Cresswell, P., Townsend, A., and DeMars, R., 1992, Presentation of viral antigen by MHC class I molecules is dependent on a putative peptide transporter heterodimer, *Nature* **355**:644–646.

Stearz, U., Karasuyama, H., and Garner, A., 1988, Cytotoxic T lymphocytes against a soluble protein, *Nature* **329**:449–450.

Tevethia, S., Flyer, D., and Tjian, R., 1980, Biology of simian virus 40 (SV40) transplantation antigen (TrAg). VI. Mechanism of induction of SV40 transplantation immunity in mice by purified SV40 T antigen (D2 protein), *Virology* **107**:13–18.

Townsend, A., 1992, A new presentation pathway? *Nature* **356**:386–387.

Townsend, A., and Skehel, J., 1982, Influenza A specific cytotoxic T-cell clones that do not recognize viral glycoproteins, *Nature* **300**:655–657.

Townsend, A., Gotch, F., and Davey, J., 1985, Cytotoxic T cells recognize fragments of influenza nucleoprotein, *Cell* **42**:457–467.

Townsend, A., Rothbard, J., Gotch, F., Bahadur, G., Wraith, D., and McMichael, A., 1986a, The epitopes of influenza nucleoprotein recognized by cytotoxic T lymphocytes can be defined with short synthetic peptides, *Cell* **44**:959–968.

Townsend, A., Bastin, J., Gould, K., and Brownlee, G., 1986b, Cytotoxic T lymphocytes recognize influenza hemagglutinin that lacks a signal sequence, *Nature* **234**:575–577.

Townsend, A., Bastin, J., Gould, K., Brownlee, G., Andrew, M., Coupar, D., Boyle, D., Chan, S., and Smith, G., 1988, Defective presentation to class I restricted cytotoxic T lymphocytes in vaccinia-infected cells is overcome by enhanced degradation of antigen, *J. Exp. Med.* **168**:1211–1218.

Townsend, A., Ohlen, C., Bastin, J., Ljunggren, H., Foster, L., and Karre, K., 1989a, Association of class I major histocompatibility heavy and light chains induced by viral peptides, *Nature* **340**:443–448.

Townsend, A., Ohlen, C., Foster, L., Bastin, J., Ljunggren, H. G., and Karre, K., 1989b, A mutant cell in which association of class I heavy and light chains is induced by viral peptides, *Cold Spring Harbor Symp. Quant. Biol.* **54**:299–308.

Townsend, A., Elliott, T., Cerundolo, V., Foster, L., Barber, B., and Tse, A., 1990, Assembly of MHC class I molecules analyzed in vitro, *Cell* **62**:195–295.

Trowsdale, J., Hanson, I., Mockridge, I., Beck, S., Townsend, A., and Kelly, A., 1990, Sequences encoded in the class II region of the MHC related to the "ABC" superfamily of transporters, *Nature* **348**:741–744.

Van Bleek, G., and Nathenson, S., 1990, Isolation of an endogenously processed immunodominant viral peptide from the class I H-2 K^b molecule, *Nature* **348**:213–216.

Vitiello, A., Potter, T., and Sherman, L., 1990, The role of β-2-microglobulin in peptide binding by class I molecules, *Science* **250**:1423–1425.

Wallny, H., and Rammensee, H. G., 1990, Identification of classical minor histocompatibility antigen as cell-derived peptide, *Nature* **343**:275–278.

Wei, M., and Cresswell, P., 1992, HLA-A2 molecules in an antigen-processing mutant cell contain signal sequence-derived peptides, *Nature* **356**:441–442.

Wraith, D., and Vessey, A., 1986, Influenza virus-specific cytotoxic T-cell recognition: Stimulation of nucleoprotein-specific clones with intact antigen, *Immunology* **59**:173–177.

Yang, Y., Fruh, K., Chambers, J., Waters, J., Wu, L., Spies, T., and Peterson, P., 1992, Major histocompatibility complex (MHC)-encoded HAM2 is necessary for antigenic peptide loading onto class I MHC molecules, *J. Biol. Chem.* **267**:11669–11672.

Yewdell, J., and Bennick, J., 1989, Brefeldin A specifically inhibits presentation of protein antigens to cytotoxic T lymphocytes, *Science* **244**:1072–1075.

Yewdell, J., Bennick, J., and Hosaka, Y., 1988, Cells process exogenous proteins for recognition by cytotoxic T lymphocytes, *Science* **239**:637.

Yide, J., Wai-Kuo, S., and Berkower, I., 1988, Human T cell response to the surface antigen of hepatitis B (HBsAg). Endosomal and nonendosomal processing pathways are accessible to both endogenous and exogenous antigen, *J. Exp. Med.* **168**:293–297.

Zinkernagel, R., and Doherty, P., 1974, Immunological surveillance against altered self components by sensitised T lymphocytes in lymphocytic choriomeningitis, *Nature* **251**:547–549.

Chapter 11

Distribution, Biosynthesis, and Function of Mevalonate Pathway Lipids

Johan Ericsson and Gustav Dallner

1. INTRODUCTION

The structures and biosynthesis of mevalonate pathway lipids, and in particular of cholesterol, have been intensively investigated throughout most of this century and a large amount of data have accumulated. In spite of this enormous effort, completely new information concerning the membrane organization, compartmentalization, function, and metabolic regulation of these lipids has recently appeared and will be summarized in this review.

The biochemistry and cell biology of mevalonate pathway lipids constitute a large volume of information, which is not possible to summarize in a single review. The established facts are summarized in a number of excellent books and reviews dealing with specific aspects of the regulation, biosynthesis, metabo-

Abbreviations used in this chapter: ACAT, acyl-CoA-cholesterol acyltransferase; CHO, Chinese hamster ovary; DMAPP, dimethylallyl pyrophosphate; ER, endoplasmic reticulum; FPP, farnesyl pyrophosphate; GPP, geranyl pyrophosphate; GGPP, geranylgeranyl pyrophosphate; HMG-CoA, 3-hydroxy-3-methylglutaryl-coenzyme A; IPP, isopentenyl pyrophosphate; LDL, low-density lipoprotein; SCP, sterol carrier protein.

Johan Ericsson and Gustav Dallner Department of Biochemistry, University of Stockholm, S-106 91 Stockholm, and Clinical Research Center at Huddinge Hospital, Karolinska Institutet, S-141 86 Huddinge, Sweden
Subcellular Biochemistry, Volume 21: Endoplasmic Reticulum, edited by N. Borgese and J. R. Harris. Plenum Press, New York, 1993.

lism, and function of these lipids (Schroepfer, 1981, 1982; Porter and Spurgeon, 1981; Olson and Rudney, 1983; Lenaz, 1985; Rip et al., 1985; Rudney and Sexton, 1986; Myant, 1990); Goldstein and Brown, 1990.

The mevalonate pathway lipids are present in all mammalian tissues and are also synthesized in all tissues. De novo synthesis may be insufficient even under normal physiological conditions. Thus, for certain lipids, in particular cholesterol, the endogenous synthesis must be complemented by dietary uptake and redistribution via the circulation. These lipids are also found and synthesized in plants and bacteria, but their structures and biosynthesis in these organisms differ from those found in mammals. These lipids have been studied in greatest detail in the liver, but there is also ample information about them in other tissues. Significant differences, specific for individual tissues, have been demonstrated, particularly with respect to function. On the other hand, many common general rules are valid for all of these lipids at various locations.

2. SEQUENCE OF THE INITIAL COMMON PATHWAY

The principles of the organization of the mevalonate pathway are shown in Figure 1. The initial substrate is acetyl-CoA and the condensation of three acetyl-CoA molecules, mediated by two enzymatic steps, gives rise to HMG-CoA. Since acetate easily penetrates practically all plasma membranes, in contrast to all of the following intermediates of the pathway, this compound is often used as a precursor in biosynthetic studies. Acetyl-CoA is, however, one of the main metabolic intermediates in a number of important cellular processes and is, consequently, utilized simultaneously in many metabolic reactions. Thus, if labeling of the mevalonate pathway lipids is performed with acetate as precursor, the rates of incorporation into the final products are highly dependent on the endogenous pool sizes for the various intermediates. Using tritiated water, it was possible to establish the absolute rates of synthesis of cholesterol in many organs and species (Dietschy and Spady, 1984). However, in general, it is not possible to apply this method for measurement of dolichol and ubiquinone biosynthesis, which occur at rates 500 times slower than that of cholesterol (Adair and Keller, 1982). This is one reason why many regulatory aspects of the metabolism of dolichol and ubiquinone remain unclear.

The initial enzymes of the pathway that transform acetyl-CoA to HMG-CoA are greatly enriched in the cytoplasm and considered to be the major supplier of HMG-CoA for polyisoprenoid biosynthesis (Rilling and Chayet, 1985). HMG-CoA is, however, also produced in mitochondria, where it is utilized for the synthesis of ketone bodies.

The next enzyme of the pathway is HMG-CoA reductase, which is not cytosolic but associated with ER membranes and, as found recently, also present in rat liver peroxisomes (Keller et al., 1985). This enzyme reduces HMG-CoA

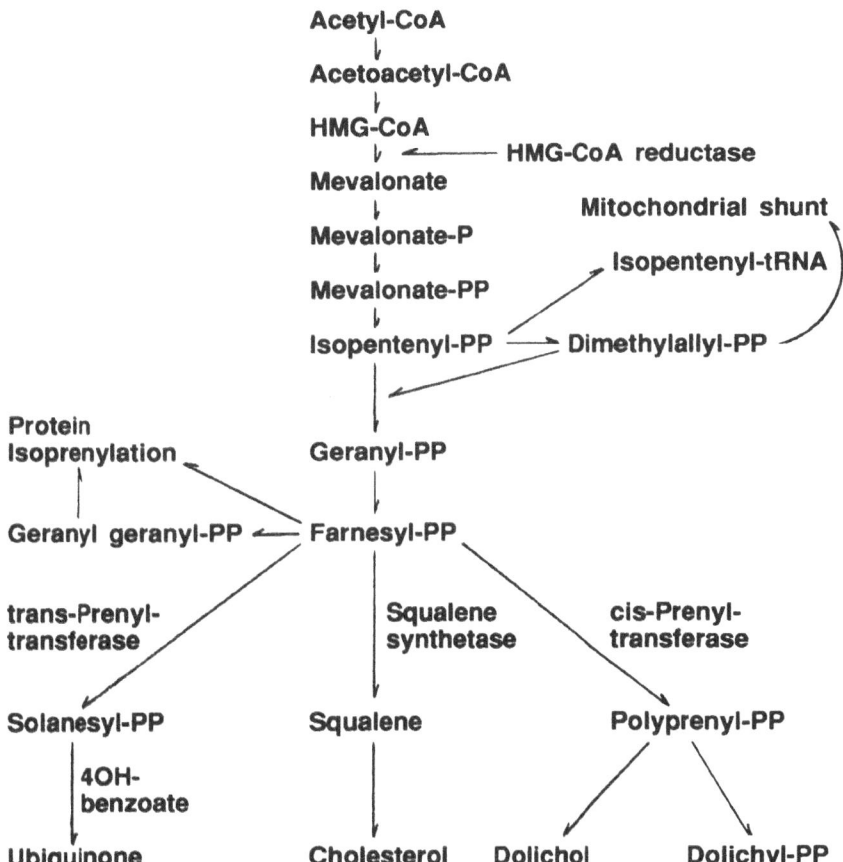

FIGURE 1. Overview of the mevalonate pathway in animal cells.

via an intermediate to mevalonate in a two-step reaction utilizing two molecules of NADPH. The intermediate involved has a structure similar to those of known inhibitors of stevol biosynthesis.

HMG-CoA reductase is one of the most studied enzymes and is considered to be the main regulatory enzyme of cholesterol synthesis. This protein is an integral component of the ER membrane and has seven transmembrane segments, with the N-terminal end facing the lumenal surface (Chin *et al.*, 1984; Liscum *et al.*, 1985) (Figure 2). One of the lumenal loops is glycosylated with a high-mannose-type residue in an *N*-glycosidic linkage. The highly conserved catalytic region is located at the C-terminal end and situated completely in the cytosol. One possible mechanism for regulation of the enzyme activity is by reversible

phosphorylation site (Ser 871)

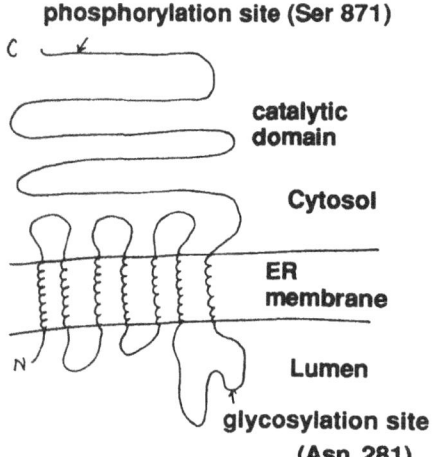

catalytic
domain

Cytosol

ER
membrane

Lumen

**glycosylation site
(Asn 281)**

FIGURE 2. The topology of HMG-CoA reductase in the ER.

phosphorylation of one or more of its 70 serine residues, most probably at a site near the C-terminus.

Mevalonate is phosphorylated by mevalonate kinase and the product so formed, mevalonate-5-phosphate, is further phosphorylated to yield mevalonate pyrophosphate. Decarboxylation of this pyrophosphate derivative results in the formation of isopentenyl pyrophosphate (IPP). IPP is the central intermediate of the pathway since it participates in several types of condensation reactions involved in polyisoprenoid biosynthesis. Isomerization leads to dimethylallyl pyrophosphate (DMAPP), which, together with IPP, is utilized as a substrate by farnesyl pyrophosphate (FPP) synthase. This enzyme condensates DMAPP with two molecules of IPP, with the intermediate formation of geranyl pyrophosphate (GPP), to give FPP. The enzymes responsible for the conversion of mevalonate to FPP are mainly present in the cytoplasm and have been studied in great detail. The two terminal double bonds in FPP are in the *trans* configuration. Many of the enzymes involved in this initial part of the mevalonate pathway have been isolated and thoroughly characterized. For example, the gene encoding FPP synthase has been identified and cloned from rat, human, yeast, and *E. coli* cells (Clarke *et al.*, 1987a; Ashby and Edwards, 1989; Anderson *et al.*, 1989; Fujisaki *et al.*, 1990).

IPP is also used in a number of other condensation reactions resulting in the synthesis of longer polyisoprenoid pyrophosphates involved in dolichol and ubiquinone biosynthesis (see Section 9). It has also been established that the tRNA species binding serine, phenylalanine, leucine, cysteine, and tryptophan contain a five-carbon isoprene unit that originates from IPP (Huneeus *et al.*, 1980; Faust *et al.*, 1980). In a reaction sequence in mitochondria, DMAPP is dephosphorylated to dimethylallyl alcohol, followed by interconversion to HMG-CoA. Further metabolism gives rise to acetyl-CoA or acetoacetate for utilization in extramitochondrial processes. This pathway is known as the mitochondrial shunt and is considered to be responsible for the removal of excess substrates (Edmond and Popjak, 1974).

FPP is the last common intermediate in the synthesis of the lipid end products, i.e., cholesterol and dolichol, and subsequent processes are associated with various organelles. FPP is also considered to be the precursor for the biosynthesis of the polyprenyl-PP side chain of ubiquinone, according to Figure 1. However, as will be discussed below, the data available indicate that DMAPP and GPP may be the preferred allylic substrates for *trans*-prenyltransferase. FPP is further elongated to all-*trans*-geranylgeranyl pyrophosphate (GGPP) in the cytosol and these two isoprenoids (FPP and GGPP) are utilized in cytosolic protein modification reactions (see Section 10).

3. CHOLESTEROL

Cholesterol is the most abundant mevalonate lipid, present in variable amounts in all cells (Figure 3). Some of its biosynthetic intermediates, i.e., squalene, lanosterol, and demethylated lanosterol derivatives, can also be found in substantial amounts in several tissues. Some tissues are very rich in cholesterol, e.g., about 30% of brain lipids are cholesterol and erythrocytes are also rich in this lipid. In membranes, only about 10% of the cholesterol present is esterified with a fatty acid, but in its nonmembranous forms, including the lipoproteins of the blood, the majority of this lipid is esterified. It is believed that esterification is required for translocation of cholesterol to the blood after *de novo* synthesis in the liver and, also, for transfer from the intestine to the blood (Johnston, 1978; Khan *et al.*, 1989). In contrast to the other mevalonate pathway lipids, a large portion of the cholesterol pool is utilized as substrate for further metabolic purposes, such as bile acid and steroid hormone production.

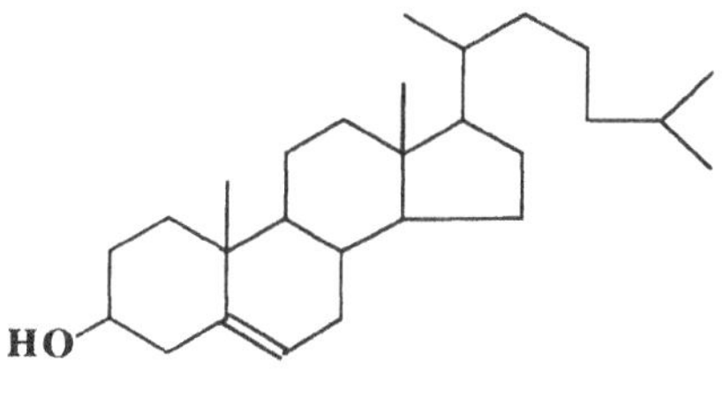

FIGURE 3. The structure of cholesterol.   **Cholesterol**

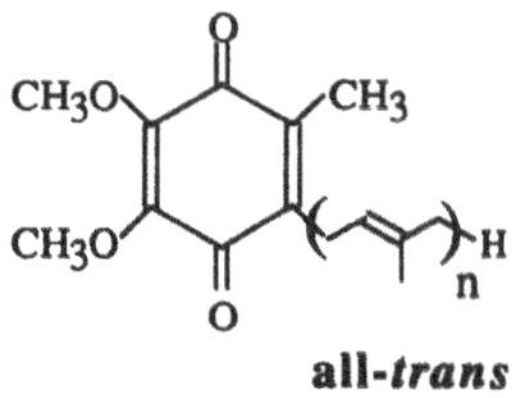

all-*trans*

Ubiquinone

n = Number of isoprene residues

FIGURE 4. The structure of ubiquinone. n is the number of isoprene units, ranging between 6 and 10.

4. UBIQUINONE

Ubiquinone (coenzyme Q) consists of a benzoquinone ring and an all-*trans* isoprenoid side chain located on C-6 of the ring (Olson and Rudney, 1983) (Figure 4). The number of isoprenoid units in this side chain is species-dependent and, furthermore, varies within a single species. The main component in rats contains nine residues but ubiquinone-10 (Åberg *et al.*, 1992) is also present. In rat brain, spleen, and intestine, as much as 30% of the total is ubiquinone-10. In human tissues, the dominating species is ubiquinone-10, but in all tissues a few percent of the total is recovered as ubiquinone-9. Under *in vivo* conditions, the majority of ubiquinone is in the reduced state.

In mitochondria, ubiquinone is a component of the respiratory chain and can thereby be reduced by various substrates (Ernster, 1977; Crane, 1986). However, it is not known which enzyme(s) is responsible for the reduction of this lipid in other cellular membranes. In inner mitochondrial membranes, ubiquinone is present in a large molar excess compared with the associated flavoprotein and cytochromes and, for as-yet-unknown reasons, this excess is required for normal respiration.

5. DOLICHOL

Dolichol in animal tissues is present primarily as a family of derivatives, containing 16–23 isoprenoid units, but smaller amounts of both shorter and longer polyisoprenoids have also been identified (Hemming, 1981; Chojnacki and Dallner, 1988) (Figure 5). The chain-length distribution varies in different species, but in the same species this distribution in various tissues is very similar. In rat tissues the two dominating components are those containing 18 and 19 residues, while in humans the dominating components have 19 and 20 residues. Under nonpathological conditions, 96–98% of these polyisoprenoids are α-saturated. The two ω-terminal isoprene residues are in the *trans* configuration,

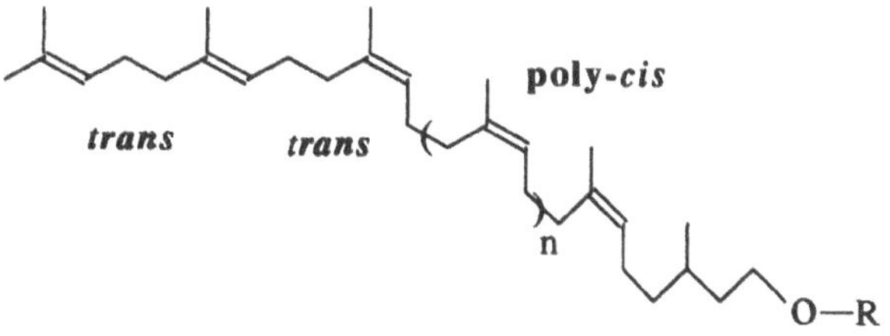

Dolichol

n =Number of isoprene units, 11-18; R = H, P, PP, FA

FIGURE 5. The structure of dolichol. n is the number of isoprene units, 11–18. R = H, P, PP, FA.

while all of the remaining units are *cis*. The center of asymmetry at C-3 allows both S and R configurations but in nature probably only the S form is present. A large portion of the total dolichol is esterified with a fatty acid and the compositions of these fatty acids in various intracellular membranes exhibit a certain specificity.

The variation in the dolichol concentrations of various tissues is considerably greater than that of cholesterol and ubiquinone. Human tissues are particularly rich in dolichol, with the highest concentrations being present in endocrine tissues. In human hypophysis, for example, the dolichol level (7 mg/g wet wt) is higher than the total amount of phospholipid (6 mg/g wet wt) (Tollbom and Dallner, 1986).

6. DOLICHYL-P

In comparison with the free alcohol, the phosphorylated form of dolichol is present in small amounts that vary to only a limited extent under different pathological conditions. One-third of this phosphorylated lipid is covalently bound to proteins in rat liver (Thelin *et al.*, 1991). A large portion of the dolichyl-P in different tissues is associated with an oligosaccharide core, which reflects the active participation of this lipid in glycoprotein synthesis, as an intermediate in *N*-glycosylation (Kornfeld and Kornfeld, 1985). Dolichyl-P is also involved in the synthesis of the oligosaccharide portion of the phosphatidylinositol (PI) anchor (Menon *et al.*, 1990; see Tartakoff, this volume).

7. SUBCELLULAR LOCALIZATION

When the lipid compositions of rat liver subfractions were investigated, it was found that both mitochondria and microsomes are poor in cholesterol, while

Table I
Mevalonate Pathway Lipids in Subcellular Organelles of Rat Liver

Organelle	Cholesterol[a]	Dolichol[b]	Ubiquinone[c]
		μg/mg protein	
Nuclei	37.5[d]	0.2	0.2
Mitochondria	2.28	0.09	1.4
Outer membranes	30.1		2.2
Inner membranes	5.0	0.18	1.9
Microsomes	27.5	0.26	0.15
Rough microsomes	15.6	0.31	
Smooth microsomes	31.3	0.21	
Lysosomes	38.0	3.8	1.9
Golgi vesicles	71.1	1.5	2.6
Peroxisomes	6.4	0.8	0.3
Plasma membranes	128.0	1.8	0.7

[a]Cholesterol + cholesteryl ester.
[b]Dolichol + dolichyl ester.
[c]Ubiquinone-9.
[d]Nuclear envelope.

the concentrations of this lipid are high in lysosomes, plasma membrane, and Golgi vesicles (Zambrano et al., 1975) (Table I). In the case of dolichol, Golgi and plasma membranes, and especially lysosomes, contain relatively high levels (Eggens et al., 1983). Both inner and outer mitochondrial membranes, as well as lysosomes and Golgi vesicles, are rich in ubiquinone (Appelkvist et al., 1991). Microsomes and lysosomes are enriched in dolichyl-P. It is clear that all cellular membranes contain all of the lipid end products of the mevalonate pathway, but that the variations in level are large. These variations may have functional explanations: membrane circulation in the Golgi–plasma membrane–lysosomal system may be promoted by the presence of dolichol, which is a fusogenic agent (van Duijn et al., 1986). Part of the lysosomal lipids may be lumenal, especially in the case of cholesterol, as a result of receptor-mediated endocytosis.

8. INTRAMEMBRANOUS LOCALIZATION

The intramembranous localization of mevalonate pathway lipids has so far been studied primarily in model membranes (Figure 6). Dolichol appears to be present in the central hydrophobic portion of the bilayer between the opposing fatty acid chains (de Ropp and Troy, 1985; Valtersson et al., 1985). This lipid is preferentially associated with phosphatidylethanolamine. A helical conformation has been proposed for dolichol in an attempt to explain the relatively high amounts of this lipid present in the bilayer.

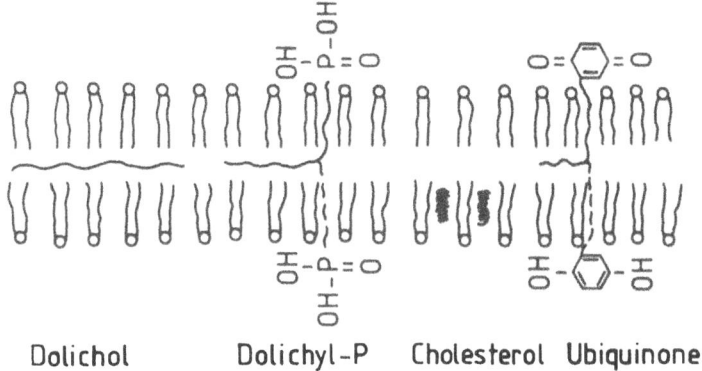

Dolichol   Dolichyl-P   Cholesterol Ubiquinone

FIGURE 6. Intramembranous distribution of the lipid products of the mevalonate pathway.

The isoprenoid chain of dolichyl-P has a localization similar to that of dolichol, but in this case the α-terminal portion makes a turn and the final isopene unit(s) is situated between neighboring phospholipids (van Duijn *et al.*, 1987). This localization allows the phosphate group to be exposed to the aqueous phase, where it can interact with the activated sugar present in the cytoplasm. It has been proposed, although not demonstrated, that the α-terminal portion of dolichyl-P rotates in such a manner that the sugar-P moiety becomes available at the inner surface of the ER, where glycoprotein synthesis takes place. In addition, intermediate dolichol-sugars may serve as carriers of substrate between different glycosyltransferases.

It is probable that ubiquinone has an intramembranous localization similar to that of dolichyl-P (Quinn and Katsikas, 1985). In this case the isoprenoid side chain is embedded in the central hydrophobic portion of the membrane, while the quinone moiety is located at the aqueous surface. Translocation of the quinone moiety (in oxidized or reduced form) to the opposite side of the membrane may occur.

The localization of cholesterol in model membranes differs from that of dolichol, which is an important factor for cholesterol's function as a membrane component. This lipid is situated between the fatty acyl chains of the phospholipids and appears to prefer phosphatidylcholine (van Dijck *et al.*, 1976). It has been proposed that cholesterol is present in the outer half of the erythrocyte bilayer (Fisher, 1976), but this requires further studies.

9.   THE BIOSYNTHETIC PROCESS

The basic requirement for the biosynthesis of mevalonate pathway lipids is the production of sufficient amounts of FPP, as illustrated in Figure 1. The

subsequent steps in cholesterol and dolichol synthesis, as well as in the biosynthesis of the side chain of ubiquinone, have been studied using microsomes
and it has been concluded that all of the enzymes involved are associated with the
ER membrane. Possible exceptions are the terminal enzymes in dolichol and
cholesterol synthesis, which are saturases recovered in the cytosolic fraction. It is
possible, however, that these enzymes are released proteolytically from fragments of the ER during subfractionation.

In spite of intensive investigations during the past decades, only a few of the
enzymes involved in these pathways have been isolated and the majority of the
biosynthetic steps have only been tentatively identified. Consequently, we can
expect modifications of the proposed biosynthetic sequences in the future. As
will be discussed below, recent investigations have demonstrated that, in addition
to the ER, peroxisomes also synthesize cholesterol and dolichol (see Section
9.3).

9.1. Cholesterol Biosynthesis

The initial, ER-bound steps in cholesterol biosynthesis, i.e., conversion of
FPP via squalene to lanosterol, are well characterized, although the enzymes
involved have not been isolated from mammalian sources (Figure 7).

The first committed step in cholesterol biosynthesis is the condensation of
two FPP molecules to form presqualene pyrophosphate, which is then reduced to
squalene in the presence of NADPH or NADH (Schroepfer, 1981, 1982; Gaylor,
1981). This two-step reaction is probably mediated by a single enzyme, namely,
squalene synthase. This enzymes is not only the first enzyme of the ER-bound
process, but may also play a regulatory role in cholesterol biosynthesis. In
addition to HMG-CoA reductase, there appear to be terminal regulatory steps in
cholesterol biosynthesis, present in the membrane-bound sequence, and squalene
synthase is a possible candidate. HMG-CoA reductase determines the availability of FPP for further metabolism to the various mevalonate pathway end products
but the flow of FPP into the sterol branch of this pathway is determined by the
activities of squalene synthase.

The first step in the cyclization of squalene is the formation of 2,3-
oxidosqualene by the microsomal squalene epoxidase. The transformation of
2,3-oxide to lanosterol is mediated by the 2,3-oxidosqualene: lanosterol cyclase,
which catalyzes a series of reactions leading to the formation of the cyclic
product.

Squalene 2,3-dioxide may also be the substrate for the formation of squal-
ene-2,3:22,23 dioxide, which may be further metabolized to other oxygenated
sterols capable of downregulating HMG-CoA reductase and, thereby, cholesterol
synthesis.

Many of the following steps have not been characterized in detail and the properties of the enzymes catalyzing these steps are not established. In the process of cholesterol biosynthesis, three steps have been shown to involve cytochrome P-450 systems, but there are an additional 14 reactions requiring NADPH or NADH. It is believed that cytochrome b_5 is involved in some of these reactions (see Borgese *et al.*, this volume), and it seems plausible that certain of these are also catalyzed by cytochrome P-450 systems.

The steps subsequent to the formation of lanosterol, which modify the sterol ring structure, involve at least 19 enzymatic reactions. These steps consist of demethylation of lanosterol at C-14, two consecutive demethylations at C-4, reduction of the double bond at C-24, isomerization of the double bond at C-8 to C-7, introduction of a double bond at C-5, reduction of the double bond at C-7, and, finally, reduction of the double bond on the side chain.

It has been demonstrated that at least two cytoplasmic proteins play a role in activation of cholesterol biosynthesis and metabolism (Ferguson and Bloch, 1977; Gavey *et al.*, 1981). Sterol carrier protein-1 (or soluble protein factor, SPF) stimulates lanosterol synthesis from squalene, while sterol carrier protein-2 (SCP2) stimulates conversion of lanosterol to cholesterol, as well as the esterification of cholesterol by ACAT. It has also been proposed that SCP2 functions as a carrier in the intracellular translocation of sterol and sterol metabolites (Billheimer and Reinhart, 1990).

It has been suggested that the oxysterols produced as intermediates in the sequence of reactions between lanosterol and cholesterol may regulate HMG-CoA reductase activity. Furthermore, it is possible that the intermediates in cholesterol biosynthesis may give rise to other oxysterols, more potent in regulating this reductase (Rudney *et al.*, 1987). It has also been proposed that oxidation of the side chain of cholesterol present in LDL induces the uptake of this lipoprotein particle into macrophages, followed by their deposition in the arterial wall (Steinberg *et al.*, 1989). This could be the start of the atherosclerotic process, and thereby be of immense medical significance.

9.2. Dolichol Biosynthesis

The proposed mechanism for dolichol biosynthesis and the modification reactions affecting the final products are based on *in vitro* and *in vivo* experiments. The sequence of events has not been clarified completely, because the enzymes involved have not been purified. The importance of local *de novo* synthesis in individual tissues is emphasized by the fact that the dietary uptake of this lipid and its redistribution via the circulation are very limited (Keller *et al.*, 1982; Chojnacki and Dallner, 1983).

The events of dolichol biosynthesis have been studied in a number of tissues, since, in contrast to cholesterol biosynthesis, the activity in liver is not so much greater than those in extrahepatic tissues. The first committed enzyme in

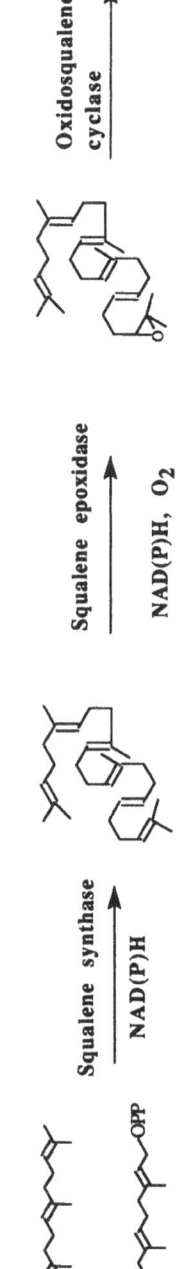

Farnesyl Diphosphate

Squalene synthase
NAD(P)H
→

Squalene

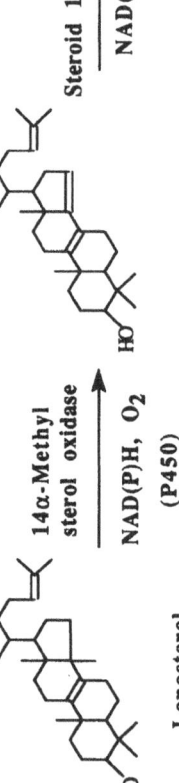

Squalene epoxidase
NAD(P)H, O$_2$
→

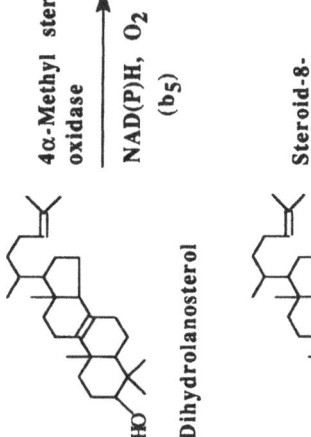

2,3-Oxidosqualene

Oxidosqualene
cyclase
→

Lanosterol

14α-Methyl
sterol oxidase
NAD(P)H, O$_2$
(P450)
→

Steroid 14-reductase
NAD(P)H
→

Dihydrolanosterol

4α-Methyl sterol
oxidase
NAD(P)H, O$_2$
(b$_5$)
→

3-Ketosterol

Steroid-3-keto-
reductase
NAD(P)H
→

4-Methyl sterol
oxidase
NAD(P)H, O$_2$, NAD
(b$_5$)
→

Zymosterol

Steroid-8-
isomerase
→

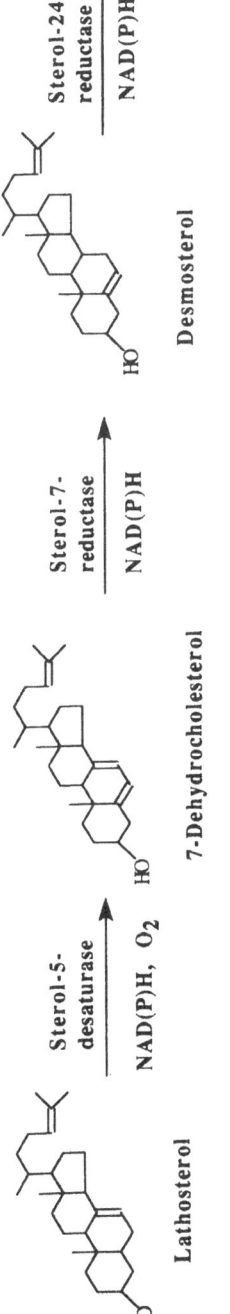

FIGURE 7. Overview of membrane-associated reactions involved in cholesterol biosynthesis.

dolichol biosynthesis is *cis*-prenyltransferase, which mediates *cis* additions of IPP units to all-*trans*-FPP, giving rise to long-chain *trans,trans*-poly-*cis*-isoprenoid pyrophosphates (Baba *et al.*, 1987; Crick *et al.*, 1991; Ericsson *et al.*, 1991a, 1992a) (Figure 8). Further modifications produce dolichyl-P, which may be converted to dolichol. However, the levels of these two substances seem to be

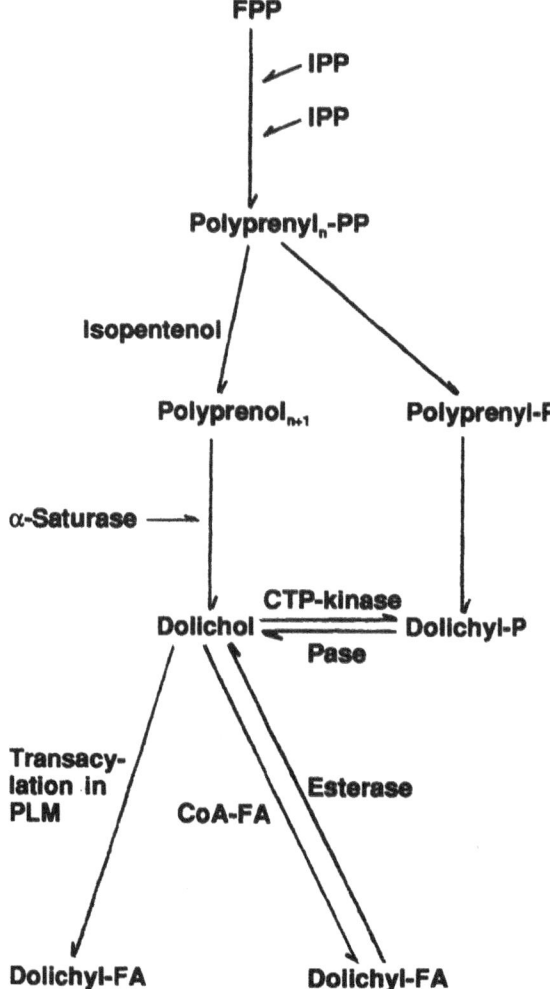

FIGURE 8. The terminal reactions in dolichol biosynthesis. PLM, plasma membrane; FA, fatty acid.

regulated in an almost independent fashion and their amounts are also modified in different fashions under pathological conditions. Therefore, it has been proposed that there are separate pathways for the biosynthesis of these two lipids (Ekström et al., 1984, 1987a). In the case of dolichyl-P, dephosphorylation of the polyprenyl-PP to polyprenyl-P, followed by α-saturation is proposed to form the final product.

In the presence of NADH, microsomal polyprenyl-PP could be converted to α-saturated dolichol upon the addition of supernatant. It has been suggested that the final condensation occurs with isopentenol, instead of IPP, and that the absence of a pyrophosphate leaving group results in termination of the sequence of condensation reactions (Ekström et al., 1987b). The presence of an α-saturase specific for NADH in rat liver has also been demonstrated. α-Saturase activity may be insufficient under a number of conditions. There are mutants of CHO cells that produce exclusively α-unsaturated polyprenols and in human hepatocellular carcinoma up to 30% of the dolichol derivatives are present in the α-unsaturated form (Stoll et al., 1988; Rosenwald and Krag, 1990; Eggens et al., 1989).

All tissues investigated to date contain an ER-bound CTP-kinase involved in the phosphorylation of dolichol, as well as a specific dolichyl-P phosphatase (Allen et al., 1978; Burton et al., 1981). A dolichyl-PP phosphatase, necessary for regeneration of the dolichyl-P in the course of protein glycosylation, has also been described. Under steady-state conditions, 90% of the dolichyl-P is produced by de novo synthesis, but the kinase may be activated under certain conditions to maintain an unchanged level of dolichyl-P, which is obviously essential for cell survival (Åstrand et al., 1986). During embryonic development, most of the dolichyl-P is synthesized by phosphorylation of the free alcohol and in human hepatocellular cancer the decreased level of de novo synthesis is compensated for by increased kinase activity (Rossignol et al., 1983; Eggens et al., 1988).

A large portion of the dolichol pool present in cells is esterified with a fatty acid, a process catalyzed by a specific ER-associated transferase utilizing CoA-activated fatty acids (Tollbom et al., 1988). It has been proposed that these fatty acids may serve as targeting signals during transport of dolichol in the cytosol. Dolichol may also be esterified through transacylation by an enzyme located on the outer surface of hepatocytes (Sindelar et al., 1992). Hydrolysis of the dolichyl esters is mediated by a specific esterase present in lysosomes and, also, to some extent in plasma membranes (Tollbom et al., 1989).

When microsomes were incubated with FPP and [³H]-IPP, not only was long-chain polyprenyl-PP produced but trans,trans,cis-geranylgeranyl-PP (GGPP) was also formed (Sagami et al., 1991; Ericsson et al., 1992a) (Figure 9). Various experimental data demonstrate the existence of a two-step reaction or, more probable, the involvement of two enzymes in this overall reaction sequence. The first reaction involves a single condensation with IPP, introducing

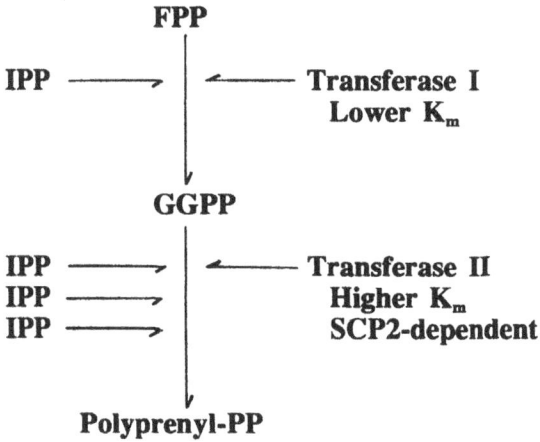

FIGURE 9. Microsomal *cis*-preyltransferase.

the first *cis* double bond. The second step involves the repetitive addition of IPP units in the *cis* configuration to *trans,trans,cis*-GGPP. The ER-associated *cis*-prenyltransferase requires the presence of detergents for activity under *in vitro* conditions. In the cell it has been proposed that a cytosolic protein factor, identified as SCP2, may play the same role as detergent *in vitro* (Ericsson *et al.*, 1991b). Most interestingly, this protein is required not only for optimal enzyme activity, but also for the production of dolichols with the same isoprenoid chain length as those occurring naturally in tissues.

9.3. ER–Peroxisomal Interactions in Cholesterol and Dolichol Biosynthesis

Investigations in recent years have demonstrated that peroxisomes are also capable of synthesizing both cholesterol and dolichol. In the presence of [^{3}H]mevalonate and cytosol, both of these lipids could be labeled in peroxisomes (Appelkvist, 1987; Thompson *et al.*, 1987; Appelkvist and Kalén, 1989). Furthermore, IPP, DMAPP, and GPP can be used as substrates for peroxisomal FPP synthesis in the absence of cytosol (Ericsson *et al.*, 1992b). These results demonstrate that peroxisomes contain both IPP isomerase and FPP synthase activities, in addition to the terminal enzymes of cholesterol and dolichol biosynthesis.

Squalene synthase, dihydrolanosterol oxidase, steroid-14-reductase, steroid-8-isomerase, and steroid-3-ketoreductase activities, involved in the synthesis of cholesterol, have also been demonstrated in rat liver peroxisomes (Appelkvist *et al.*, 1990). The peroxisomal *cis*-prenyltransferase is also active in the absence of detergents, since this organelle contains high levels of SCP2

(Ericsson *et al.*, 1992b). Both peroxisomal *cis*-prenyltransferase and squalene synthase activities have properties differing from those of their microsomal counterparts. Inducers affect these enzymes in the different organelles differently, e.g., mevinolin or cholestyramine treatment induces the peroxisomal much more than the microsomal activities.

Peroxisomes also contain acetoacetyl-CoA thiolase, HMG-CoA reductase, and mevalonate kinase activities (Thompson and Krisans, 1990; Stamellos *et al.*, 1992) and it is highly possible that the whole initial portion of the mevalonate pathway, commencing from acetyl-CoA, is present in this organelle. In this case it will be a future task to determine suitable conditions for obtaining the complete biosynthetic process with this organelle. Furthermore, the most active peroxisomal metabolic process, i.e., β-oxidation of fatty acids, produces substantial quantities of acetyl-CoA and NADH, both of which are consumed during cholesterol and dolichol biosynthesis.

One may speculate to what extent peroxisomes contribute to the total cellular synthesis of these two lipids. As shown in Figure 10, the specific activities of squalene synthase in peroxisomes and microsomes are equally high, while the peroxisomal *cis*-prenyltransferase has at least a four-fold higher specific activity than its microsomal counterpart. If these enzymes represent the rate-limiting steps in the synthetic processes, one can calculate that approximately 50% of cellular dolichol and 20% of cholesterol is synthesized in peroxisomes under normal conditions. Under conditions when peroxisomes are selectively induced, this organelle may account for an even larger portion of the total cellular production of these lipids.

9.4. Ubiquinone Biosynthesis

Ubiquinone biosynthesis has been studied almost exclusively in bacteria and yeast, so that our present concept of the biosynthetic sequence is based on these organisms (Casey and Threlfall, 1978; Ramasarma, 1985). In the first portion of this sequence, the ring structure, 4-hydroxybenzoate, condensates with the polyisoprenoid pyrophosphate side chain (Figure 11). 4-Hydroxybenzoate is derived from tyrosine in animal tissues and is believed to be present in excess. Originally it was proposed that the synthesis of the polyisoprenoid moiety and condensation with the ring structure, as well as the subsequent modifications of this structure, take place in the inner mitochondrial membrane. It appears, however, that the distribution of this biosynthetic process within the cell is a more complex question that is difficult to study because of the low specific radioactivity of most of the precursors available.

Solanesyl-PP in rat, decaprenyl-PP in human, and hexaprenyl-PP in yeast are synthesized by the *trans*-prenyltransferase, which catalyzes *trans* addition of IPP units to the allylic substrate. In this respect the isoprenoid moiety of ubiqui-

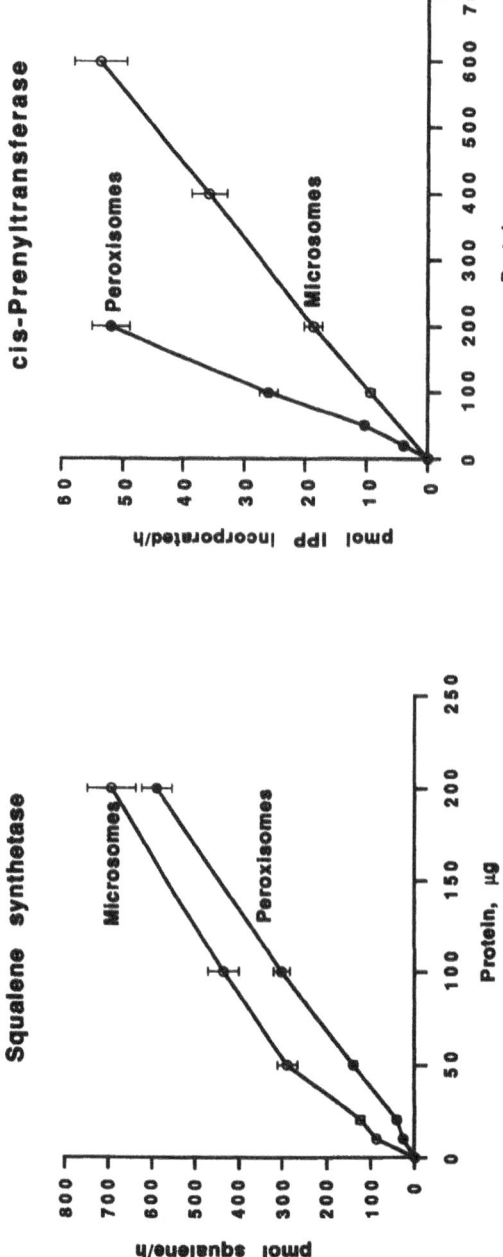

FIGURE 10. Comparison of microsomal and peroxisomal *cis*-prenyltransferase and squalene synthetase activities. Data from Ericsson *et al.* (1992b).

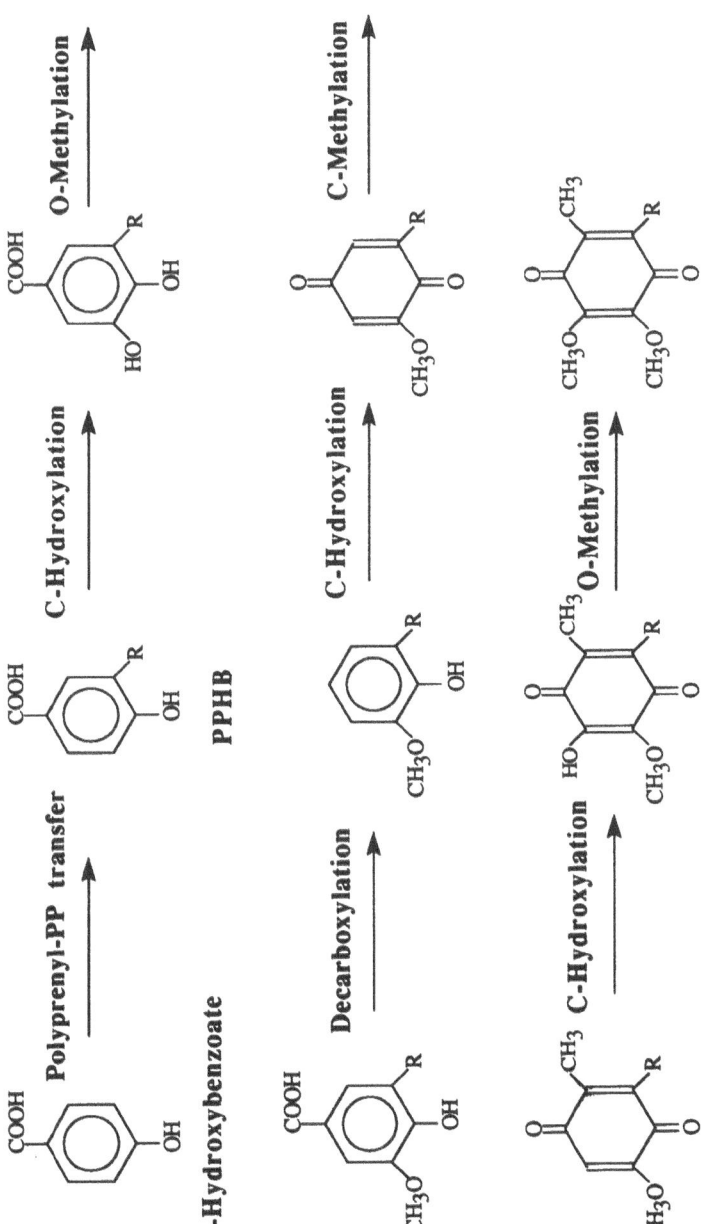

FIGURE 11. Biosynthesis of ubiquinone. R = polyisoprenoid side chain; PPHB = polyprenyl-4-hydroxybenzoate.

none differs significantly from dolichol, which is a poly-*cis* isoprenoid. It appears that *trans*-prenyltransferase is unable to use FPP as substrate and that the initial substrate in rat liver microsomes and *Saccharomyces cerevisiae* mitochrondria, as well as in spinach microsomes is GPP (Ashby and Edwards, 1990; Teclebrahan *et al.*, 1993; Swiezewska *et al.*, 1992). In rat liver the site of solanesyl-PP biosynthesis is the ER, which is logical in light of the fact that the biosynthesis of other isoprenoid lipids also occurs at this localization. Concerning other tissues and organisms, more detailed subfractionation studies will be required in the future.

The transfer of solanesyl-PP to 4-hydroxybenzoate, mediated by 4-hydroxybenzoate : polyprenyl-PP transferase, has been studied in rat liver subfractions and the highest activities were detected in Golgi fractions and in a microsomal subfraction (smooth II) most probably representing Golgi membranes (Kalén *et al.*, 1987, 1990a). This intermediate must undergo modification of the ring structure in order to form the final product.

These reactions are *C*-hydroxylations, *O*-methylations, *C*-methylation, and decarboxylation. The exact sequence of events is, however, unclear and considerable modification of our present view of the pathway may be required in the future. Respiratory yeast mutants, defective in ubiquinone biosynthesis, may be very useful in attempts to identify the various enzymatic steps (Clarke *et al.*, 1991). Recently, the genes coding for the enzymes responsible for the assembly of the polyisoprenoid side chain and its subsequent transfer to 4-hydroxybenzoate have been identified in such mutants (Ashby and Edwards, 1990; Ashby *et al.*, 1992).

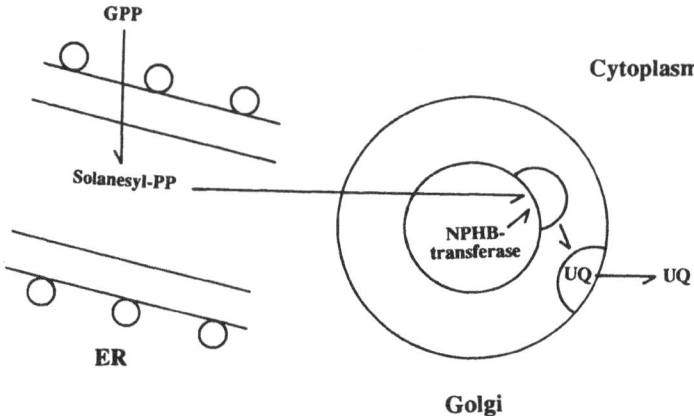

FIGURE 12. Localization of ubiquinone biosynthesis in rat liver. UQ = ubiquinone; NPHB = nonaprenyl-4-hydroxybenzoate.

The proposed pathway for ubiquinone synthesis in rat hepatocytes is depicted in Figure 12. Solanesyl-PP synthesized in the ER is transferred to the Golgi region and final synthesis of ubiquinone is probably associated with a specific, as-yet-unidentified compartment of this membrane system. Experiments both *in vitro* and *in vivo* indicate that the transferase reaction is associated with the inner surface of vesicles, with the final product being translocated during completion to the cytoplasmic surface. This localization requires an active cytoplasmic mechanism for redistribution of the lipid to various cellular membranes. In perfused liver, newly synthesized ubiquinone is discharged to the circulation in association with lipoproteins, a finding that further supports the localization of its biosynthesis to the ER–Golgi region (Elmberger *et al.*, 1989). Future studies designed to determine whether other organelles are also involved in the biosynthesis of this lipid are required.

10. PROTEIN ISOPRENYLATION

Covalent binding of isoprenoids to proteins was described by Glomset and his co-workers in 1984, but interest in this process increased dramatically in 1989, when it was demonstrated that *ras*-oncogenic products require isoprenylation for activation (Schmidt *et al.*, 1984; Hancock *et al.*, 1989). The isoprenoid moiety was identified as farnesol or geranylgeraniol and it has now been established that a large number of proteins in all cells are isoprenylated (Farnsworth *et al.*, 1990; Rilling *et al.*, 1990). There are considerable tissue differences in protein isoprenylation, brain being the most active tissue in this respect.

The sequence of events is shown in Figure 13, which illustrates the reactions involved in *ras*-protein activation. Isoprenylation is started by the establishment of a thioether linkage between a cysteine residue on the protein and the isoprenoid residue, which is a cytoplasmic process. The cysteine participating in this reaction is located close to the C-terminal end of the protein and is followed by three amino acids, two aliphatic and an additional residue that, it was originally believed, could be any amino acid (CAAX). It has, however, now been established that this terminal amino acid is of decisive importance: when it is leucine, geranylgeraniol is added to the cysteine, and when this amino acid is serine or methionine, the isoprenoid moiety added is farnesol.

The next step in isoprenylation is removal of the last three amino acids, which is followed by methylation of the cysteine residue, now at the C-terminal. The final reaction in the case of some *ras* proteins is the palmitoylation of a cysteine farther upstream. The protein is now activated and associates with a receptor at the inner surface of the plasma membrane. As many as 40–60 different proteins per cell may be isoprenylated, but only a few of these have been identified (Maltese, 1990).

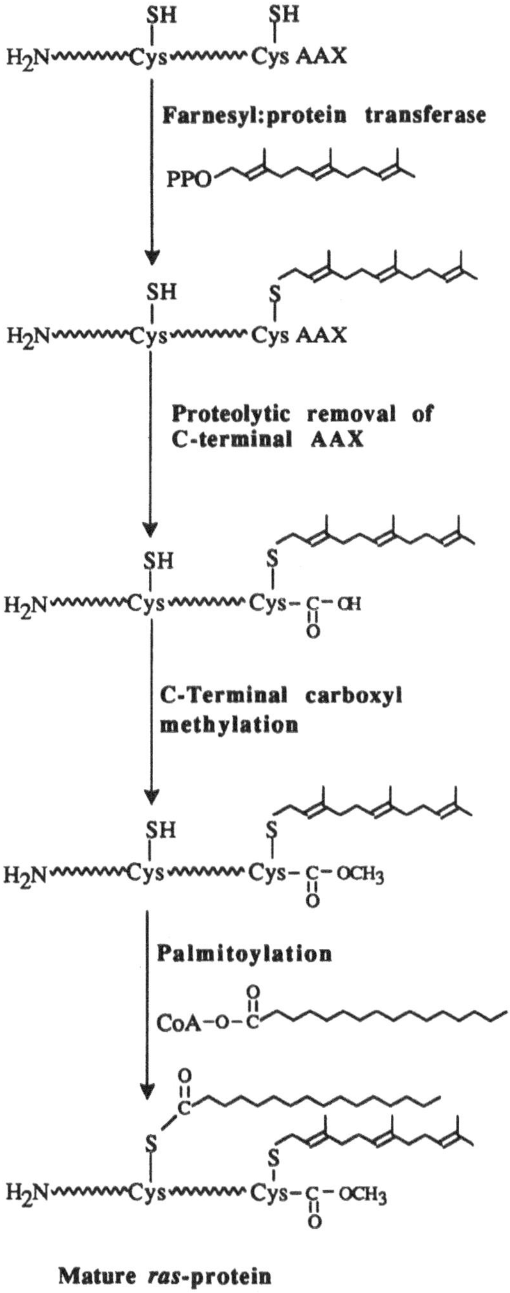

Mature *ras*-protein

FIGURE 13. Isoprenylation of *ras* proteins.

11. CENTRAL AND TERMINAL REGULATION

Cholesterol, dolichol, and ubiquinone are continuously synthesized and removed from tissues and blood, and their cellular levels vary broadly. The half lives of these lipids in liver are around 100 h (Turley, and Dietschy, 1988; Edlund *et al.*, 1988; Andersson *et al.*, 1990).

Obviously, cellular functions require two types of regulation in this connection. First, it is important for the cell to have adequate amounts of different substrates for various functions. Second, the differential requirements for the end products of the mevalonate pathway do not allow a common, central regulation of the overall pathway. An increase or decrease in cholesterol synthesis, metabolism, or amount is not usually paralleled by similar changes with respect to the other end products. A large body of data clearly establish that the major central regulatory step in the biosynthesis of cholesterol is at the level of HMG-CoA reductase. A large number of investigations have demonstrated that the reductase is highly regulated and at several levels.

1. Transcriptional regulation is regarded as the most important mechanism for influencing the enzyme activity and has been intensively investigated (Clarke *et al.*, 1987b; Edwards, 1991). Several sequences within the nontranscribed region of the gene have been shown to be involved in gene activation or inhibition by different agents, for which the direct mediator(s) seems to be sterols or sterol metabolites, e.g., oxysterols (Osborne *et al.*, 1987, 1988). Transcriptional control is involved in the diurnal rhythm in reductase activity and in the increase in activity observed after administration of cholestyramine or mevinolin. It has also been established that coordinated transcriptional control of HMG-CoA synthase, HMG-CoA reductase, FPP-synthase, and the LDL receptor occurs (Rosser *et al.*, 1989; Goldstein and Brown, 1990).

2. HMG-CoA reductase can also be regulated at the translational level (Peffley and Sinensky, 1985; Nakanishi *et al.*, 1988; Simonet and Ness, 1989). Most investigations in this field are restricted to analyses of mRNA levels and corresponding enzyme amounts. It has been concluded that changes in translational efficiency of mRNA probably are involved in enzyme modulation during the diurnal cycle, after administration of cholestyramine, compactin, or mevinolin, and after administration of LDL-cholesterol. In these cases, however, it may be difficult to differentiate between several possible alternatives, such as increased synthesis of mRNA, stabilization of mRNA, or changes in translational efficiency.

3. An increased amount of protein is observed when protein synthesis is activated. On the other hand, the amount of protein may also be modulated by changing the rate of protein degradation (Edwards *et al.*, 1983; Chun and Simoni, 1992). It was found that degradation of the reductase is decreased upon

treatment with mevinolin or compactin, whereas the rate of degradation is stimulated by the administration of oxysterols, LDL-cholesterol, and mevalonate. The seven transmembrane sequences of HMG-CoA reductase are considered to be involved in regulating the degradation process.

4. Phosphorylation/dephosphorylation is considered to be a possible mechanism for modulating the level of active reductase enzyme, where the phosphorylated protein is inactive (Beg et al., 1987; Clarke and Hardie, 1990). The physiological significance of this regulatory mechanism is not known, but the general opinion is that initial, rapid responses can be obtained in this manner, whereas long-term effects are elicited by regulating the amount of enzyme protein. It has been suggested that mevalonate and glucagon treatment initiate phosphorylation, while insulin administration causes dephosphorylation of the reductase.

5. It has also been proposed that the sterol content of ER membranes influences reductase activity (Mitropoulos and Venkatesan, 1985). Increasing the cholesterol concentration could cause inhibition of the reductase. It is at present difficult to decide whether this effect of cholesterol reflects primarily effects on membrane stability and permeability or a direct interaction with the enzyme.

6. It has also been proposed that a nonsterol product derived from mevalonate may be involved in repression of the reductase activity, probably by inhibiting the translational process (Popjak et al., 1985; Panini et al., 1989). Such nonsterol regulatory factors might be isoprenylated peptides, proteins, and/or nucleotides.

HMG-CoA reductase obviously plays a central role in the regulation of cholesterol biosynthesis. Total inhibition of this enzyme also inhibits the formation of other products of the mevalonate pathway, such as dolichol, ubiquinone, isoprenylated tRNA, and isoprenylated proteins. In in vivo systems, however, total inhibition is never obtained and these experiments also demonstrate that HMG-CoA reductase does not participate in the regulation of lipid products other than cholesterol in vivo (James and Kandutsch, 1980; Panini et al., 1985). After treatment with various compounds, including cholesterol, drugs, and inducers, it is commonly observed that cholesterol synthesis is decreased or increased, while the synthesis of other mevalonate lipids is unchanged or regulated in the opposite direction (Kalén et al., 1990b). Most studies have concluded that the first committed enzymes in dolichol and ubiquinone biosynthesis, as well as in protein prenylation, have much higher affinities for the branch-point intermediate, i.e., FPP, than squalene synthase (Faust et al., 1979; Reiss et al., 1991; Ericsson et al., 1992a) (Figure 14). Even if the FPP concentration is decreased to a very large extent, far below the K_m for squalene synthase, this concentration is still sufficient for saturation of the other FPP-utilizing enzymes at this branch point. These findings indicate the necessity for peripheral regulation of the synthesis of mevalonate pathway lipids other than cholesterol. We have, however, no information as to what the rate-limiting steps in the individual branches may be.

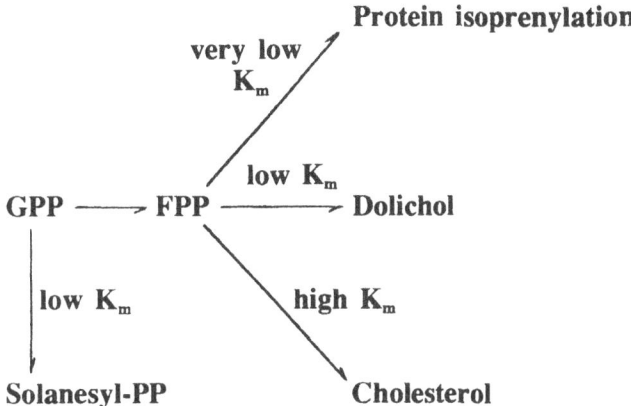

FIGURE 14. Affinities of branch-point enzymes in the mevalonate pathway for their substrates.

12.   THE EFFECTS OF INHIBITORS

One approach to studying biosynthesis and metabolic regulation is the use of enzyme inhibitors. In the case of the mevalonate pathway, only a few well-defined inhibitors, affecting HMG-CoA reductase and cholesterol synthesis, are known but no inhibitors of ubiquinone and dolichol biosynthesis or protein-isoprenylation are available (Figure 15). The two most widely used inhibitors of HMG-CoA reductase, i.e., compactin (mevastatin) and mevinolin (lovastatin), are of fungal origin (Brown *et al.*, 1978; Alberts, 1988). These compounds differ only by a methyl group on the hexahydronaphthalene ring (Figure 16). Microbial metabolism of compactin gives rise to pravastatin, which is employed to reduce

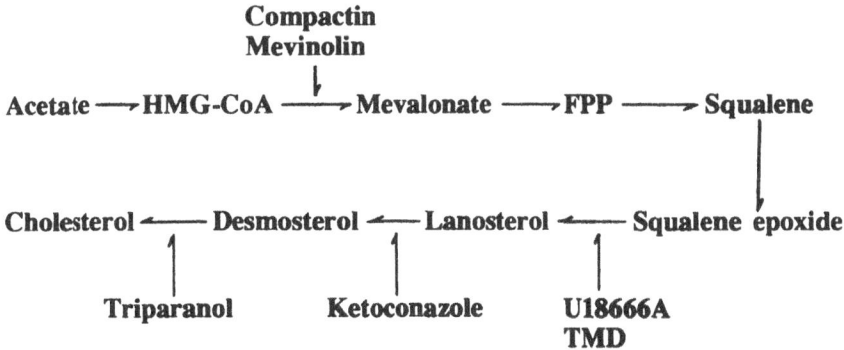

FIGURE 15. Inhibitors of the mevalonate pathway.

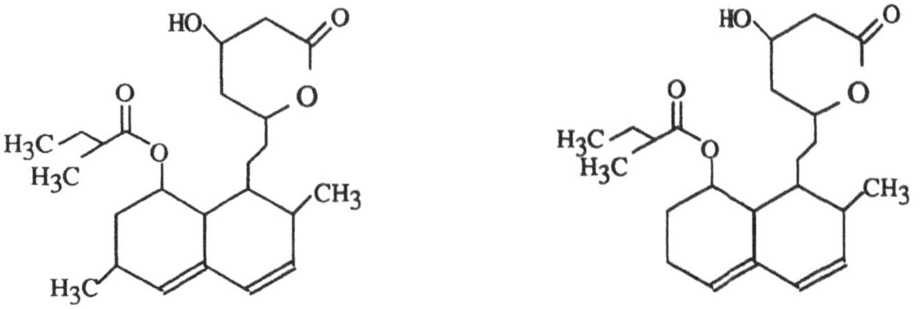

Lovastatin   Compactin
(Mevinolin)   (Mevastatin)

FIGURE 16. The structures of compactin and mevinolin.

high blood levels of cholesterol. Similarly, the modification of mevinolin to
simvastatin increases the efficiency of this inhibitor. In order to act as inhibitors
of the HMG-CoA reductase, both compactin and mevinolin must be converted,
chemically or enzymatically, from the inactive lactone form to the open acid
form. The active forms of these inhibitors resemble the structure of the half-
reduced intermediate form of HMG-CoA. These compounds are very potent
competitive inhibitors of the HMG-CoA reductase. Lovastatin has an inhibition
constant of 6.4×10^{-10} M, which is much lower than the K_m value for HMG-
CoA.

Inhibition of HMG-CoA reductase activity results in a cascade of events: the
decreased cholesterol synthesis results in an increase in LDL-receptor activity
and, at the same time, the synthesis of HMG-CoA reductase is upregulated (Chin
et al., 1982; Li et al., 1988). This latter event has no direct functional consequences as long as the inhibitor is present in the system. It has also been
established that other enzymes of the mevalonate pathway are also affected to
some extent and in an experimental system the dolichol level in the liver increased, while that of ubiquinone decreased following mevolin treatment (Löw
et al., 1992).

Beyond the branching point at the level of FPP, cholesterol synthesis can be
inhibited at the level of 2,3-squalene oxide : lanosterol cyclase with the compounds U18666A and TMD (Panini et al., 1984; Rudney and Sexton, 1986).
Ketoconazole is an inhibitor of cytochrome P-450-mediated reactions and inhibits cholesterol synthesis at the level of lanosterol demethylase, since this is the
first step requiring this cytochrome (Borgers, 1980; van den Bossche et al.,
1980). The last step in cholesterol synthesis is the reduction of desmosterol and
the enzyme involved here can be inhibited by triparanol (Steinberg et al., 1961).
Because of the toxic effects of the other compounds in humans, only HMG-CoA

reductase inhibitors have so far been used as clinical drugs. Development of drugs affecting cholesterol biosynthesis subsequent to FPP for human use is an important task, in order to avoid undesirable side effects on the synthesis of other end products of the mevalonate pathway. These future inhibitors must also inhibit oxysterol production in order to avoid secondary downregulation of the reductase. The effects of inhibitors described above concern primarily the microsomal mevalonate pathway. Both cholesterol and dolichol are also synthesized in peroxisomes by separate enzyme systems, and it is quite possible that the peroxisomal processes are influenced differently by these same compounds.

13. FUNCTIONAL ASPECTS

13.1. Cholesterol

In most membranes neutral lipids are minor components in comparison with phospholipids. Cholesterol, however, is one of the major lipids in several membranes, such as in erythrocytes, myelin, and plasma membranes (Myant, 1981). Our understanding of the role of cholesterol in determining membrane structure is mainly based on experiments with model membranes and cholesterol is considered to be one of the major factors giving the membrane form and stability (Table II). The stabilizing effect of this lipid also involves decreased fluidity of phospholipid fatty acids and decreased membrane permeability (Gibbons *et al.,* 1982). However, in spite of the intensive investigations in this field, it is still difficult to evaluate and quantitate the effects of cholesterol on membranes.

Certain integral membrane proteins, which are dependent on the presence of phospholipids for activity, exhibit modified activity in the presence of cholesterol. Na^+/K^+-ATPase, Ca^{2+}-ATPase, and HMG-CoA reductase activities are inhibited by cholesterol, whereas ATP/ADP exchange and microsomal cholesterol esterification are activated by this lipid. These effects are not necessarily caused by direct interaction with the enzymes, but changes in membrane structure may be the factor leading to modified enzyme function.

Table II
Functions of Cholesterol and Dolichol in Membranes

Cholesterol	Dolichol
Decreases fluidity and permeability	Increases fluidity and permeability
Increases stability	Decreases stability
Enzyme regulation	Fusogenic agent
Precursor for steroid hormones	
Precursor for bile acids	

Cholesterol is the common precursor for the formation of numerous compounds participating in normal physiological functions. Steroid hormones are produced from cholesterol by endocrine cells at various locations. Cholesterol is also the substrate for bile acid synthesis in the liver, a synthesis that supplements the bile acid pool which is recirculated. Furthermore, cholesterol is one of the major components of lipoproteins involved in lipid transport and redistribution. The cellular requirement for cholesterol is high and in humans as much as one-third of the cholesterol supply originates from dietary uptake.

13.2. Dolichol

Our knowledge about the function of dolichol is limited. Experiments with model membranes have led to the conclusion that the effects of dolichol on membranes are opposite those of cholesterol, probably because of the localization of the former lipid in the central, hydrophobic region of the bilayer (de Ropp and Troy, 1985; Valtersson et al., 1985; Schutzbach et al., 1987). In model membranes, dolichol decreases stability and increases fluidity and membrane permeability for small molecules. It was also found that longer-chain dolichols have more pronounced effects on membranes than the shorter species. Studies on liposomes and liposomal–microsomal systems have established that dolichols have a fusogenic effect.

The exact biological significance of these findings is not known, but it is remarkable that the highest dolichol concentrations are found in lysosomal, Golgi, and plasma membranes. It is now well established that continuous fusion and membrane exchange occur in these compartments as part of cellular transport processes. The extremely high content of dolichol in human endocrine organs may perhaps be explained by the high rates of membrane fusion and circulation associated with hormone transport and secretion.

In contrast to cholesterol, no metabolic function has been assigned to dolichol. The presence of this lipid in all biological membranes, its de novo biosynthesis in all tissues, its relatively high rate of turnover, and the absence of dolichol excretion to any large extent suggest that as-yet-unidentified, metabolically active products are formed from dolichol (Elmberger et al., 1988). The isolation of mutants that are defective in dolichol biosynthesis or that overexpress dolichol metabolites will be useful in identifying the metabolic function(s) of dolichol in the future (Stoll et al., 1988).

Upon analyzing bile, one can identify bile acids, phosphatidylcholine, unesterified cholesterol and dolichol (Figure 17). There are two pools of bile acids in the liver: one is synthesized de novo and the second is the recirculated pool, which is taken up from the blood by the liver and transported to the bile ducts (Björkhem, 1985; Coleman and Rahman, 1992). The bile acids taken up from the circulation are conjugated in the ER before secretion. It has been proposed

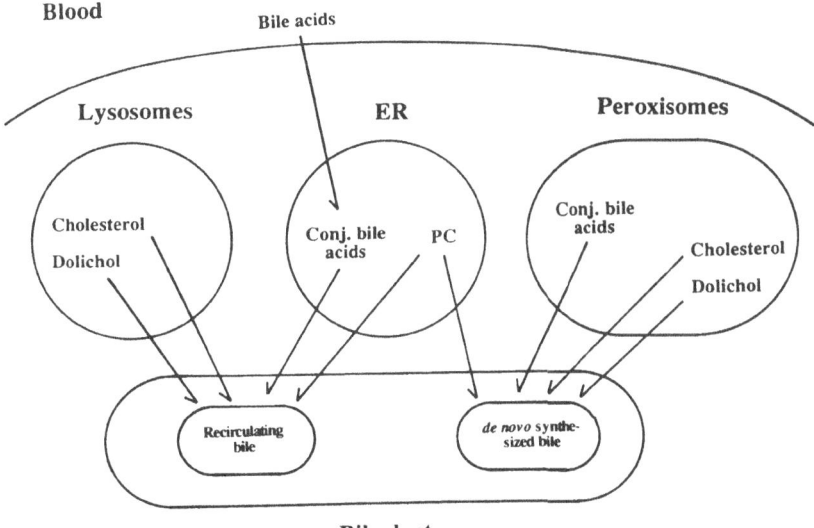

FIGURE 17. Proposed function for peroxisomal cholesterol and dolichol. Conj., conjugated; PC, phosphatidylcholine.

that a part of the cholesterol in the bile originates from lipoproteins taken up and degraded by lysosomes. Since lysosomes contain a large amount of dolichol, it is possible that a part of this lipid is also discharged to the bile.

The bile also contains newly synthesized bile acids, produced by biosynthetic pathways initiated in the ER membrane and terminating in peroxisomes. It has also been demonstrated that bile acids may be converted into their taurine and glycine conjugates by transferases present in peroxisomes (Kase and Björkhem, 1989). *In vivo* experiments have shown that unesterified cholesterol and dolichol, synthesized in peroxisomes, disappear rapidly from this organelle (Appelkvist and Kalén, 1989). These events suggest that bile acids synthesized *de novo* and then conjugated, are associated with newly synthesized cholesterol and dolichol in peroxisomes and these lipids are then transported to the bile ducts for secretion. A detailed understanding of the exact mechanism of bile formation and, especially, the role of peroxisomes in this secretion does, however, require further investigation.

13.3.   Dolichyl Monophosphate

The main or exclusive function of dolichyl-P is considered to be its participation in the biosynthesis of oligosaccharide chains linked N-glycosidically to proteins (Struck and Lennarz, 1980; Kornfeld and Kornfeld, 1985; Hirschberg and Snider, 1987). This type of protein glycosylation is the major form in most cells and is of vital importance for the transport and function of many proteins. Dolichyl-P is present in both the ER and lysosomes, but oligosaccharide synthesis involving dolichyl-P appears to occur exclusively in the ER. The oligosaccharide chain assembled on dolichyl-PP is transferred to an Asn (in the sequence Asn-X-Thr/Ser) in the polypeptide cotranslationally within the ER membrane. The protein-bound core oligosaccharides are processed and completed during transport in the ER–Golgi system, but none of these latter processes involves the participation of dolichyl-P.

The process of oligosaccharide assembly has been established in detail (Figure 18). One dolichyl-P molecule accumulates an oligosaccharide chain composed of 14 sugars, with the involvement of a number of ER-bound glycosyltransferases. The substrates for these transferases are either nucleotide-activated monosaccharides or monosaccharides bound to dolichyl-P.

Dolichyl-P also participates in glycosylation reactions not related to N-glycosylation, namely, the biosynthesis of Pl anchors (Menon *et al.*, 1990; see Tartakoff, this volume). It is well established that a number of proteins in cells are bound to membranes through Pl anchors, whose structure has been characterized (Low, 1989) (Figure 19). This structure is composed of membranous phosphatidylinositol bound to an oligosaccharide moiety consisting, usually, of three mannose residues and one nonacetylated glucosamine residue. This core is associated, through an ethanolamine phosphate moiety, with the C-terminus of the protein. There are a number of known variations with respect to the structure of the oligosaccharide core. Furthermore, the lipid moiety is specific for the individual protein, and variations in the types and binding of the fatty acids, giving rise to lipids of the alkenyl, alkyl, or acyl types, have been identified. Dolichyl-P-mannose is the sugar donor in the assembly of the oligosaccharide

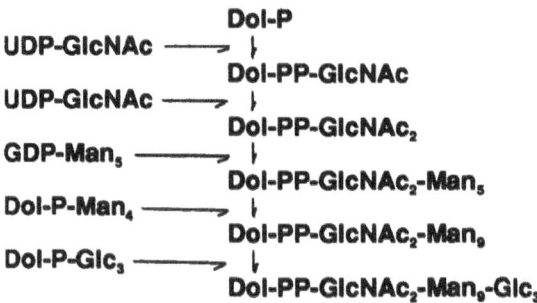

FIGURE 18. Biosynthesis of core oligosaccharides associated with dolichyl-P. Dol-P = dolichyl-P; GlcNAc = N-acetylglucosamine; Man = mannose; Glc = glucose.

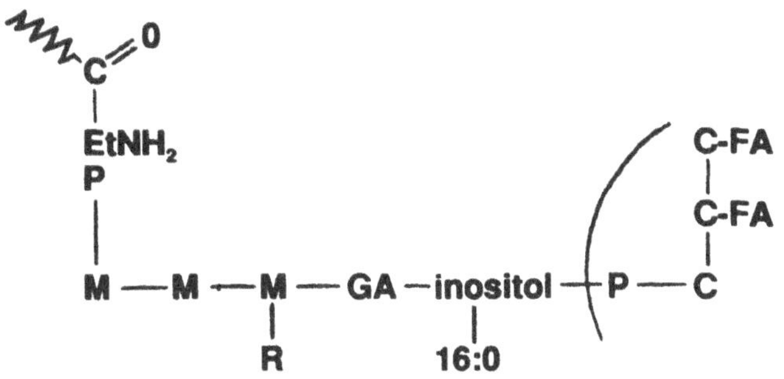

FIGURE 19. Structure of phosphatidylinositol anchor. EtNH$_2$-P = ethanolamine phosphate; M = mannose; GA = glucosamine; FA = fatty acid; R = substituents such as ethanolamine phosphate, galactose (1 to 4) or more complex oligosaccharides.

core and, consequently, an obligatory intermediate in the biosynthesis of this membrane anchor.

The role of dolichyl-P in glycosylation processes may be twofold: first, the binding of substrate to the membrane close to the active site of the glycosyltransferase involved and, second, translocation of the sugar from the outside (cytosolic) to the inside (lumenal) surface of the ER membrane, where a major part of the oligosaccharide synthesis takes place.

The nature and amount of dolichyl-P present in the membrane also influence the glycosylation process (Table III). α-Unsaturation greatly decreases or completely eliminates glycosylation in eukaryotic cells, whereas the length of the dolichyl-P molecule seems to be of no importance. Only dolichyl-P in the S form is utilized and the number of *cis* and *trans* double bonds present is probably of little importance. Part of the dolichyl-P is covalently bound to proteins, but it is not known how this linkage influences the glycosylation process. The level of

Table III
Influence of Dolichyl Phosphate
on Glycoprotein Synthesis

Parameter	Effect on glycoprotein synthesis
α-Unsaturated	Not utilized
Isoprene length (17–22)	None
S and R forms	S form is preferred
Number of *cis* and *trans* residues	Probably not decisive
Covalent linkage	Not investigated
Amount	Rate-limiting

dolichyl-P in the membrane is probably an important regulator of the glycosylation process.

13.4. Ubiquinone

Ubiquinone is a well-known component of the mitochondrial respiratory chain and its role in electron transport has been well established. This lipid is present in great excess relative to the cytochromes and the flavoprotein. Interestingly and unexpectedly, complete restoration of respiration in submitochondrial particles depleted of neutral lipids, including ubiquinone, requires replacement of the original amount of ubiquinone.

In recent years the role of ubiquinone as an antioxidant has received both new interest and experimental support. It is now clear that this lipid in reduced form (ubiquinol) functions as an antioxidant, inhibiting lipid peroxidation in model systems, as well as in biological membranes *in vitro* and *in vivo* (Beyer *et al.*, 1987; Frei *et al.*, 1990). Concerning its mode of action in this regard, there are several suggestions, including the hypothesis that it may act by reducing the superoxide radical ($O_2^-\bullet$) or perferryl radical ($Fe^{3+}-O_2^-\bullet$). It is also possible that ubiquinol may quench carbon-centered lipid radicals (L•) or lipid peroxyl radicals (LOO•). Additionally, its action may be explained in part by its involvement in reduction of the α-tocopheroxyl radical (vitamin E–O•).

The uniqueness of ubiquinone as an antioxidant resides in the fact that it is synthesized endogenously in all mammalian tissues and, consequently, in contrast to all other antioxidants, is not dependent on dietary supply (Table IV). In human and rat tissues this lipid is present mostly in the reduced form, required for antioxidant function. Its reduction in mitochondria is mediated by the respiratory chain but it is not clear what mechanism is responsible for this reduction in other intracellular membranes. It has a high scavenger capacity and its action is facilitated by its solubility in membranes. The living cell uses ubiquinone as an antioxidant preferentially to α-tocopherol and ubiquinone is also present at tenfold higher levels than is vitamin E.

Table IV
The Antioxidant Properties of Ubiquinone

Endogenously produced
Endogenous reactivation
Membrane soluble
Scavenger capacity: 1.1 radical/molecule
Preferential utilization
Tenfold more UQ than α-tocopherol

Table V
Effect of Drugs and Chemicals on the Synthesis
of Mevalonate Pathway Lipids

Treatment	Cholesterol	Dolichol	Ubiquinone
Cholesterol	Decreased	Unchanged	Increased
Cholestyramine	Increased	Decreased	Decreased
Clofibrate	Decreased	Increased	Increased
Diethylhexylphthalate	Decreased	Increased	Increased
Phenobarbital	Increased	Increased	Increased
Nitrosodiethylamine	Decreased	Increased	Increased

14. EFFECTS OF INDUCERS AND DRUGS

Biosynthesis, turnover, and levels of mevalonate lipids can be influenced to a great extent by treatment of experimental animals with various drugs and membrane inducers such as clofibrate (peroxisomes) and phenobarbital (ER). This approach can be used to obtain important information about the regulation of the metabolism of these lipids and such treatments may also be used to selectively increase or decrease the biosynthesis of individual lipids.

Table V presents a few examples where [^{3}H]mevalonate incorporation into liver lipids of treated rats was analyzed (Potter and Kandutsch, 1982; Keller and Nellis, 1986; Edlund et al., 1987; Kalén et al., 1990b). These types of experiments do not provide information about individual biosynthetic enzymes, but serve as an estimate of the overall biosynthetic capacity. The treatments listed are well established as being lipid modulators or membrane inducers. It is strikingly clear from these studies, as well as from a number of others, that the biosynthesis of cholesterol is regulated independently from that of dolichol and ubiquinone and that these processes are often affected in opposite manners by the same treatment. Obviously, peripheral regulation of the biosynthesis of the mevalonate pathway lipids is an important complement to the central regulation of HMG-CoA reductase.

15. AGING AND DISEASE

During the aging process, both in experimental animals and in humans, membrane lipid compositions display characteristic patterns of change. Table VI shows the lipid content of human pancreas as a function of age, but the pattern is similar in other human organs and animal tissues (Pullarkat and Reha, 1982; Kalén et al., 1989; Daniels and Hemming, 1990). The levels of cholesterol and

Table VI
Distribution of Lipids in Human Pancreas during Aging[a]

Lipid	Age				
	2 days	2 years	20 years	41 years	79 years
Cholesterol[b]	3	2	2	2	2
Phospholipid[b]	7	9	10	11	12
Ubiquinone[c]	9	38	21	19	6
Dolichol[c]	21	18	1117	2033	3058
Dolichyl-P[c]	5	7	8	21	23

[a]Modified from Kalén et al. (1989).
[b]mg/g wet wt.
[c]μg/g wet wt.

phospholipid are largely unaltered during aging. The amount of ubiquinone increases extensively after birth, in humans up to the age of 20–30 years, followed by a continuous decrease, so that at around 80 years of age the amount is as low as in the newborn.

During aging there is an extensive accumulation of free dolichol in all tissues, which can be as much as 150-fold during a lifetime of 80 years. Significantly, the amount of dolichyl-P is elevated to only a limited extent. The reason for this increase in dolichol concentration is not known but the increase and subsequent decrease in ubiquinone concentration is probably associated with the capacity of the cell to detoxify free radicals and reactive oxygen metabolites.

The involvement of cholesterol in many human pathological conditions is well known. Disturbances in uptake, synthesis, and discharge of cholesterol are characteristic of several familiar lipid metabolic disturbances. These types of disturbances are also considered to be the main factors inducing atherosclerotic diseases and chronic heart failure. The decrease in ubiquinone concentration observed in several heart and skeletal muscle diseases is considered to be an important factor in the etiology of these conditions (Yamamura, 1985; Folkers, 1990).

In experimentally induced preneoplastic noduli in rodent liver, the levels of dolichol and ubiquinone increase, that of dolichol-P is unchanged, while the cholesterol amount is decreased (Olsson et al., 1991) (Table VII). In developed hepatocellular cancer, on the other hand, the level of dolichol is decreased drastically, while that of dolichyl-P remains constant. In this same process, the ubiquinone concentration is decreased by 50%, while the cholesterol concentration is increased almost threefold (Eggens et al., 1989).

In Alzheimer's disease of the brain, the dolichol concentration is decreased, while that of dolichyl-P is increased (Söderberg et al., 1992). In this case no large changes are observed in the concentrations of cholesterol or ubiquinone.

Table VII

Mevalonate Pathway Lipids in Pathological Conditions

	% of control				
Lipid	Preneoplastic noduli[a]	Hepatocellular cancer[b]	Alzheimer's disease[c]	Niemann–Pick disease, type C[d]	Prion disease[f]
Dolichol	168	15	56	115	67
Dolichyl-P	105	110	172	250	158
Ubiquinone	216	51	130	65	180
Cholesterol	62	260	102	400	103

[a]Rat liver. [b]Human liver. [c]Human brain. [d]Mouse liver. [f]Mouse brain.

The main characteristic of Niemann–Pick type C and D diseases is an elevation of cholesterol and dolichyl-P concentrations and a decrease in the level of ubiquinone (Pentchev et al., 1987; Schedin, Ericsson, Pentchev, and Dallner, unpublished observations). In prion disease the brain dolichol content is decreased by 50% while the level of dolichyl-P is elevated.

Analysis of such data may be useful in understanding the etiology of such conditions, as well as in establishing the functions of these lipids under normal conditions. Increasing protection against free radicals may be required in preneoplastic noduli, whereas in developed cancer the antioxidant capacity of ubiquinone is exhausted. Similarly, the extensive production of β-amyloid (pathological glycoprotein) in Alzheimer's disease is associated with an increase in the obligatory intermediate of glycoprotein synthesis, dolichyl-P.

16. FUTURE DIRECTIONS

The metabolism and function of products of the mevalonate pathway are of great importance at several levels of cellular life and, consequently, one may raise a number of important questions concerning these processes. Taking into consideration contemporary research in this field, the problems discussed, and the main lines of interest, it is possible to pose the following questions.

The ER was previously considered to be the exclusive or main site of lipid synthesis. Recent investigations have shown that a considerable portion of such synthetic processes are associated with peroxisomes and Golgi vesicles (Figure 20). The study of cooperation and interaction of the ER with other cellular compartments in connection with lipid biosynthesis will be an important task for the future. It appears that the biosynthetic systems at different locations have different organizations and may also be regulated independently. It will probably turn out that the same lipids synthesized in different compartments have different functions.

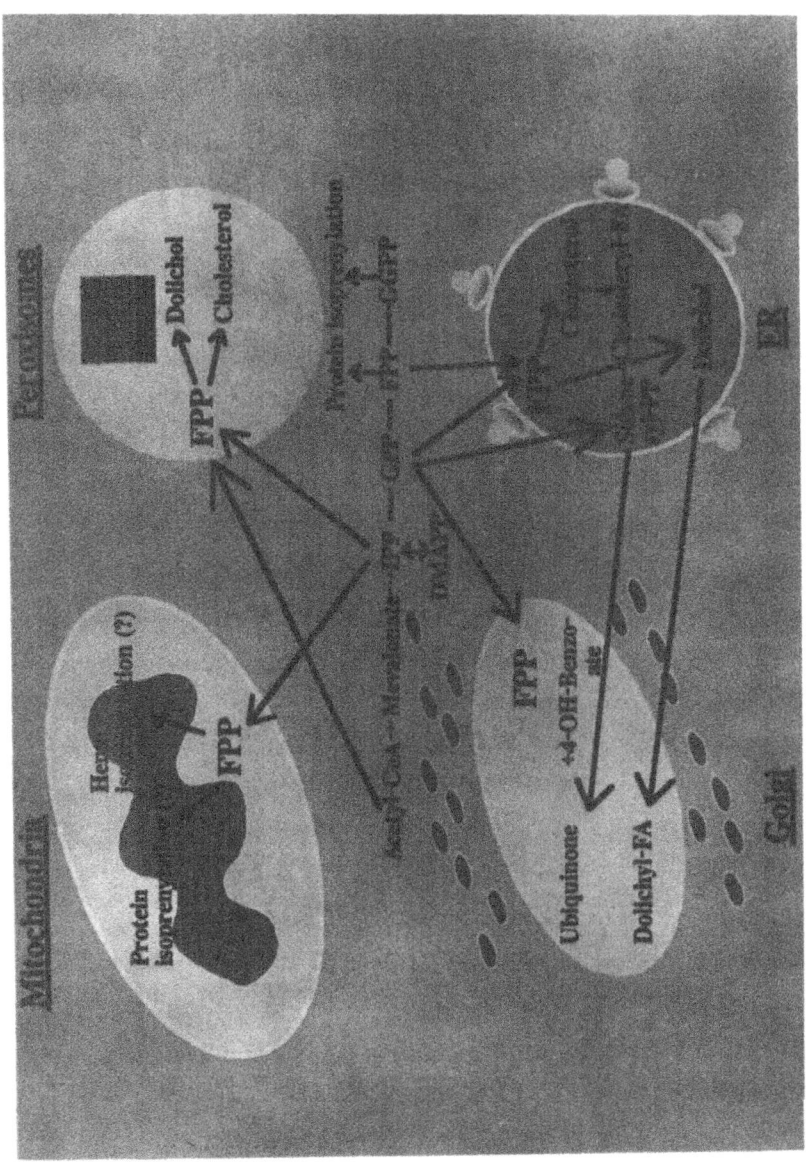

FIGURE 20. Biosynthesis of mevalonate pathway lipids—interactions among organelles.

According to established concepts, HMG-CoA reductase is the main regulatory enzyme in the biosynthesis of mevalonate pathway lipids. It becomes, however, more and more clear that peripheral and individual regulation of the levels of these lipids is indispensable for normal cellular life. Since the various products have different properties, as well as functions, it is quite probable that a general increase or decrease in the biosynthesis of all mevalonate pathway products may have negative consequences for the cell. Drugs used to reduce high cholesterol levels must act selectively by inhibiting cholesterol synthesis, without influencing cellular dolichol or ubiquinone content. It appears that selective regulatory mechanisms are operating, but our knowledge concerning this selectivity is very limited. The understanding of these regulatory processes will require a large amount of effort in the future, involving *in vivo* systems and biosynthetic mutants, as well as enzyme purification.

In spite of intensive investigations in this century, the functional role(s) of these lipids is only partly understood. Protein isoprenylation and the antioxidant role of ubiquinone are very recent findings. The generally accepted regulation of membrane stability and fluidity by these lipids is based on studies with model membranes, without actually demonstrating their influence on biological membranes. In spite of the high turnover of dolichol, no breakdown products of polyisoprenes have been identified and it is very possible that important active metabolites or hormones formed from dolichol will be identified in the future. We can expect to identify a number of new functions for these lipids when new techniques have been developed and our biological concepts become more sophisticated.

17. REFERENCES

Åberg, F., Appelkvist, E.-L., Dallner, G., and Ernster, L., 1992, Distribution and redox state of ubiquinones in rat and human tissues, *Arch. Biochem. Biophys.* **295**:230–234.

Adair, W. L., and Keller, R. K., 1982, Dolichol metabolism in rat liver: Determination of the subcellular distribution of dolichyl phosphate and its site and rate of de novo biosynthesis, *J. Biol. Chem.* **257**:8990–8996.

Alberts, A. W., 1988, Discovery, biochemistry and biology of lovastatin, *Am. J. Cardio.* **62**:10J–15J.

Allen, C. M., Kalin, J. R., Sack, J., and Verizzo, D., 1978, CTP-dependent dolichol phosphorylation by mammalian cell homogenates, *Biochemistry* **17**:5020–5026.

Anderson, M. S., Yarger, J. G., Burcke, C. L., and Poulter, C. D., 1989, Farnesyl diphosphate synthetase: Molecular cloning, sequence, and expression of an essential gene from Saccharomyces cerevisiae, *J. Biol. Chem.* **264**:19176–19184.

Andersson, M., Elmberger, P. G., Edlund, C., Kristensson, K., and Dallner, G., 1990, Rates of cholesterol, dolichol and dolichyl-P biosynthesis in rat brain slices, *FEBS Lett.* **269**:15–18.

Appelkvist, E.-L., 1987, In vitro labeling of peroxisomal cholesterol with radioactive precursors, *Biosci. Rep.* **7**:853–858.

Appelkvist, E.-L., and Kalén, A., 1989, Biosynthesis of dolichol by rat liver peroxisomes, *Eur. J. Biochem.* **185**:503–509.

Appelkvist, E.-L., Reinhart, M., Fischer, R., Billheimer, J., and Dallner, G., 1990, Presence of individual enzymes of cholesterol biosynthesis in rat liver peroxisomes, *Arch. Biochem. Biophys.* **282**:318–325.

Appelkvist, E.-L., Kalén, A., and Dallner, G., 1991, Biosynthesis and regulation of coenzyme Q, in: *Biomedical and Clinical Aspects of Coenzyme Q* (K. Folkers, G. P. Littarru, and T. Yamagami, eds.), pp. 141–150, Elsevier, Amsterdam.

Ashby, M. N., and Edwards, P. A., 1989, Identification and regulation of a rat liver cDNA encoding farnesyl pyrophosphate synthetase, *J. Biol. Chem.* **264**:635–640.

Ashby, M. N., and Edwards, P. A., 1990, Elucidation of the deficiency in two yeast coenzyme Q mutants: Characterization of the structural gene encoding hexaprenyl pyrophosphate synthetase, *J. Biol. Chem.* **265**:13157–13164.

Ashby, M. N., Kutsunai, S. Y., Ackerman, S., Tzagoloff, A., and Edwards, P. A., 1992, COQ2 is a candidate for the structural gene encoding parahydroxybenzoate:polyprenyltransferase, *J. Biol. Chem.* **267**:4128–4136.

Åstrand, I.-M., Fries, E., Chojnacki, T., and Dallner, G., 1986, Inhibition of dolichyl phosphate biosynthesis by compactin in cultured rat hepatocytes. *Eur. J. Biochem.* **155**:447–452.

Baba, T., Morris, C., and Allen, C. M., 1987, Dehydrodolichyl diphosphate synthetase from rat seminiferous tubules, *Arch. Biochem. Biophys.* **252**:440–450.

Beg, Z. H., Stonik, J. A., and Brewer, H. B., Jr., 1987, Phosphorylation and modulation of the enzymatic activity of native and protease-cleaved purified hepatic 3-hydroxy-3-methylglutaryl-coenzyme A reductase by a calcium/calmodulin-dependent protein kinase, *J. Biol. Chem.* **262**:13228–13240.

Beyer, R. E., Nordenbrand, K., and Ernster, L., 1987, The function of coenzyme Q in free radical production and as an antioxidant: A review, *Chem. Scr.* **27**:145–153.

Billheimer, J. T., and Reinhart, M. P., 1990, Intracellular trafficking of sterols, in: *Intracellular Transfer of Lipid Molecules* (H. J. Hilderson, ed.), pp. 301–331, Plenum Press, New York.

Björkhem, I., 1985, Mechanism of bile acid biosynthesis in mammalian liver, in: *Sterols and Bile Acids* (H. Danielsson and J. Sjövall, eds.), pp. 231–278, Elsevier, Amsterdam.

Borgers, M., 1980, Mechanism of action of antifungal drugs with special reference to the imidazole derivatives, *Rev. Infect. Dis.* **2**:520–534.

Brown, M. S., Faust, J. R., Goldstein, J. L., Kaneko, I., and Endo, A., 1978, Induction of 3-hydroxy-3-methylglutaryl coenzyme A reductase activity in human fibroblasts incubated with compactin (ML-236B), a competitive inhibitor of the reductase, *J. Biol. Chem.* **253**:1121–1128.

Burton, W. A., Lucas, J. J., and Waechter, C. J., 1981, Enhanced chick oviduct dolichol kinase activity during estrogen-induced differentiation, *J. Biol. Chem.* **256**:632–635.

Casey, J., and Threlfall, D. R., 1978, Formation of 3-hexaprenyl-4-hydroxybenzoate by matrix-free mitochondrial membrane-rich preparations of yeast, *Biochim. Biophys. Acta* **530**:487–502.

Chin, D. J., Luskey, K. L., Anderson, R. G. W., Faust, J. R., Goldstein, J. L., and Brown, M. S., 1982, Appearance of crystalloid endoplasmic reticulum in compactin-resistant Chinese hamster cells with a 500-fold increase in 3-hydroxy-3-methylglutaryl coenzyme A reductase, *Proc. Natl. Acad. Sci. USA* **79**:1185–1189.

Chin, D. J., Gil, G., Russell, D. W., Liscum, L., Luskey, K. L., Basu, S. K., Okayama, H., Berg, P., Goldstein, J. L., and Brown, M. S., 1984, Nucleotide sequence of 3-hydroxy-3-methylglutaryl coenzyme A reductase, a glycoprotein of endoplasmic reticulum, *Nature* **308**:613–617.

Chojnacki, T., and Dallner, G., 1983, The uptake of dietary polyprenols and their modification to active dolichols by the rat liver, *J. Biol. Chem.* **258**:916–922.

Chojnacki, T., and Dallner, G., 1988, The biological role of dolichol, *Biochem. J.* **251**:1–9.

Chun, K. T., and Simoni, R. D., 1992, The role of the membrane domain in the regulated degradation of 3-hydroxy-3-methylglutaryl coenzyme A reductase, *J. Biol. Chem.* **267**:4236–4246.

Clarke, C. F., Tanaka, R. D., Svenson, K., Wamsley, M., Fogelman, A. M., and Edwards, P. A., 1987a, Molecular cloning and sequence of a cholesterol-repressible enzyme related to prenyltransferase in the isoprene biosynthetic pathway, *Mol. Cell. Biol.* **7**:3138–3146.

Clarke, C. F., Edwards, P. A., and Fogelman, A. M., 1987b, Cellular regulation of cholesterol metabolism, in: *Plasma Lipoproteins* (A. M. Gotto, Jr., ed.), pp. 261–276, Elsevier, Amsterdam.

Clarke, C. F., Williams, W., and Teruya, J., 1991, Ubiquinone biosynthesis in Saccharomyces cerevisiae: Isolation and sequence of COQ3, the 3,4-dihydroxy-5-hexaprenylbenzoate methyltransferase gene, *J. Biol. Chem.* **266**:16636–16644.

Clarke, P. R., and Hardie, D. G., 1990, Regulation of HMG-CoA reductase: Identification of the site phosphorylated by the AMP-activated protein kinase in vitro and in intact rat liver, *EMBO J.* **9**:2439–2446.

Coleman, R., and Rahman, K., 1992, Lipid flow in bile formation, *Biochim. Biophys. Acta* **1125**:113–133.

Crane, F. L., 1986, Physiological coenzyme Q function and pharmacological relations, in: *Biomedical and Clinical Aspects of Coenzyme Q* (K. Folkers and Y. Yamamura, eds.), pp. 3–14, Elsevier, Amsterdam.

Crick, D. C., Rush, J. S., and Waechter, C. J., 1991, Characterization and localization of a long-chain isoprenyltransferase activity in porcine brain: Proposed role in the biosynthesis of dolichyl phosphate, *J. Neurochem.* **57**:1354–1362.

Daniels, I., and Hemming, F. W., 1990, Changes in murine tissue concentrations of dolichol and dolichol derivatives associated with age, *Lipids* **25**:586–593.

de Ropp, J. S., and Troy, F. A., 1985, ^{2}H NMR investigation of the organization and dynamics of polyisoprenols in membranes, *J. Biol. Chem.* **260**:15669–15674.

Dietschy, J. M., and Spady, D. K., 1984, Measurement of rates of cholesterol synthesis using tritiated water, *J. Lipid Res.* **25**:1469–1476.

Edlund, C., Ericsson, J., and Dallner, G., 1987, Changes in hepatic dolichol and dolichyl monophosphate caused by treatment of rats with inducers of the endoplasmic reticulum and peroxisomes and during ontogeny, *Chem. Biol. Interact.* **62**: 191–208.

Edlund, C., Brunk, U., Chojnacki, T., and Dallner, G., 1988, The half-lives of dolichol and dolichyl phosphate in rat liver, *Biosci. Rep.* **8**:139–146.

Edmond, J., and Popjak, G., 1974, Transfer of carbon atoms from mevalonate to n-fatty acids, *J. Biol. Chem.* **249**:66–71.

Edwards, P. A., 1991, Regulation of sterol biosynthesis and isoprenylation of proteins, in: *Biochemistry of Lipids, Lipoproteins and Membranes* (D. E. Vance and J. Vance, eds.), pp. 383–401, Elsevier, Amsterdam.

Edwards, P. A., Lan, S. F., and Fogelman, A. M., 1983, Alterations in the rates of synthesis and degradation of rat liver 3-hydroxy-3-methylglutaryl coenzyme A reductase produced by cholestyramine and mevinoline, *J. Biol. Chem.* **258**:10219–10222.

Eggens, I., Chojnacki, T., Kenne, L., and Dallner, G., 1983, Separation, quantitation and distribution of dolichol and dolichyl phosphate in rat and human tissues, *Biochim. Biophys. Acta* **751**:355–368.

Eggens, I., Ericsson, J., and Tollbom, Ö., 1988, Cytidine 5′-triphosphate-dependent dolichol kinase and dolichol phosphatase activities and levels of dolichyl phosphate in microsomal fractions from highly differentiated human hepatomas, *Cancer Res.* **48**:3418–3424.

Eggens, I., Elmberger, P. G., and Löw, P., 1989, Polyisoprenoid, cholesterol and ubiquinone levels in human hepatocellular carcinomas, *Br. J. Exp. Pathol.* **70**:83–92.

Ekström, T. J., Chojnacki, T., and Dallner, G., 1984, Metabolic labeling of dolichol and dolichyl phosphate in isolated hepatocytes, *J. Biol. Chem.* **259**:10460–10468.

Ekström, T. J., Chojnacki, T., and Dallner, G., 1987a, The α-saturation and terminal events in dolichol biosynthesis, *J. Biol. Chem.* **262**:4090–4097.

Ekström, T. J., Ericsson, J., and Chojnacki, T., 1987b, Localization and terminal reactions of dolichol biosynthesis, *Chem. Scr.* **27**:39–47.

Elmberger, P. G., Engfeldt, P., and Dallner, G., 1988, Presence of dolichol and its derivatives in human blood, *J. Lipid Res.* **29**:1651–1662.

Elmberger, P. G., Kalén, A., Brunk, U. T., and Dallner, G., 1989, Discharge of newly-synthesized dolichol and ubiquinone with lipoproteins to rat liver perfusate and to the bile, *Lipids* **24**:919–930.

Ericsson, J., Thelin, A., Chojnacki, T., and Dallner, G., 1991a, Characterization and distribution of cis-prenyl transferase participating in liver microsomal polyisoprenoid biosynthesis, *Eur. J. Biochem.* **202**:789–796.

Ericsson, J., Scallen, T. J., Chojnacki, T., and Dallner, G., 1991b, Involvement of sterol carrier protein-2 in dolichol biosynthesis, *J. Biol. Chem.* **266**:10602–10607.

Ericsson, J., Thelin, A., Chojnacki, T., and Dallner, G., 1992a, Substrate specificity of cis-prenyltransferase in rat liver microsomes, *J. Biol. Chem.* **267**:19730–19735.

Ericsson, J., Appelkvist, E.-L., Thelin, A., Chojnacki, T., and Dallner, G., 1992b, Isoprenoid biosynthesis in rat liver peroxisomes: Characterization of cis-prenyltransferase and squalene synthetase, *J. Biol. Chem.* **267**:18708–18714.

Ernster, L., 1977, Facts and ideas about the functions of coenzyme Q in mitochondria, in: *Biomedical and Clinical Aspects of Coenzyme A* (K. Folkers and Y. Yamamura, eds.), pp. 15–21, Elsevier, Amsterdam.

Farnsworth, C. C., Gelb, M. H., and Glomset, J. A., 1990, Identification of geranylgeranyl-modified proteins in HeLa cells, *Science* **247**:320–322.

Faust, J. R., Goldstein, J. L., and Brown, M. S., 1979, Synthesis of ubiquinone and cholesterol in human fibroblasts: Regulation of a branched pathway, *Arch. Biochem. Biophys.* **192**:86–99.

Faust, J. R., Brown, M. S., and Goldstein, J. L., 1980, Synthesis of Δ^2-isopentenyl tRNA from mevalonate in cultured human fibroblasts, *J. Biol. Chem.* **255**:6546–6548.

Ferguson, J. B., and Bloch, K., 1977, Purification and properties of a soluble protein activator of rat liver squalene epoxidase, *J. Biol. Chem.* **252**:5381–5385.

Fisher, K. A., 1976, Analysis of membrane halves: Cholesterol, *Proc. Natl. Acad. Sci. USA* **73**:173–177.

Folkers, K., 1990, Progress in biochemical approaches to clinical therapy with coenzyme Q10, in: *Highlights in Ubiquinone Research* (G. Lenaz, O. Barnabei, A. Rabbi and M. Battino, eds.), pp. 309–322, Taylor & Francis, London.

Frei, B., Kim, M. C., and Ames, B. N., 1990, Ubiquinol-10 is an effective lipid-soluble antioxidant at physiological concentrations, *Proc. Natl. Acad. Sci. USA* **87**:4879–4883.

Fujisaki, S., Hara, H., Nishimura, Y., Horiuchi, K., and Nishino, T., 1990, Cloning and nucleotide sequence of the ispA gene responsible for farnesyl diphosphate synthase activity in Escherichia coli, *J. Biochem.* **108**:995–1000.

Gavey, K. L., Noland, B. J., and Scallen, T. J., 1981, The participation of sterol carrier protein 2 in the conversion of cholesterol to cholesterol ester by rat liver microsomes, *J. Biol. Chem.* **256**:2993–2999.

Gaylor, J. A. L., 1981, Formation of sterols in animals, in: *Biosynthesis of Isoprenoid Compounds* (S. A. L. Spurgeon and J. O. W. Porter, eds.), Vol. 1, pp. 481–544, Wiley, New York.

Gibbons, G. F., Mitropoulos, K. A., and Myant, N. B., 1982, *Biochemistry of Cholesterol*, Elsevier, Amsterdam.

Goldstein, J. L., and Brown, M. S., 1990, Regulation of the mevalonate pathway, *Nature* **343**:425–430.

Hancock, J. F., Magee, A. I., Childs, J. E., and Marshall, C. J., 1989, All ras proteins are polyisoprenylated but only some are palmitoylated, *Cell* **57**:1167–1177.

Hemming, F. W., 1981, Biosynthesis of dolichol and related compounds, in: *Biosynthesis of Isoprenoid Compounds* (S. A. L. Spurgeon and J. O. W. Porter, eds.), Vol. 2, pp. 305–354, Wiley, New York.

Hirschberg, C. B., and Snider, M. D., 1987, Topography of glycosylation in the rough endoplasmic reticulum and Golgi apparatus, *Annu. Rev. Biochem.* **56**:63–87.

Huneeus, V. Q., Wiley, M. H., and Siperstein, M. D., 1980, Isopentenyladenine as a mediator of mevalonate-regulated DNA replication, *Proc. Natl. Acad. Sci. USA* **77**:5842–5846.

James, M. J., and Kandutsch, A. A., 1980, Regulation of hepatic dolichol synthesis by β-hydroxy-β-methylglutaryl coenzyme A reductase, *J. Biol. Chem.* **255**:8618–8622.

Johnston, J. M., 1978, Esterification reactions in the intestinal mucosa and lipid adsorption, in: *Disturbance in Lipid and Lipoprotein Metabolism* (J. M. Dietshy, A. M. Gotto, Jr. and J. A. Ontko, eds.), pp. 57–68, American Physiological Society, Bethesda.

Kalén, A., Norling, B., Appelkvist, E.-L., and Dallner, G., 1987, Ubiquinone biosynthesis by the microsomal fraction from rat liver, *Biochim. Biophys. Acta* **926**:70–78.

Kalén, A., Appelkvist, E.-L., and Dallner, G., 1989, Age-related changes in the lipid composition of rat and human tissues, *Lipids* **24**:579–584.

Kalén, A., Appelkvist, E.-L., Chojnacki, T., and Dallner, G., 1990a, Nonaprenyl-4-hydroxybenzoate transferase, an enzyme involved in ubiquinone biosynthesis, in the endoplasmic reticulum–Golgi system of rat liver, *J. Biol. Chem.* **265**:1158–1164.

Kalén, A., Appelkvist, E.-L., and Dallner, G., 1990b, The effects of inducers of the endoplasmic reticulum, peroxisomes and mitochondria on the amounts and synthesis of ubiquinone in rat liver subcellular membranes, *Chem. Biol. Interact.* **73**:221–234.

Kase, B. F., and Björkhem, I., 1989, Peroxisomal bile acid–CoA:amino acid N-acyltransferase in rat liver, *J. Biol. Chem.* **264**:9220–9223.

Keller, G. A., Barton, M. C., Shapiro, D. J., and Singer, S. J., 1985, 3-Hydroxy-methylglutaryl-coenzyme A reductase is present in peroxisomes in normal rat liver cells, *Proc. Natl. Acad. Sci. USA* **82**:770–774.

Keller, R. K., and Nellis, S. W., 1986, Quantitation of dolichyl phosphate and dolichol in major organs of the rat as a function of age, *Lipids* **21**:353–355.

Keller, R. K., Jehle, E., and Adair, W. L., 1982, The origin of dolichol in the liver of the rat: Determination of the dietary contribution, *J. Biol. Chem.* **257**:8985–8989.

Khan, B., Wilcox, H. G., and Heimberg, M., 1989, Cholesterol is required for secretion of very-low-density lipoprotein by rat liver, *Biochem. J.* **258**:807–816.

Kornfeld, R., and Kornfeld, S., 1985, Assembly of asparagine-linked oligosaccharides, *Annu. Rev. Biochem.* **54**:631–664.

Lenaz, G. (ed.) 1985, *Biochemistry, Bioenergetics and Clinical Applications of Ubiquinone*, Wiley, New York.

Li, A. C., Tanaka, R. D., Callaway, K., Fogelman, A. M., and Edwards, P. A., 1988, Localization of 3-hydroxy-3-methylglutaryl CoA reductase and 3-hydroxy-3-methylglutaryl CoA synthase in the rat liver and intestine is affected by cholestyramine and mevinolin, *J. Lipid Res.* **29**:781–796.

Liscum, L., Finer-Moore, J., Stroud, R. M., Luskey, K. L., Brown, M. S., and Goldstein, J. L., 1985, Domain structure of 3-hydroxy-3-methylglutaryl coenzyme A reductase, a glycoprotein of the endoplasmic reticulum, *J. Biol. Chem.* **260**:522–530.

Low, M. G., 1989, The glycosyl-phosphatidylinositol anchor of membrane proteins, *Biochim. Biophys. Acta* **988**:427–454.

Löw, P., Andersson, M., Edlund, C., and Dallner, G., 1992, Effects of mevinolin treatment on tissue dolichol and ubiquinone levels in the rat, *Biochim. Biophys. Acta* **1165**:102–109.

Maltese, W. A., 1990, Posttranslational modification of proteins by isoprenoids in mammalian cells, *FASEB J.* **4**:3319–3328.

Menon, A. K., Mayor, S., and Schwarz, R. T., 1990, Biosynthesis of glycosylphosphatidylinositol lipids in Trypanosoma brucei: Involvement of mannosylphosphoryldolichol as the mannose donor, *EMBO J.* **9**:4249–4258.

Mitropoulos, K. A., and Venkatesan, S., 1985, Membrane-mediated control of reductase activity, in: *Regulation of HMG-CoA Reductase* (B. Preiss, ed.), pp. 1–48, Academic Press, New York.

Myant, N. B., 1981, *The Biology of Cholesterol and Related Steroids*, Heinemann, London.

Myant, N. B., 1990, *Cholesterol Metabolism, LDL and the LDL Receptor*, Academic Press, New York.

Nakanishi, M., Goldstein, J. L., and Brown, M. S., 1988, Multivalent control of 3-hydroxy-3-methylglutaryl coenzyme A reductase. Mevalonate-derived product inhibits translation of mRNA and accelerates degradation of enzyme, *J. Biol. Chem.* **263**:8929–8937.

Olson, R. E., and Rudney, H., 1983, Biosynthesis of ubiquinone, *Vitam. Horm. (N.Y.)* **40**:1–42.

Olsson, J., Eriksson, L. C., and Dallner, G., 1991, Lipid compositions of intracellular membranes isolated from liver nodules in Wistar rats, *Cancer Res.* **51**:3774–3780.

Osborne, T. F., Gil, G., Brown, M. S., Kowal, R. C., and Goldstein, J. L., 1987, Identification of promoter elements required for in vitro transcription of hamster 3-hydroxy-3-methylglutaryl coenzyme A reductase gene, *Proc. Natl. Acad. Sci. USA* **84**:3614–3618.

Osborne, T. F., Gil, G., Goldstein, J. L., and Brown, M. S., 1988, Operator constitutive mutation of 3-hydroxy-3-methylglutaryl coenzyme A reductase promoter abolishes protein binding to sterol regulatory element, *J. Biol. Chem.* **263**:3380–3387.

Panini, S. R., Sexton, R. C., and Rudney, H., 1984, Regulation of HMG-CoA reductase by oxysterol by-products of cholesterol biosynthesis. Possible mediators of low density lipoprotein action, *J. Biol. Chem.* **259**:7767–7771.

Panini, S. R., Rogers, D. H., and Rudney, H., 1985, Regulation of HMG-CoA reductase and the biosynthesis of nonsteroid prenyl derivatives, in: *Regulation of HMG-CoA Reductase* (B. Preiss, ed.), pp. 149–181, Academic Press, New York.

Panini, S. R., Schnitzer-Polokoff, R., Spencer, T. A., and Sinensky, M., 1989, Sterol-independent regulation of 3-hydroxy-3-methylglutaryl coenzyme A reductase by mevalonate in Chinese hamster ovary cells: Magnitude and specificity, *J. Biol. Chem.* **264**:11044–11052.

Peffley, D., and Sinensky, M., 1985, Regulation of 3-hydroxy-3-methylglutaryl coenzyme A reductase synthesis by a non-sterol mevalonate-derived product in Mev-1 cells: Apparent translational control, *J. Biol. Chem.* **260**:9949–9952.

Pentchev, P. G., Comly, M. E., Kruth, H. S., Tokoro, T., Butler, J., Sokol, J., Filling-Katz, M., Quirk, J. M., Marshall, D. C., Patel, S., Vanier, M. T., and Brady, R. O., 1987, Group C Niemann–Pick disease: Faulty regulation of low-density lipoprotein uptake and cholesterol storage in cultured fibroblasts, *FASEB J.* **1**:40–45.

Popjak, G., Clarke, C. F., Hadley, C., and Meenan, A., 1985, Role of mevalonate in regulation of cholesterol synthesis and 3-hydroxy-3-methylglutaryl coenzyme A reductase in cultured cells and their cytoplasts, *J. Lipid Res.* **26**:831–841.

Porter, J. W., and Spurgeon, S. L., (eds.), 1981, *Biosynthesis of Isoprenoid Compounds*, Vols. 1 and 2, Wiley, New York.

Potter, J. E., and Kandutsch, A. A., 1982, Increased synthesis and concentration of dolichyl phosphate in mouse spleens during phenylhydrazine-induced erythropoiesis, *Biochem. Biophys. Res. Commun.* **106**:691–696.

Pullarkat, R., and Reha, H., 1982, Accumulation of dolichols in brains of elderly, *J. Biol. Chem.* **257**:5991–5993.

Quinn, P. J., and Katsikas, H., 1985, Thermal characteristics of coenzyme Q and its interaction with model membrane systems, in: *Biochemistry, Bioenergetics and Clinical Applications of Ubiquinone* (G. Lenaz, ed.), pp. 107–130, Wiley, New York.

Ramasarma, T., 1985, Metabolism of coenzyme Q, in: *Coenzyme Q: Biochemistry, Bioenergetics and Clinical Applications of Ubiquinone* (G. Lenaz, ed.), pp. 131–142, Wiley, New York.

Reiss, Y., Seabra, M. C., Armstrong, S. A., Slaughter, C. A., Goldstein, J. L., and Brown, M. S., 1991, Nonidentical subunits of p21 H-ras farnesyltransferase, *J. Biol. Chem.* **266:**10672–10677.

Rilling, H. C., and Chayet, L. T., 1985, Biosynthesis of cholesterol, in: *Sterols and Bile Acids* (H. Danielsson and J. Sjövall, eds.), pp. 1–39, Elsevier, Amsterdam.

Rilling, H. C., Bruenger, E., Epstein, W. W., and Crain, P. F., 1990, Prenylated proteins: The structure of the isoprenoid group, *Science* **247:**318–320.

Rip, J. W., Rupar, A. C., Ravi, K., and Carroll, K., 1985, Distribution, metabolism and function of dolichol and polyprenols, *Prog. Lipid Res.* **24:**269–309.

Rosenwald, A. G., and Krag, S. S., 1990, Lec9 CHO glycosylation mutants are defective in the biosynthesis of dolichol, *J. Lipid Res.* **31:**523–533.

Rosser, D. S. E., Ashby, M. N., Ellis, J. L., and Edwards, P. A., 1989, Coordinate regulation of 3-hydroxy-3-methylglutaryl-coenzyme A synthase, 3-hydroxy-3-methylglutaryl-coenzyme A reductase, and prenyltransferase synthesis but not degradation in HepG2 cells, *J. Biol. Chem.* **264:**12653–12656.

Rossignol, D. P., Scher, M., Waechter, C. J., and Lennarz, W. J., 1983, Metabolic interconversion of dolichol and dolichyl phosphate during development of the sea urchin embryo, *J. Biol. Chem.* **258:**9122–9127.

Rudney, H., and Sexton, R. C., 1986, Regulation of cholesterol biosynthesis, *Annu. Rev. Nutr.* **6:**245–272.

Rudney, H., Sexton, R. C., Gupta, A. K., and Panini, S. R., 1987, Regulation of isoprenoid biosynthesis: Oxygenated sterols as modulators of HMG-CoA reductase activity, *Chem. Scr.* **27:**57–62.

Sagami, H., Matsuoka, S., and Ogura, K., 1991, Formation of Z,Z,E-geranylgeranyl diphosphate by rat liver microsomes, *J. Biol. Chem.* **266:**3458–3463.

Schmidt, R. A., Schneider, C. J., and Glomset, J. A., 1984, Evidence for posttranslational incorporation of a product of mevalonic acid into Swiss 3T3 cell proteins, *J. Biol. Chem.* **259:**10175–10180.

Schroepfer, G. J., 1981, Sterol biosynthesis, *Annu. Rev. Biochem.* **50:**585–621.

Schroepfer, G. J., 1982, Sterol biosynthesis, *Annu. Rev. Biochem.* **51:**555–585.

Schutzbach, J. S., Jensen, J. W., Lai, C. S., and Monti, J. A., 1987, Membrane structure and mannosyltransferase activities: The effects of dolichols on membranes, *Chem. Scr.* **27:**109–118.

Simonet, W. S., and Ness, G. C., 1989, Post-transcriptional regulation of 3-hydroxy-3-methylglutaryl-CoA reductase mRNA in rat liver: Glucocorticoids block the stabilization caused by thyroid hormones, *J. Biol. Chem.* **264:**569–573.

Sindelar, P., Chojnacki, T., and Valtersson, C., 1992, Phosphatidylethanolamine: dolichol acyltransferase: Characterization and partial purification of a novel rat liver enzyme, *J. Biol. Chem.* **267:**20594–20599.

Söderberg, M., Edlund, C., Alafuzoff, I., Kristensson, K., and Dallner, G., 1992, Lipid composition in different regions of the brain in Alzheimer's disease/senile dementia of Alzheimer's type, *J. Neurochem.* **59:**1646–1653.

Stamellos, K. D., Shackelford, J. E., Tanaka, R. D., and Krisans, S. K., 1992, Mevalonate kinase is localized in rat liver peroxisomes, *J. Biol. Chem.* **267:**5560–5568.

Steinberg, D., Avigan, J., and Feigelson, E. B., 1961, Effects of triparanol (MER-29) on cholesterol biosynthesis and the blood sterol levels in man, *J. Clin. Invest.* **40:**884–893.

Steinberg, D., Parthasarathy, S., Carew, T. E., Khoo, J. C., and Witztum, J. L., 1989, Beyond cholesterol. Modifications of low-density lipoprotein that increase its atherogenicity, *N. Engl. J. Med.* **320**:915–924.

Stoll, J., Rosenwald, A. G., and Krag, S. S., 1988, A Chinese hamster ovary cell mutant F2A8 utilizes polyprenol rather than dolichol for its lipid-dependent asparagine-linked glycosylation reactions, *J. Biol. Chem.* **263**:10774–10782.

Struck, D. O. K., and Lennarz, W. I. J., 1980, The function of saccharide-lipids in synthesis of glycoproteins, in: *The Biochemistry of Glycoproteins and Proteoglycans* (W. I. J. Lennarz, ed.), pp. 35–84, Plenum Press, New York.

Swiezewska, E., Dallner, G., Andersson, B., and Ernster, L., 1992, Biosynthesis of ubiquinone and plastoquinone in the endoplasmic reticulum–Golgi membranes of spinach leaves, *J. Biol. Chem.* **268**:1494–1499.

Teclebrahan, M., Olsson, M. J., Swiezewska, E., and Dallner, G., 1993, Biosynthesis of the side chain of ubiquinone: *trans*-prenyltransferase in rat liver microsomes, *J. Biol. Chem.*, (in press).

Thelin, A., Löw, P., Chojnacki, T., and Dallner, G., 1991, Covalent binding of dolichyl phosphate to proteins in rat liver, *Eur. J. Biochem.* **195**:755–761.

Thompson, S. L., and Krisans, S. K., 1990, Rat liver peroxisomes catalyze the initial step in cholesterol synthesis: The condensation of acetyl-CoA units into acetoacetyl-CoA, *J. Biol. Chem.* **265**:5731–5735.

Thompson, S. L., Burrows, R., Laub, R. J., and Krisans, S. K., 1987, Cholesterol synthesis in rat liver peroxisomes. Conversion of mevalonic acid to cholesterol, *J. Biol. Chem.* **262**:17420–17425.

Tollbom, Ö., and Dallner, G., 1986, Dolichol and dolichyl phosphates in human tissues, *Br. J. Exp. Pathol.* **67**:757–764.

Tollbom, Ö., Valtersson, C., Chojnacki, T., and Dallner, G., 1988, Esterification of dolichol in rat liver, *J. Biol. Chem.* **263**:1347–1352.

Tollbom, Ö, Chojnacki, T., and Dallner, G., 1989, Hydrolysis of dolichyl esters in rat liver lysosomes, *J. Biol. Chem.* **264**:9836–9841.

Turley, S. D., and Dietschy, J. M., 1988, The metabolism and excretion of cholesterol by the liver, in: *The Liver, Biology and Pathobiology* (I. M. Arias, W. B. Jakoby, H. Popper, and D. Schachter, eds.), pp. 617–641, Raven Press, New York.

Valtersson, C., van Duijn, G., Verkleij, A. J., Chojnacki, T., de Kruijff, B., and Dallner, G., 1985, The influence of dolichol, dolichol esters, and dolichyl phosphate on phospholipid polymorphism and fluidity in model membranes, *J. Biol. Chem.* **260**:2742–2751.

van den Bossche, H., Willemsen, G., Cools, W., Cornellisen, F., Lauwers, W. F., and van Cutsem, J. M., 1980, In vitro and in vivo effects of the antimycotic drug ketoconazole on sterol synthesis, *Antimicrob. Agents Chemother.* **17**:922–928.

van Dijck, P. W., de Kruijff, B., van Deenen, L. L., de Gier, J., and Demel, R. A., 1976, The preference of cholesterol for phosphatidylcholine in mixed phosphatidylcholine–phosphatidylethanolamine bilayers, *Biochim. Biophys. Acta* **455**:576–587.

van Duijn, G., Valtersson, C., Chojnacki, T., Verkleif, A. J., Dallner, G., and de Kruijff, B., 1986, Dolichyl phosphate induces non-bilayer structures, vesicle fusion and transbilayer movement of lipids: A model membrane study, *Biochim. Biophys. Acta* **861**:211–223.

van Duijn, G., Verkleij, A. J., de Kruijff, B., Valtersson, C., Dallner, G., and Chojacki T., 1987, Influence of dolichols on lipid polymorphism in model membranes and the consequences for phospholipid flip–flop and vesicle fusion, *Chem. Scri.* **27**:95–100.

Yamamura, Y., 1985, A survey of the therapeutic uses of coenzyme Q, in: *Biochemistry, Bioenergetics and Clinical Applications of Ubiquinone* (G. Lenaz, ed.), pp. 479–505, Wiley, New York.

Zambrano, F., Fleischer, S., and Fleischer, B., 1975, Lipid composition of the Golgi apparatus of rat kidney and liver in comparison with other subcellular organelles, *Biochim. Biophys. Acta* **380**:357–369.

Chapter 12

Phospholipid Translocation in the Endoplasmic Reticulum

Philippe F. Devaux

1. INTRODUCTION

Phospholipid topology in biomembranes as well as the rate of lipid translocation (or lipid flip-flop) has been thoroughly investigated during the last 20 years. The conclusions appear relatively consistent in the case of the plasma membrane of eukaryotes, for which it is admitted that phospholipids are organized in an asymmetrical bilayer. The existence of the bilayer was inferred from X-ray diffraction studies as well as ^{31}P-NMR and freeze-fracture electron microscopy carried out with many different cell membranes. The chemical distribution of lipids between inner and outer leaflets was obtained by selective enzymatic attack, by chemical labeling, or by exchange of phospholipids catalyzed by soluble exchange proteins. For early reviews including critical appraisal of the techniques, see Op den Kamp (1979) and Etemadi (1980). More recent data can be found in Devaux (1991) and Schroit and Zwaal (1991).

Abbreviations used in this chapter: CDP, cytidine diphosphate; DiC$_4$PC, *sn*-1,2[^{32}P]dibutyroyl-PC; ER, endoplasmic reticulum; NMR, nuclear magnetic resonance; PA, phosphatidic acid; PC, phosphatidylcholine; PE, phosphatidylethanolamine; PS, phosphatidylserine; SM, sphingomyelin; TNBS, trinitrobenzenesulfonic acid.

Philippe F. Devaux   Institut de Biologie Physico-Chimique, F-75005 Paris, France.

Subcellular Biochemistry, Volume 21: Endoplasmic Reticulum, edited by N. Borgese and J. R. Harris. Plenum Press, New York, 1993.

In the case of organelle membranes, the situation is more complex. Although the bilayer structure is generally not questioned, results from the literature concerning phospholipid asymmetry in organelles led to numerous conflicting conclusions. In many respects, the endoplasmic reticulum (ER) seems to be the worst case. Indeed, microsomal membranes from liver or brain cells, which correspond to partially purified ER, have been investigated by several laboratories, yet conclusions are difficult to draw. The present review was undertaken by the author in a hope to untangle the diverging views put forth in the literature over the last 15 years. Unfortunately, the data do not allow one to infer any definitive conclusion about lipid topology in microsomal membranes. On the other hand, phospholipid translocation in the ER appears to be fast, much faster than in most plasma membranes. There are strong indications that specific microsomal proteins are responsible for this fast lipid flip-flop.

2. OVERVIEW OF LIPID ASYMMETRY AND LIPID FLIP-FLOP IN BIOMEMBRANES

In the plasma membrane of human erythrocytes, the best documented system, different techniques and different laboratories have come to the same conclusion: phosphatidylserine (PS), phosphatidylethanolamine (PE), and phosphatidylinositol (PI) are located mainly on the inner monolayer and, thus, are facing the cytosolic side, while phosphatidylcholine (PC) and sphingomyelin (SM) are essentially on the outer monolayer. The minor divergences that appear in the literature concern less than 10% of each lipid species and may reflect slight sample variations as well as technical limitations. In addition to the asymmetry of the head group distribution, it has been reported that the average fatty acid composition of PS and PE shows more unsaturation than that of PC and SM (Myher et al., 1989); furthermore, within the same class of phospholipids (SM or PE) acyl chains from the outer monolayer differ from those of the inner monolayer (Boegheim et al., 1983; Hullin et al., 1991). Consequently, the viscosity of the outer monolayer exceeds that of the inner monolayer (Morrot et al., 1986). Erythrocytes from other mammals have yielded similar results. The lipid asymmetry in the plasma membrane of other eukaryotic cells such as platelets, brush borders, heart sarcolemna, and fibroblasts shows comparable asymmetry, with aminophospholipids principally on the inner monolayer and the choline-containing lipids on the outer monolayer (Op den Kamp, 1979; Devaux, 1991, and references therein).

As shown originally in McConnell's laboratory, transmembrane phospholipid diffusion, or flip-flop, is a slow process, the typical time for phospholipid translocation being several hours in a lipid vesicle (Kornberg and McConnell,

1971). In red cells, the half-time of PC flip-flop varies between 3 h and 26 h depending on the nature of the acyl chains (Middlekoop *et al.*, 1986). On the other hand, neutral lipids such as diacylglycerol, fatty esters, and probably cholesterol traverse a phospholipid bilayer in less than a second. Molecules that are in equilibrium between a charged and an uncharged form, such as free fatty acids, can flip rapidly in the neutral form. Their equilibrium distribution between inner and outer leaflets depends on the local pH. Thus, as shown by Cullis's group, a pH gradient through a membrane allows one to create an asymmetrical distribution of fatty acids, phosphatidic acid (PA), or phosphatidylglycerol (PG) (Hope and Cullis, 1987).

The origin or the asymmetrical distribution of the main phospholipids in plasma membranes is not the pH because PS, PE, PC, and SM cannot be driven by a pH gradient. It has been shown that a specific protein—*aminophospholipid translocase*—transports PS and PE from the outer to the inner monolayer of the erythrocyte membrane and therefore accumulates aminophospholipids on the cytosolic leaflet. This protein is a Mg^{2+}-ATPase, inhibited by vanadate, cytosolic calcium, and all SH reagents (Seigneuret and Devaux, 1984; Daleke and Huestis, 1985; Zachowski *et al.*, 1986; Tilley *et al.*, 1986). Its affinity is higher for PS than for PE. Originally discovered in erythrocytes, there is now compelling evidence of a similar mechanism operating in many eukaryotic plasma membranes, in particular in platelets, lymphocytes, fibroblasts, and synaptosomes (Devaux, 1991, and references therein; Schroit and Zwaal, 1991). The existence of an aminophospholipid translocase in chromaffin granules has also been reported (Zachowski *et al.*, 1989); in the latter case, the pump accumulates aminophospholipids on the outer monolayer, which for this organelle corresponds to the cytosolic face. There is no evidence of a similar active mechanism for the transportation of PC and SM. Yet, these choline-containing phospholipids eventually accumulate on the outer monolayer of plasma membranes. *A priori*, PC and SM segregation could be achieved by simple diffusion of the latter lipids if one assumes that "inner sites" are occupied by aminophospholipids. However, the difference in kinetics between inward transport of aminophospholipids and outward diffusion of choline-containing lipids makes it impossible to propose such a scheme for the *establishment* of phospholipid asymmetry in the case of a totally scrambled membrane. Indeed, the difference in lipid flux would generate a difference in surface areas between inner and outer monolayers so important that large membrane curvature and eventually membrane collapse would be created. The origin of the lipid asymmetry in all biomembranes is likely to be the asymmetrical lipid synthesis which takes place for the most part in organelles: ER, Golgi system, and mitochondria (see the reviews by Bishop and Bell, 1988, and Voelker, 1991). Membrane trafficking from these organelles to the plasma membrane establishes plasma membrane asymmetry; the aminophospholipid translocase would be principally devoted to the *maintenance* of this asymmetry, particularly in erythrocytes where there is no lipid synthesis and thus no trafficking.

3. ATTEMPTS TO DETERMINE PHOSPHOLIPID TOPOLOGY IN THE ER

One prerequisite for the study of lipid asymmetry in the ER is purified membranes; a direct *in vivo* approach is not feasible. Obviously this constitutes the first obstacle. The microsomal fraction may contain different proportions of light and heavy ER or contaminations from mitochondria and Golgi system. Determining lipid topology after organelle fractionation implies that lipid asymmetry is stable, which is probably not the case for microsomes where lipid flip-flop is fast (see below). Furthermore, the ER is normally the site of lipid synthesis and lipid exportation. Thus, a static view of this heterogeneous membrane is likely to represent a distribution different from that of the real *in vivo* steady state.

Table I summarizes the results gathered from the publications of seven laboratories. It is immediately apparent that the results are not consistent. For example, in 1977, four different groups used exogenous phospholipase A_2 in order to determine the topology of rat liver microsomal phospholipids and came to different conclusions. According to Higgins and Dawson (1977), complete hydrolysis of the membrane phospholipids takes place. According to Nilsson and Dallner (1977) and van den Besselaar *et al.* (1978), 80% to 100% of PE but only 55% of PC is hydrolyzed. According to Sundler *et al.* (1977), the same phospholipase A_2 reveals that 50% of *all* phospholipids are outwardly accessible. Later, it was reported that phospholipase A_2 from *Vipera russeli* hydrolyzes 100% of all glycerophospholipids in brain microsomal membranes (Dominski *et al.*, 1983). Thus, this technique of lipid localization, which has been successful with the erythrocyte plasma membrane, appears unreliable with more fragile membranes.

Two groups used phospholipase C and reported consistent results in liver and in brain microsomal membranes (Higgins and Dawson, 1977; Higgins and Pigott, 1982; Dominski *et al.*, 1983; Freysz *et al.*, 1985). Both groups claimed that the majority of PC is exposed on the outer monolayer while the majority of PE would be on the inner monolayer (see Table I). A control of the membrane, impermeability to sucrose, was carried out to ensure integrity of the membrane after phospholipase C treatment. However, as pointed out by van Meer in a critical review-note published in 1986, it was demonstrated that diglycerides, produced in microsomes by phospholipase C, were responsible for important membrane reorganization and that this technique can be used reliably for lipid localization only if the substrate is present as a minor component (van Meer,

Table I
Results on the Transmembrane Orientation of Phospholipids
in Microsomal Membranes

References[a]	Technique	% of each phospholipid exposed on cytosolic leaflet		Comments
Nilsson and Dallner (1977)	PL A$_2$[b]	PE, PS	100	Incubation at 0°C
		PC	55	
		PI, SM	0	
Sundler et al. (1977)	PL A$_2$	PC, PE, PI	50	Incubation at 0°C
Higgins and Dawson (1977)	PL C	PC	75	
		PE	17	
		SM	64	
		PS	40	
	PL A$_2$		?	Complete hydrolysis with PL A$_2$
Zilversmit and Hughes (1977)	TP		?	85% of all lipids (PC, PE, PS,PI) exchange rapidly at 30°C
van den Besselaar et al. (1978)	PL A$_2$	PC	55–60	Incubation at 0°C (if incubation at 37°C, higher percentage of hydrolysis)
		PE	80	
	TP		?	Rapid exchange
Higgins and Pigott (1982)	TNBS	PE	33	Incubation at 7°C
	PL C	PE	30	Incubation at 37°C
Dominski et al. (1983)	TNBS	PE	35	Incubation at 20°C
		Plasmalog-PE	29	
		PS	18	
	PL C	PC	90	Incubation at 37°C
		PE	32	
		SM	60	
van Duijn et al. (1986)	TP	PC	45	Incubation at 4°C (if incubated at higher temperature, total exchange)

[a]Microsomal membranes are purified from rat liver with the exception of Dominski and collaborators, who work with brain microsomes.
[b]PL A$_2$, phospholipase A$_2$; PL C, phospholipase C; TP, transfer protein; TNBS, trinitrobenzenesulpfhonic acid.

1986). Furthermore, the experiments with phospholipase C were carried at 37°C, a temperature at which lipids traverse rapidly the microsomal membrane (see below).

The results from Freysz's and Higgins's laboratories with phospholipase C were confirmed, however, by experiments carried out at low temperature with trinitrobenzene sulfonic acid (TNBS), a probe that reacts with amino groups. They found again one-third of PE outside and two-thirds inside. But the latter technique is also subject to criticisms: first, the probe slowly permeates through bilayers; second, when microsomes are labeled with TNBS, the flip-flop of PC is accelerated which reveals a destabilization of the lipid topology (van Duijn et al., 1986). To add to the confusion, using the same techniques that Freysz's and Higgins's groups used, Butler and Morell (1982) found in brain microsomal membranes, in the steady state, a 1 : 1 ratio of outer to inner PE.

In summary, it is very difficult to draw definitive conclusions on the lipid asymmetry in the ER. The probable reason is the dynamic character of the phospholipid distribution in this membrane.

4. FAST REDISTRIBUTION OF ENDOGENOUS PHOSPHOLIPIDS IN THE ER

Evidence of fast phospholipid flip-flop in ER, at least at temperatures that are close to physiological, was presented in 1977 by two groups using phospholipid exchange proteins to determine the pool of exchangeable radioactive lipids. In Zilversmit and Hughes's report, liver microsomal fractions were prepared from rats injected with a single dose of choline [^{14}C]methylchloride or with single or multiple doses of $^{32}P_i$. Labeled PC, PE, PS, and PI were found to act as a single pool and were 85–95% exchangeable in 1–2 h. The upper limit for the half-time of PC translocation at 30°C was calculated to be 45 min or less while values of several hours or days were reported in erythrocytes (Zilversmit and Hughes, 1977). In another laboratory, a purified PC-exchange protein was used to assay the exchangeability of $^{32}P_i$-PC in rat liver microsomal membranes. At 25 and 37°C, PC was completely and rapidly available for exchange with a half-time of 1 h at 25°C (van den Besselaar et al., 1978).

Another approach to the determination of lipid flip-flop in ER is to directly follow the redistribution of newly synthesized phospholipids. The synthesis de novo of PC and PE in animal cells is mainly through the Kennedy pathway. The enzymes in ER responsible for the terminal steps of PC or PE synthesis, ethanolamine- or choline-transferase, utilize diacylglycerol and CDP-choline or CDP-ethanolamine as substrate (Bishop and Bell, 1988, Voelker, 1991). When cells are labeled with [^{14}C]choline or [^{14}C]ethanolamine, the location of newly synthesized phospholipids can be determined. Using phospholipase C as a probe, Hutson and Higgins (1982) observed that labeled PE was initially (1–2 min) concentrated in the outer leaflet of the membrane bilayer of liver microsomes. On longer incubation, up to 30 min, the specific activity of the outer leaflet PE approached that of the inner leaflet. These observations suggest that PE is synthesized at the cytoplasmic leaflet of the ER and subsequently transferred across the membrane to the cisternal leaflet of the bilayer. A half-time of less than 15 min was observed at 37°C. Transmembrane movement is apparently temperature dependent and independent of continued synthesis of PE. The same conclusion was reached by Freysz and collaborators who investigated brain microsomes (Dominski *et al.*, 1982; Corazzi *et al.*, 1983; Freysz *et al.*, 1985). Using either TNBS or phospholipase C under nonpenetrating conditions, these authors found that PE, plasmalogen-PE, and PC are synthesized on the outer leaflet; PE and plasmalogen-PE are rapidly translocated ($t_{1/2}$ = 15 min), while PC would diffuse more slowly to the inner layer. They also reported that the synthesis of phospholipids through the base exchange pathway takes place on the outer leaflet of microsomes and is accompanied by a more rapid transfer of PE compared with that of PC. Note that the same group reported in 1990 that the translocation of PE is energy dependent but did not show any data to prove this important statement (Erhardt *et al.*, 1990). Finally, Butler and Morell came to a comparable conclusion concerning the sidedness of phospholipid synthesis in brain microsomes; however, they observed a rapid equilibration of both phospholipids, PE and PC, with a half-time of 30 min (Butler and Morell, 1982).

In an effort to understand the contradictory results concerning the equilibration of newly synthesized PC in microsomes, Hutson *et al.* (1985) repeated the localization experiments with phospholipase C and with a PC exchange protein and concluded that there are at least two pools of PC in rat liver microsomes. One of these is preferentially labeled with [^{14}C]choline and does not equilibrate across the bilayer. The second pool is labeled with [^{3}H]glycerol and does equilibrate across the bilayer. Thus, two subpopulations of microsomal vesicles exist.

In concluding this section, it seems that the enzymes responsible for the terminal steps of glycerophospholipid synthesis in animal cells appear to have their catalytic sites at the cytosol-facing aspect of the ER; this was inferred several years ago (Vance *et al.*, 1977); rapid flip-flop of newly synthesized lipids seem to take place, for most if not all glycerophospholipids. Note that differential rates of lipid synthesis and flip-flop would permit the establishment of an asymmetrical membrane even if the enzymes that synthesize PC and PE are located on the cytoplasmic side of the ER.

5. THE PHOSPHOLIPID FLIPPASE

In 1973, Bretscher suggested the existence of "flip-flop enzymes" to explain the transmembrane distribution of lipid in eukaryotic membranes. Later, De Oliveira Filgueiras and collaborators attempted to purify a proteolipid from rat liver microsomes but found no catalytic activity (De Oliveira Filgueiras et al., 1981). Nevertheless, Barsukov et al. postulated that intrinsic proteins were involved in the rapid phospholipid flip-flop encountered in microsomes (Barsukov et al., 1982). Finally, Bishop and Bell (1985) proved the existence of a PC-specific flippase capable of catalyzing PC translocation in the ER membrane. They used the water-soluble molecule sn-1,2[^{32}P]dibutyroyl-PC (DiC$_4$PC) to measure PC translocation. This homologue retains the polar head group, the portion of the phospholipid unable to undergo spontaneous transmembrane movement in vesicles, and its water solubility permits the application of standard transport methods. DiC$_4$PC entered the lumenal compartment of microsomal vesicles. Transport was saturable and was dependent on time, amount of microsomes, and an intact permeability barrier. DiC$_4$PC was inhibited by structural analogues (but not sn-2,3-diC$_4$PC) and by treatment of microsomes with proteases, N-ethylmaleimide, and TNBS. The latter result, however, is at variance with the results of van Duijn and collaborators who reported an *acceleration* of PC translocation upon incubation with TNBS (van Duijn et al., 1986). DiC$_4$PC was not transported across PC vesicles or red cell membranes, where PC translocation is slow. The possible metabolism of DiC$_4$PC was investigated; time course indicated that hydrolysis lagged behind uptake, raising the possibility of lumenal metabolism but excluding lipid metabolism as being directly involved in the translocation.

The existence of the ER flippase was confirmed by reconstitution experiments using nonpurified microsomal proteins (Backer and Dawidowicz, 1987). However, attempts to purify the flippase up to now have been unsuccessful. Using a short-chain derivative (sn-1-monobutyroyl PC), Kawashima and Bell (1987) found that lyso-PC transport through microsomal membranes is also protein mediated. The data are consistent with diC$_4$PC and monoC$_4$PC transport activities being dual functions of a single transporter. On the other hand, glycerophosphocholine is transported by a different protein (Kawashima and Bell, 1987).

The broad lipid specificity of the microsomal flippase was confirmed by experiments carried out with amphiphilic spin-labeled phospholipid analogues. Three classes of spin labels were synthesized. The first class possessed a short β chain bearing a nitroxide (Figure 1a). Four different head groups were used: choline, serine, ethanolamine, and monomethylated ethanolamine. The second class of spin labels had a ceramide backbone (Figure 1b). Two different head groups were used for this class of lipids: choline and serine. The third class of spin labels had a nitroxide on the α chain and corresponded to lyso compounds (Figure 1c); the head groups used were again choline and serine. All of these molecules were transported at approximately the same rate to an equilibrium distribution where they were equally distributed between two layers. The half-time for equilibration corresponded to approximately 20 min at 37°C. The veloc-

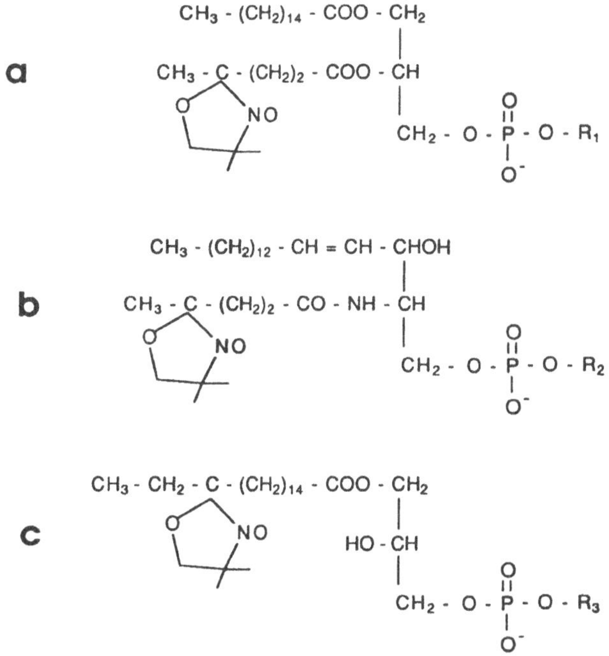

FIGURE 1. Spin-labeled phospholipids used to determine translocation rates in microsomal vesicles. (a) Glycerophospholipids; (b) sphingolipids; (c) lyso compounds.

ity of diffusion was reduced by the action of N-ethylmaleimide on the microsomal proteins. This protein-mediated (facilitated) diffusion was saturable with respect to the substrate available and exhibited a low lipid specificity (Figure 2). Direct competition experiments, using simultaneously two spin-labeled lipids, showed that the same protein transports PC and lyso-PS (Herrmann *et al.*, 1990). Note that the transmembrane diffusion of the spin-labeled PC with one long chain and one short chain was not influenced by the presence of up to 20 mM unlabeled diC_4PC. Thus, water-soluble molecules do not compete efficiently with more hydrophobic lipids.

While the influence of one or several proteins on the translocation of phospholipids in the ER seems to be well established, the mechanism of this facilitated diffusion process is far from being understood. It has been suggested that local nonbilayer structures would be present in microsomal membranes. Indeed, while ^{31}P-NMR, with most biological membranes, gives a typical line shape associated with the bilayer structure, rat microsomal membranes give a different line shape indicative of a more isotropic motion, at least for a fraction of the lipids (de Kruijff *et al.*, 1978). Although there was some excitement after this

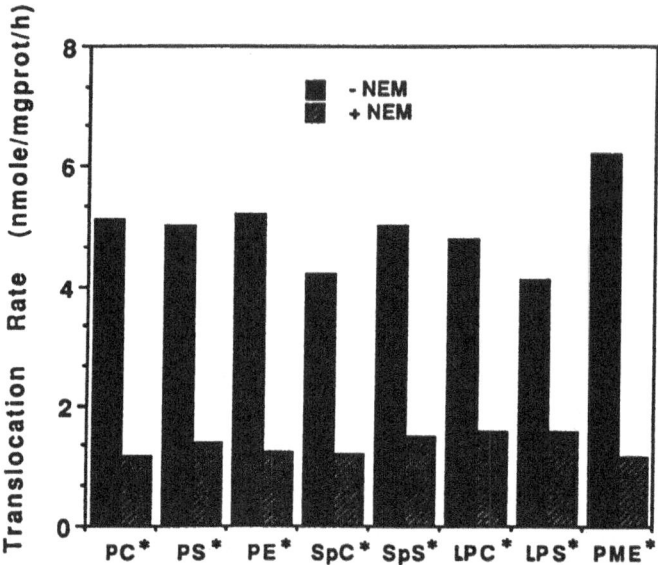

FIGURE 2. Initial outside–inside diffusion rate of various spin-labeled analogues in intact rat liver microsomal vesicles (solid columns) and in microsomes treated with N-ethylmaleimide (hatched columns). The analogues were introduced on the microsomal outer leaflet as 1% of the endogenous lipids; their incorporation is achieved in 1 or 2 min because of the partial water-solubility of these lipids. The reorientation kinetics is followed by back-exchange using fatty acid-free bovine serum albumin to reextract the spin labels. From Herrmann *et al.* (1990). PC*, 1-palmitoyl-2-(4-doxyl-pentanoyl) phosphatidylcholine; PS*, 1-palmitoyl-2-(4-doxylpentanoyl) phosphatidylserine; PE*, 1-palmitoyl-2-(4-doxylpentanoyl) phosphatidylethanolamine; SpC*, N-(4-doxylpentanoyl)-*trans*-4-sphingenyl-1-phosphocholine; SpS*, N-(4-doxylpentanoyl)-*trans*-4-sphingenyl-1-phosphoserine; LPC*, 1-(16-doxylstearoyl)lyso-phosphatidylcholine; LPS*, 1-(16-doxylstearoyl)lyso-phosphatidyl-serine; PME*, 1-palmitoyl-2-(4-doxylpentanoyl) phosphatidyl-(N-monomethyl) ethanolamine.

discovery (de Kruijff *et al.*, 1980), it was later suggested that artifacts such as local membrane bending could be responsible for such NMR line shapes (Burnell *et al.*, 1980).

6. CONCLUSIONS

We have seen that the phospholipid topology in microsomal membranes is difficult to establish. We can say that phospholipids are synthesized in this membrane in an asymmetrical fashion and the membrane has proteins associated with it that catalyze PL redistribution. Depending on temperature, the amount of substrate and the metabolic state of the cell, specific lipid synthesis does or does not take place, and hence the lipid asymmetry may vary. It appears relatively clear that the microsomal membranes are unique because of the rapid relocation of phospholipids. This must be contrasted with plasma membranes and liposomes where translocation either does not take place or takes place in a highly selective fashion through the activity of the aminophospholipid translocase. The latter is an ATP-requiring protein whose presence in microsomes was not found.

ACKNOWLEDGMENTS. Work suported by grants from the Centre National de la Recherche Scientifique (UA 526), the Université Paris VII, the Institut National de la Santé et de la Recherche Médicale (No. 900104) and the "Fondation pour la Recherche Médicale."

7. REFERENCES

Backer, J. M., and Dawidowicz, E. A., 1987, Reconstitution of a phospholipid flippase from rat liver microsomes, *Nature* **327**:341–343.

Barsukov, L. I., Kulikov, V. I., Ivanova, V. P., and Bergelson, L. D., 1982, Phospholipid dynamics and transbilayer distribution in rat liver and hepatoma microsomes, *Stud. Biophys.* **90**:147–148.

Bishop, W. R., and Bell, R. M., 1985, Assembly of the endoplasmic reticulum phospholipid bilayer: The phosphatidylcholine transporter, *Cell* **45**:51–60.

Bishop, W. R., and Bell, R. M., 1988, Assembly of phospholipids into cellular membranes: Biosynthesis, transmembrane movement and intracellular translocation, *Annu. Rev. Cell Biol.* **4**:579–610.

Boegheim, J. P. J., Jr., van Linde, M., Op den Kamp, J. A. F., and Roelofsen, B., 1983, The sphingomyelin pools in the outer and inner layer of the human erythrocyte membrane are composed of different molecular species, *Biochim. Biophys. Acta* **735**:438–442.

Bretscher, M. S., 1973, Membrane structure: Some general principles, *Science* **181**:622–629.

Burnell, E. E., Cullis, P. R., and de Kruijff, B., 1980, Effects of tumbling and lateral diffusion on phosphatidylcholine model membrane ^{31}P-NMR lineshapes, *Biochim. Biophys. Acta* **603**:63–69.

Butler, M., and Morell, P., 1982, Sidedness of phospholipid synthesis on brain membranes, *J. Neurochem.* **39**:155–164.

Corazzi, L., Binaglia, L., Roberti, R., Freysz, L., Arienti, G., and Porcellati, G., 1983, Compartmentation of membrane phosphatidylethanolamine formed by base-exchange reaction in rat brain microsomes, *Biochim. Biophys. Acta* **730**:104–110.

Daleke, D. L., and Huestis, W. H., 1985, Incorporation and translocation of aminophospholipids in human erythrocytes, *Biochemistry* **24**:5406–5416.

de Kruijff, B., van den Besselaar, A.M.H.P., Cullis, P. R., van den Bosch, H., and van Deenen, L. L. M., 1978, Evidence for isotropic motion of phospholipids in liver microsomal membranes, *Biochim. Biophys. Acta* **514**:1–8.

de Kruijff, B., Cullis, P. R., and Verkleij, A. J., 1980, Non-bilayer lipid structures in model and biological membranes, *Trends Biochem. Sci.* **5**:79–81.

De Oliveira Filgueiras, O. M., de Winter, J. M., and van den Bosch, H., 1981, Phosphatidylcholine accessibility in single bilayer vesicles prepared from rat liver microsomal lipids containing proteolipids, *Biochem. Biophys. Res. Commun.* **100**:800–806.

Devaux, P. F., 1991, Static and dynamic lipid asymmetry in cell membranes, *Biochemistry* **30**:1163–1173.

Dominski, J., Binaglia, L., Porcellati, G., and Freysz, L., 1982, Asymmetric synthesis of eth-
    anolamine phospholipids in chicken brain microsomes, through the cytidine pathway, *FEBS
    Lett.* **147:**153–155.
Dominski, J., Binaglia, L., Dreyfus, H., Massarelli, R., Mersel, M., and Freysz, L., 1983, A study
    on the topological distribution of phospholipids in microsomal membranes of chick brain using
    phospholipase C and trinitrobenzenesulfonic acid, *Biochim. Biophys. Acta* **734:**257–266.
Erhardt, A., Leray, C., Binaglia, L., Roberti, R., Dreyfus, H., Massarelli, R., and Freysz, L., 1990,
    In vitro synthesis and transbilayer movement of phosphatidylethanolamine molecules labelled
    with different fatty acids in chick brain microsomes, *Biochim. Biophys. Acta* **1021:**126–132.
Etemadi, A. H., 1980, Membrane asymmetry. A survey and critical appraisal of methodology. II.
    Methods for assessing the unequal distribution of lipids, *Biochim. Biophys. Acta* **604:**423–475.
Freysz, L., Binaglia, L., Dominski, J., and Porcellati, G., 1985, Topological biosynthesis of phos-
    pholipids in brain microsomes, in: *Phospholipids in the Nervous System* (L. A. Horrocks, J. N.
    Kanfer, and G. Porcelli, eds.), pp. 279–288, Raven Press, New York.
Herrmann, A., Zachowski, A., and Devaux, P. F., 1990, Protein-mediated phospholipid transloca-
    tion in the endoplasmic reticulum with a low lipid specificity, Biochemistry **29:**2023–2027.
Higgins, J. A., and Dawson, R. M. C., 1977, Asymmetry of the phospholipid bilayer of rat liver
    endoplasmic reticulum, *Biochim. Biophys. Acta* **470:**342–356.
Higgins, J. A., and Pigott, C. A., 1982, Asymmetric distribution of phosphatidylethanolamine in the
    endoplasmic reticulum demonstrated using trinitrobenzenesulphonic acid as a probe, *Biochim.
    Biophys. Acta* **693:**151–158.
Hope, M. J., and Cullis, P. R., 1987, Lipid asymmetry induced by transmembrane pH gradients in
    large unilamellar vesicles, *J. Biol. Chem.* **262:**4360–4366.
Hullin, F., Bossant, M.-J., and Salem, N., Jr., 1991, Aminophospholipid molecular species asym-
    metry in the human erythrocyte plasma membrane, *Biochim. Biophys. Acta* **1061:**15–25.
Hutson, J. L., and Higgins, J. A., 1982, Asymmetric synthesis followed by transmembrane move-
    ment of phosphatidylethanolamine in rat liver endoplasmic reticulum, *Biochim. Biophys. Acta*
    687:247–256.
Hutson, J. L., Higgins, J. A., and Wirtz, K. W. A., 1985, Microsomal membranes contain phospha-
    tidylcholine that equilibrates across the bilayer, and phosphatidylcholine that does not, *FEBS
    Lett.* **183:**145–150.
Kawashima, Y., and Bell, R. M., 1987, Assembly of the endoplasmic reticulum phospholipid
    bilayer. Transporters for phosphatidylcholine and metabolites, *J. Biol. Chem.* **262:**14495–
    14502.
Kornberg, R. D., and McConnell, H. M., 1971, Inside–outside transitions of phospholipids in
    vesicle membranes, *Biochemistry* **10:**1111–1120.
Middelkoop, E., Lubin, B. H., Op den Kamp, J. A. F., and Roelofsen, B., 1986, Flip-flop rates of
    individual molecular species of phosphatidylcholine in the human red cell membrane, *Biochim.
    Biophys. Acta* **855:**421–424.
Morrot, G., Cribier, S., Devaux, P. F., Geldwerth, D., Davoust, J., Bureau, J. F., Fellmann, P.,
    Hervé, P., and Frilley, B., 1986, Asymmetric lateral mobility of phospholipids in the human
    erythrocyte membrane, *Proc. Natl. Acad. Sci. USA* **83:**6863–6867.
Myher, J. J., Kuksis, A., and Pinder, S., 1989, Molecular species of glycerophospholipids and
    sphingomyelins of human erythrocytes: Improved method of analysis, *Lipids* **24:**396–407.
Nilsson, O. S., and Dallner, G., 1977, Enzyme and phospholipid asymmetry in liver microsomal
    membranes, *J. Cell Biol.* **72:**568–583.
Op den Kamp, J. A. F., 1979, Lipid asymmetry in membranes, *Annu. Rev. Biochem.* **48:**47–71.
Schroit, A. J., and Zwaal, R. F. A., 1991, Transbilayer movement of phospholipids in red cell and
    platelet membranes, *Biochim. Biophys. Acta* **1071:**313–329.
Seigneuret, M., and Devaux, P. F., 1984, ATP-dependent asymmetric distribution of spin-labeled

phospholipids in the erythrocyte membranes: Relation to shape change, *Proc. Natl. Acad. Sci. USA* **81**:3751–3755.

Sundler, R., Sarcione, S. L., Alberts, A. W., and Vagelos, P. R., 1977, Evidence against phospholipid asymmetry in intracellular membranes from liver, *Proc. Natl. Acad. Sci. USA* **74**:3350–3354.

Tilley, L., Cribier, S., Roelofsen, B., Op den Kamp, J. A. F., and van Deenen, L. L. M., 1986, ATP-dependent translocation of aminophospholipids across the human erythrocyte membrane, *FEBS Lett.* **194**:21–27.

Vance, D. E., Choy, P. C., Farren, S. B., Lim, P. H., and Schneider, W. J., 1977, Asymmetry of phospholipid biosynthesis, *Nature* **270**:268–269.

van den Besselaar, A. M. H. P., de Kruijff, B., van den Bosch, H., and van Deenen, L. L. M., 1978, Phosphatidylcholine mobility in liver microsomal membranes, *Biochim. Biophys. Acta* **510**:242–255.

van Duijn, G., Luiken, J., Verkleij, A. J., and de Kruijff, B., 1986, Relation between lipid polymorphism and transbilayer movement of lipids in rat liver microsomes, *Biochim. Biophys. Acta* **863**:193–204.

van Meer, G., 1986, The lipid bilayer of the ER, *Trends Biochem. Sci.* **11**:194–195.

Voelker, D. R., 1991, Organelle biogenesis and intracellular lipid transport in eukaryotes, *Microbiol. Rev.* **55**:543–560.

Zachowski, A., Favre, E., Cribier, S., Hervé, P., and Devaux, P. F., 1986, Outside–inside translocation of aminophospholipids in the human erythrocyte membrane is mediated by a specific enzyme, *Biochemistry* **25**:2585–2590.

Zachowski, A., Henry, J. P., and Devaux, P. F., 1989, Control of transmembrane lipid asymmetry in chromaffin granules by an ATP-dependent protein, *Nature* **340**:75–76.

Zilversmit, D. B., and Hughes, M. E., 1977, Extensive exchange of rat liver microsomal phospholipids, *Biochim. Biophys. Acta* **469**:99–110.

Chapter 13

Cytochrome P-450 in the Endoplasmic Reticulum

Biosynthesis, Distribution, Induction, and Degradation

Yutaka Tashiro, Ryuichi Masaki, and Akitsugu Yamamoto

1. INTRODUCTION

Microsomal cytochrome P-450 (microsomal P-450) represents a family of hemoprotein monoxygenases that are localized in endoplasmic reticulum (ER) membranes of various cells and function in the metabolism of a wide variety of exogenous compounds such as drugs and chemical carcinogens and endogenous ones such as steroids, fatty acids, and retinoids (Porter and Coon, 1991). Various P-450 forms have been characterized by their specific, but overlapping, substrate specificities and by their inducibility and have been grouped into several families based on structural homology (Nebert and Gonzales, 1987; Gonzales, 1988). The ER membrane provides a framework that facilitates the interaction of P-450

Abbreviations used in this chapter: ER, endoplasmic reticulum; MC, methylcholanthrene; PB, phenobarbital; SRP, signal recognition particle.

Yutaka Tashiro, Ryuichi Masaki, and Akitsugu Yamamoto   Department of Physiology and Department of Cell Biology, Liver Research Center, Kansai Medical University, Moriguchi, Osaka 570, Japan.
Subcellular Biochemistry, Volume 21: Endoplasmic Reticulum, edited by N. Borgese and J. R. Harris. Plenum Press, New York, 1993.

molecules with a specific NADPH cytochrome P-450 reductase that transfers electrons to the cytochrome P-450 heme moiety.

Rat liver microsomes, which are mostly derived from ER membranes during homogenization, contain 0.5–1.0 nmol P-450/mg protein and this P-450 content increases approximately two- to threefold after treatment with xenobiotics such as phenobarbital (PB), so that P-450 comprises ~5 and ~10% or more of the proteins of noninduced and PB-induced liver microsomes, respectively (Sato and Omura, 1978). Thus, microsomal P-450 is certainly the most abundant ER membrane protein of the hepatocytes and could be a good marker protein to investigate the static and dynamic aspects of ER membrane proteins.

In this review, we describe cell biological aspects of microsomal P-450 in rat hepatocytes such as membrane topology, biosynthesis, intracellular distribution, induction, turnover, and degradation. The molecular biological aspects of P-450 are described in detail in recent reviews (Fujii-Kuriyama *et al.*, 1992; Gonzales, 1988; Nebert and Gonzales, 1987).

In order to carry out cell biological experiments on microsomal P-450, it is essential to prepare specific antibody. Because it was very difficult to solubilize and purify microsomal P-450, reliable anti P-450 antibody was not available until the mid 1970s. In 1974, Imai and Sato succeeded in purifying P-450 from liver microsomes of PB-treated rabbits. In collaboration with Imai, we purified P-450(PB) from PB-treated rat livers and prepared antibody (Negishi *et al.*, 1976), which immunoreacted mostly with P-450IIB1 and IIB2, two major molecular forms of PB-inducible P-450, but also bound to a small amount of constitutive species of P-450s. Most of the early experiments on the microsomal P-450 in this laboratory were carried out using this antibody, which will be called anti-P-450(PB) antibody 1 herein.

Later, Masaki *et al.* (1984) raised antibodies against P-450(PB) prepared by a different procedure. This antibody showed immunochemical properties similar to antibody 1 and will be called anti-P-450(PB) antibody 2. We also purified microsomal P-450 [microsomal P-450(MC)] from 3-methylcholanthrene-treated rat livers (Masuda-Mikawa *et al.*, 1979) and prepared anti-P-450(MC) antibody.

2. MEMBRANE TOPOLOGY

Information concerning the spatial arrangement and localization of P-450 in the microsomal membrane is essential for an understanding of its function. Localization of microsomal P-450 at the cytoplasmic surface of microsomes was suggested by enzymatic iodination (Welton and Aust, 1974) and by inhibition of its function with antibodies (Welton *et al.*, 1975; Thomas *et al.*, 1978). Direct evidence, obtained by visualization of the distribution of P-450(PB) in rat liver microsomes by ferritin immunoelectron microscopy, was presented in 1978

(Matsuura *et al.*, 1978). As shown in Figures 1 and 2, the outer surfaces of the microsomes were heavily labeled with ferritin particles. Since all of the ribosomes are attached to the outer surface of the microsomes (Figure 1), it is certain that the outer surface corresponds to the cytoplasmic surface of the ER.

Another organelle that was heavily labeled with ferritin was the nuclear enveloped. As shown in Figure 3, also in this case, the cytoplasmic surface was labeled.

These results, however, did not exclude the possibility that some of the epitopic sites of P-450(PB) also exist on the lumenal surface of the ER membranes, because ferritin–antibody conjugates cannot cross the ER or outer nuclear membranes. Although it is very difficult to prepare inside-out microsomal

FIGURES 1–5. Rat liver microsomes, nuclei, and nuclear envelopes labeled with ferritin anti-P-450(PB) antibody conjugates by preembedding method. Bars = 0.2 μm.

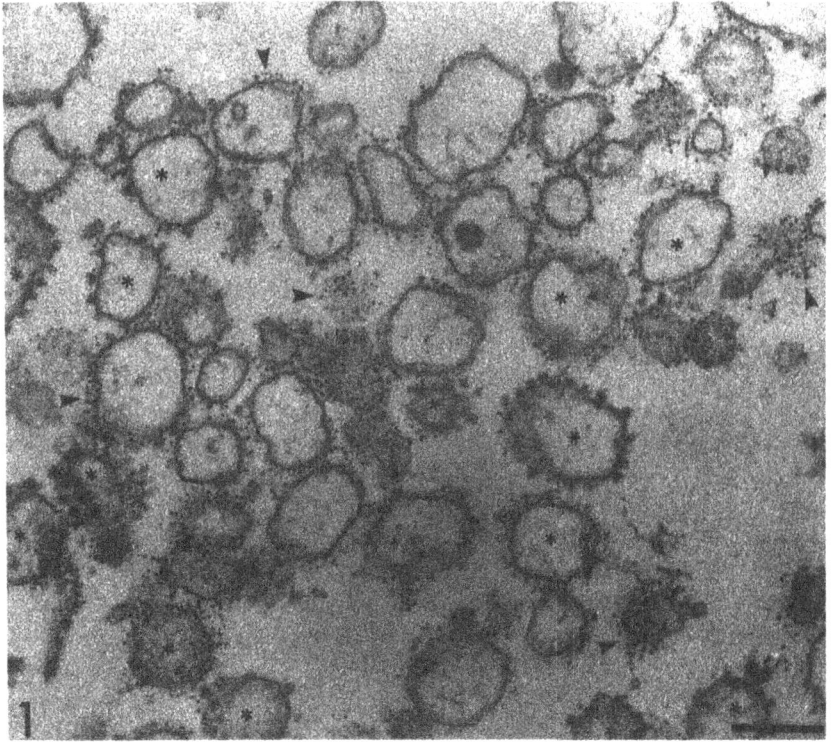

FIGURE 1. Total microsomes: Rough microsomes (asterisks), especially the domains densely loaded with ribosomes, are less densely labeled than smooth microsomes. In the smooth microsomes, ferritin particles frequently show clustering (arrowheads).

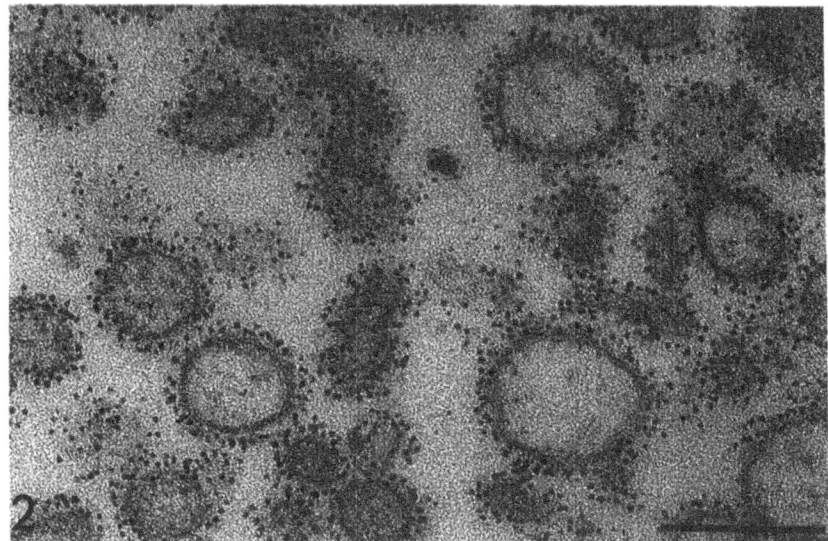

FIGURE 2. Smooth microsomes prepared from PB-treated rat. The cytoplasmic surfaces are completely covered with ferritin particles.

vesicles, it is possible to prepare nuclei from which the outer nuclear membranes are partially detached as shown in Figure 4. When such a preparation was incubated with ferritin–antibody conjugates, only the cytoplasmic surface of the outer nuclear membrane was labeled; the lumenal surfaces of the outer and the inner nuclear membranes were not labeled (Matsuura *et al.*, 1983).

We also studied whether the nucleoplasmic surface of the inner nuclear membranes is labeled or not. Nuclear envelopes were prepared from MC-treated livers and incubated with ferritin-antibody conjugates. It is evident from Figure 5 that the cytoplasmic surface of the nuclear envelopes, which is studded with ribosomes, is heavily labeled with ferritin, while the nucleoplasmic surface is barely labeled.

The complete absence of ferritin particles on the lumenal surface of the outer nuclear membrane suggests that most, if not all, of the epitopic sites of the P-450(PB) molecules are localized on the cytoplasmic surface of the outer nuclear membranes and presumably also of the ER membranes. This finding has important implications for the membrane topology of P-450(PB) as discussed later.

Later, the complete amino acid sequence of two microsomal P-450s was determined biochemically and their membrane topology in the ER membrane was estimated from their hydropathy profiles (Ozols *et al.*, 1985; Tarr *et al.*, 1983). A multitopic model of P-450 spanning the ER membranes eight or ten

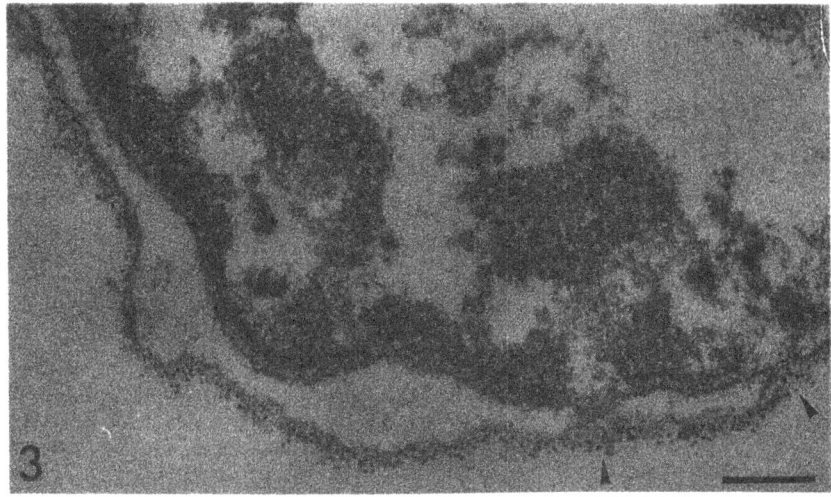

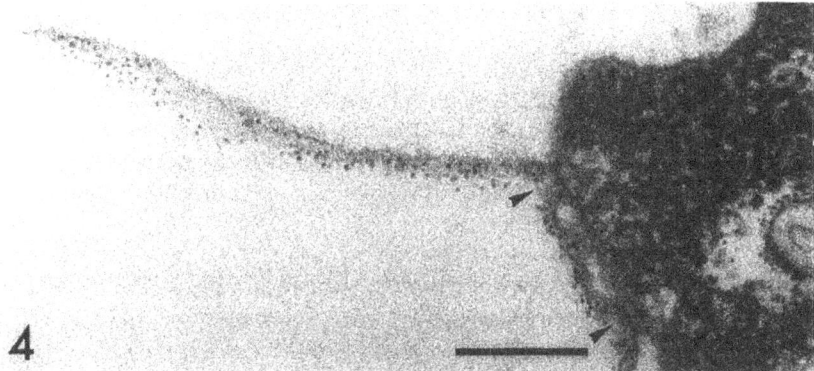

FIGURES 3 and 4. Nuclei. Rat liver nuclei incubated with ferritin anti-P-450(PB) antibody conjugates. P-450(PB) was maximally induced by PB treatment. In Figure 4, the outer membranes of the nuclear envelope were detached from the nucleus by homogenization. Nuclear pores are indicated by arrowheads. Reproduced with permission from Matsuura *et al.* (1978, 1983).

times was proposed. This model was not consistent with our immunocytochemical results.

In order to solve this problem, De Lemos-Chiarandini *et al.* (1987) prepared site-specific antibodies against 15 synthetic peptides that correspond to selected domains of rat liver P-450IIB1, and showed that many of the antibodies bound to the cytoplasmic surface of the membrane but none bound to the lumenal surface of ruptured or inverted microsomal vesicles. They further showed that an antibody raised against the first 31 residues of P-450 bound very poorly to rough

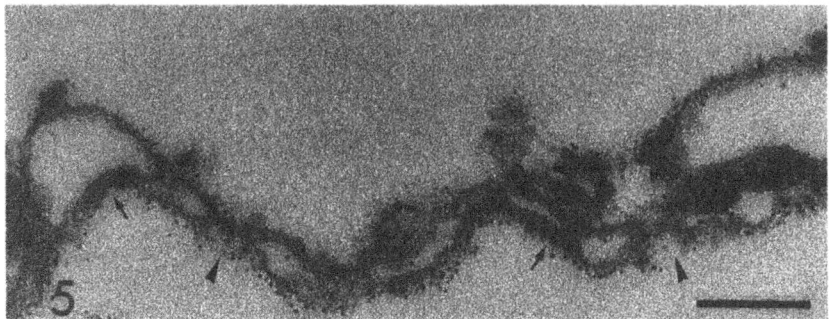

FIGURE 5. Rat liver nuclear envelopes. Nuclear envelopes were isolated from methylcholanthrene (MC)-treated rats and incubated with anti-P-450(MC) antibody conjugates. Nuclear pores are indicated by arrowheads. Only the cytoplasmic surface of the outer nuclear membrane loaded with ribosomes (arrows) is heavily labeled with ferritin particles. Reproduced with permission from Matsuura *et al.* (1981).

microsomes, whereas an antibody to a peptide comprising residues 24–38 showed relatively strong binding to intact microsomes. These results are not consistent with the multitopic model of microsomal P-450 but are consistent with the proposal that the amino-terminal segment of P-450 extending approximately to residue 20 is embedded in the ER membranes and the following segment is exposed to the cytoplasmic surface.

The membrane topology of hepatic microsomal P-450 was also studied by trypsinolysis of the PB-treated microsomes (Brown and Black, 1989) and it was suggested that the P-450 is bound to the ER membrane by only one or two transmembrane segments located at the N-terminus.

These results are consistent with the model of Nelson and Strobel (1988), according to which P-450 is anchored to the membrane only by the hydrophobic N-terminus. These authors proposed that the protein spans the membrane twice, the *N*-terminal methionine facing the cytosol (Figure 6B).

It has not yet been settled, however, whether the N-terminal segment of microsomal P-450 spans the membrane once or twice. In fact, localization of the N-terminal methionine was estimated by a simple determination of the sidedness with the site-specific reagent fluorescein isothiocyanate. Vergeres *et al.* (1991) concluded that the N-terminus faces the lumen of the ER and proposed that P-450 spans the membrane only once with amino acid residues 1 to 21 (Figure 6A). This conclusion is consistent with the membrane topology of microsomal P-450 deduced from translational studies by Monier *et al.* (1988) and Sakaguchi *et al.* (1992) as described in the next section.

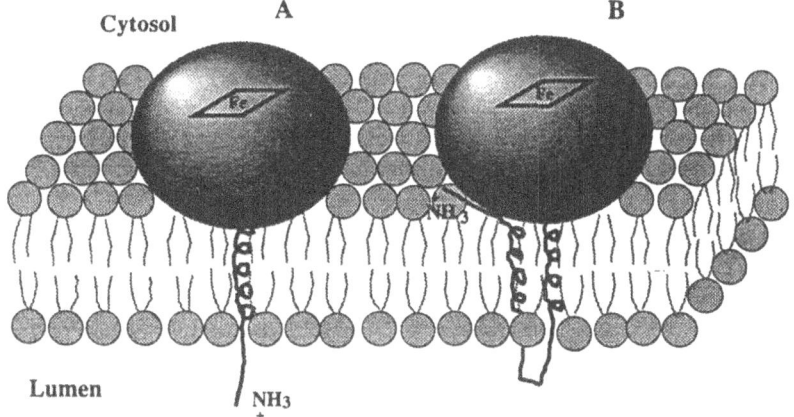

FIGURE 6. Two models of microsomal P-450 molecule in the ER membranes. The N-terminal region spans once in model A and twice in model B. The N-terminus, therefore, faces the lumen of the ER in model A and the cytoplasmic surface of the ER in model B.

3. BIOSYNTHESIS

In the previous section we have shown that microsomal P-450 is localized on the cytoplasmic surface of ER and nuclear envelope membranes, anchored to the membranes only by the hydrophobic N-terminal segment. Now we discuss whether microsomal P-450 is synthesized on the free or membrane-bound ribosomes in rat hepatocytes, how it is incorporated into the ER membranes, and how it is equilibrated between the rough and smooth ER membranes.

Figure 7 shows binding experiments of [125]-labeled anti-P-450(PB) antibody to rat liver membrane-bound and free ribosomes (panel A) and immunoprecipitation of [3H]puromycin-labeled nascent P-450(PB) polypeptides that were released from membrane-bound and free ribosomes (panel B) (Negishi *et al.*, 1976). It is evident that P-450(PB) is synthesized exclusively in the membrane-bound ribosomes.

We next examined the fate of the nascent P-450(PB) peptides after release from bound ribosomes by using protease digestion (Fujii-Kuriyama *et al.*, 1979). It was revealed that most of the nascent peptides (~90%) were recovered in the microsomal membrane fraction after the puromycin treatment but were easily digested by subsequent protease digestion (Table I). This result suggests that nascent P-450 peptides are incorporated directly from the bound ribosomes onto or into the outer surface of the ER membranes.

Figure 8A shows the time course of [14C]leucine incorporation into microsomal P-450(PB) in rat liver rough and smooth microsomes after intravenous

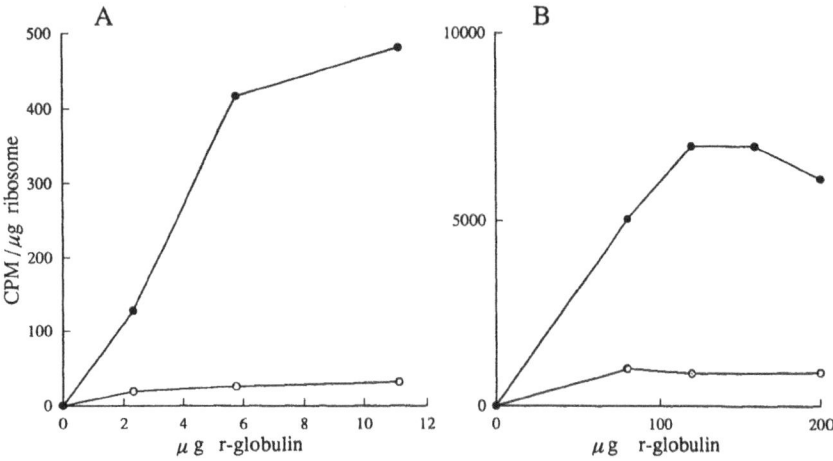

FIGURE 7. Binding of ^{125}I-labeled anti-P-450 antibody with free and membrane-bound ribosomes (A) and immunoprecipitation of [^{3}H]puromycin-labeled nascent P-450 chains released from free and membrane-bound ribosomes (B). ●, membrane-bound ribosomes, ○, free ribosomes.

injection of [^{14}C]leucine. At early time points, the specific radioactivity of P-450(PB) was clearly higher in the rough microsomes than in the smooth microsomes. The radioactivity in the rough microsomes reached a peak value around 10 min, then decreased, while that in the smooth microsomes continued to increase until these two values became equal at ~ 60 min after the labeling. This result indicates that the newly synthesized P-450(PB) directly incorporated into the cytoplasmic surface of the rough ER membranes was translocated to the smooth ER membranes to be evenly distributed in both types of the ER in ~1 h.

The movement of P-450 molecules from rough to smooth ER could be brought about by either attachment–detachment of ribosomes to and from ER membranes accompanying the ribosomal cycle for protein synthesis, lateral

Table I
Protease Digestion of Nascent Peptides in Microsomal Vesicles

	^{3}H-amino acid label *in vivo*		^{3}H-puromycin label *in vitro*	
Protease treatment	−	+	−	+
Total membrane protein	17,200	9430	4.50×10^5	2.19×10^5
P-450	246	25	9.25×10^3	1.36×10^3
Albumin	2,900	2350	—	—

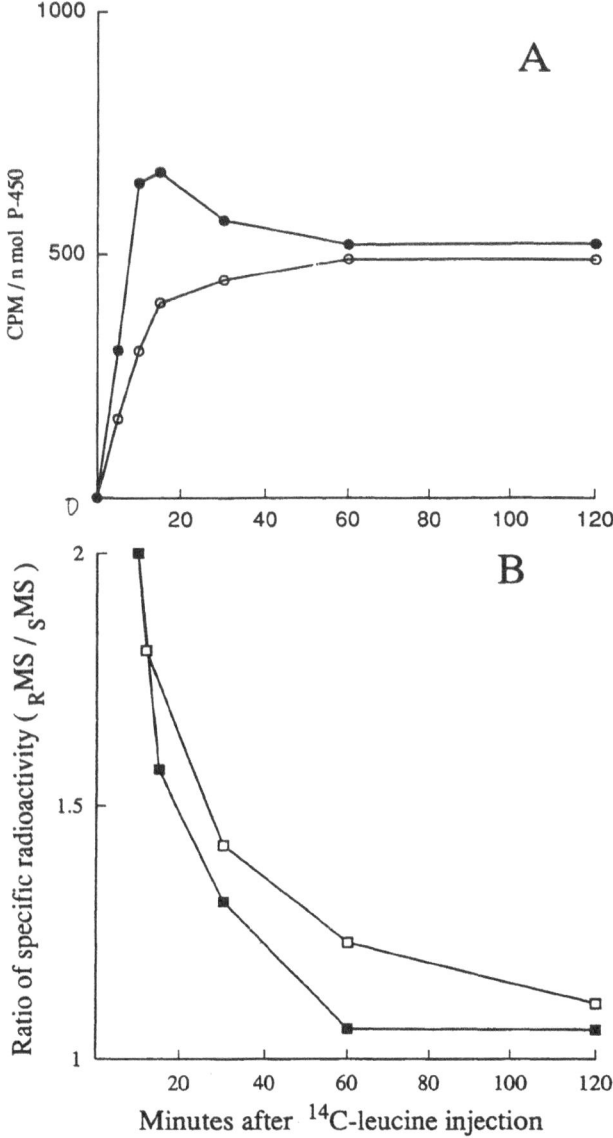

FIGURE 8. (A) Incorporation *in vivo* of [14C]leucine into P-450(PB) in rat liver rough (●) and smooth (○) microsomes. (B) Change in the ratios of the specific radioactivity in the immunoprecipitates of P-450(PB) from rough to smooth microsomes without (■) and with (□)cycloheximide treatment. Reproduced with permission from Fujii-Kuriyama *et al.* (1979).

movement of the protein molecules *per se* on the membranes, or both. To test these alternatives, we followed the change in the radioactivity of P-450 in the rough and the smooth microsomes after complete inhibition of protein synthesis by the injection of cycloheximide. As shown in Figure 8B, the ratios of the specific radioactivities of P-450 in the rough and the smooth microsomes decreased rapidly and approached a value of 1, with or without drug treatment. The time course in the treated rats was essentially similar to that in the untreated rats, except that the rate was somewhat slowed. This result indicates that the newly synthesized P-450 molecules are rapidly translocated from the rough to the smooth ER membrane and distributed evenly within 1 h mainly by the lateral movement of these molecules in the membrane rather than the attachment–detachment of ribosomes during the ribosomal cycle for protein synthesis (Fujii-Kuriyama *et al.*, 1979).

Our results were confirmed by Bar-Nun *et al.* (1980), who showed that P-450 mRNA prepared from PB-treated rat livers is primarily associated with membrane-bound polysomes. They also suggested that microsomal P-450 is directly inserted into the ER membranes because it was not released by treatment with low concentrations of detergent that release albumin and other microsomal content proteins. Furthermore, they pointed out that the N-terminal amino acid sequence of P-450(PB) synthesized in an mRNA-dependent system resembles the signal sequence of presecretory proteins and suggested that it may serve to insert the polypeptide into the membrane.

In 1984, Sakaguchi *et al.* reported the important finding that the insertion of P-450 into microsomal membranes requires signal recognition particle (SRP). Later, they showed (Sakaguchi *et al.*, 1987) that chimeric proteins containing N-terminal P-450 segments consisting of ≥ 29 amino acid residues were cotranslationally inserted into the ER membranes in an SRP-dependent fashion, and concluded that a short N-terminal segment (≥ 29 residues) of P-450 molecules functions not only as an insertion signal but also as a stop-transfer sequence.

Monier *et al.* (1988) also constructed chimeric proteins and showed that the N-terminal 20 residues of P-450IIB1 function as a combined insertion–stop-transfer signal. They also obtained evidence suggesting that during the early stage of insertion, the signal enters the membrane in a loop configuration and reorients within the membrane so that the N-terminus of the signal becomes translocated into the microsomal lumen (Figure 6A).

Table II shows the N-terminal regions of various rat liver microsomal P-450s. In contrast to typical eukaryotic secretory signal sequences, the N-terminal region preceding the hydrophobic sequence lacks a positively charged amino acid residue(s) but has a negatively charged amino acid residue(s), and the hydrophobic sequence is usually longer than the secretory signal sequences, as pointed out by Sato *et al.* (1990).

Substitutions of the N-terminal acidic amino acid with basic amino acid(s)

Table II

Comparison of the N-Terminal Sequences of Rat Microsomal P450s

I A1	MPSVYGFPAFTSATELLLAVTTFCLGFWVVRVTRTWVPKGLKSPPGPWGL	Sogawa et al. (1984)
II A1	MLDTGLLLVVILASLSVMLLVSLWQQKIRGRLPPGPTPLPFIGNYLQLNT	Nagata et al. (1987)
II B1	MEPSILLLLALLVGFLLLLVRGHPKSRGNFPPGPRPLPLLGNLLQLDRGG	Fujii-Kuriyama et al. (1982)
II C7	MDLVTFLVLTLSLILLLSLWRQSSRRRKLPPGPTPLPIIGNFLQIDVKNI	Gonzales et al. (1986)
II C11	MDPVLVLVLTLSSLLLLSLWRQSFGRGKLPPGPTPLPIIGNTLQIYMKDI	Yoshioka et al. (1987)
II C12	MDPFVVLVLSLSFLLLLYLWRPSPGRGKLPPGPTPLPIFGNFLQIDMKDI	Zaphiropoulos et al. (1988)
II C13	MDPVVVLLSLFFLLFLSLWRPSSGRGKLPPGPTPLPIIGNFFQVDMKDI	McClellan-Green et al. (1989)
II D9	MDPVLVLVLTLSSLLLLSLWRQSFGRGKLPPGPTPLPIIGNTLQIYMKDI	Strom et al. (1988)
II E1	MAVLGITIALLVWVATLLVISIWKKIYNSWNLPPGPFPLPILGNIFQLDL	Song et al. (1986)
III A1	MDLLSALTLETWVLLAVVLVLLYGFGTRTHGLFKKQGIPGPKPLPFFGTV	Gonzales et al. (1985)
VII	MMTISLIWGIAVLVSCCIWFIVGIRRRKAGEPPLENGLIPYLGCALKFGS	Li et al. (1990)
X VII A1	MWELVGLLLLILAYFFWVKSKTPGAKLPRSLPSLPLVGSLPFLPRRGHMH	Namiki et al. (1988)

converted the signal-anchor peptide of P-450 to a secretory signal peptide (Szezesna-Skorupa *et al.*, 1988; Szcesna-Skorupa and Kemper, 1989; Sato *et al.*, 1990). Shortening of the hydrophobic core sequences showed similar effect (Sato *et al.*, 1990).

Sakaguchi *et al.* (1992) analyzed the functions of signal and signal-anchor sequences systematically by changing the charge and the hydrophobicity and showed that their functions are determined by the balance between the N-terminal charge and the hydrophobicity. According to their interpretation, the longer hydrophobic segment appears to pull the negatively charged N-terminal portion into the membrane, resulting in the translocation of the N-terminus into the ER lumen, as suggested by Monier *et al.* (1988).

4. INTRACELLULAR DISTRIBUTION

We have shown in the previous sections that microsomal P-450 is present in high concentration on the cytoplasmic surfaces of the rough and smooth microsomes and of the outer nuclear membranes. Since it is well established that liver microsomes are mostly derived from the ER membranes, this strongly suggests that the cytoplasmic surfaces of ER membranes in hepatocytes are heavily loaded with P-450. In order to demonstrate this directly, we visualized the distribution of P-450(PB) in rat hepatocytes by using cryoimmunogold electron microscopy. As shown in Figure 9, both the rough and smooth ER as well as the nuclear envelopes are heavily labeled.

In this section we discuss (1) whether microsomal P-450 is distributed equally between rough and smooth ER membranes; (2) whether it is distributed homogeneously within each ER membrane; and (3) whether it is present on other organelles such as the Golgi apparatus, plasma membrane, and outer mitochondrial membrane.

Table III shows the amount of microsomal P-450s determined spectrophotometrically from the carbon monoxide difference spectra. The specific content of microsomal P-450s in the smooth microsomes from control rat livers was always ~ twice that in the rough microsomes. These data are consistent with the morphological observations of Figure 1, in which it can be seen that smooth microsomes are more densely labeled than rough microsomes. The higher concentration of P-450 in the smooth ER might be explained by the exclusion of P-450 molecules from the protein biosynthetic domains of the rough microsomes.

As can be seen in Figure 1, the distribution of ferritin particles on each microsome is not always homogeneous, occasionally forming clusters or patches. The heterogeneous distribution of ferritin particles was confirmed by statistical analyses of the labeled microsomes (Matsuura *et al.*, 1979). An oli-

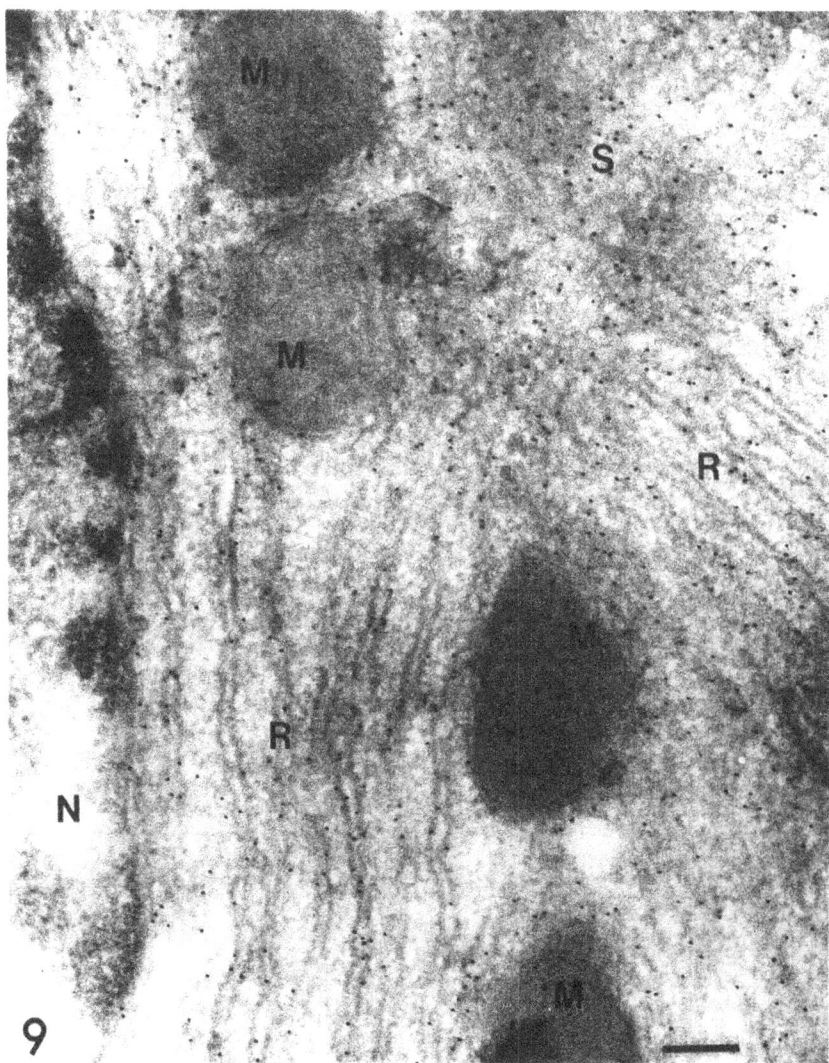

FIGURE 9. Immunogold localization of P-450IIB in frozen ultrathin sections of rat hepatocytes maximally induced by PB treatment (72 h treatment). M, mitochondria; N, nucleus; R, rough ER region; S, smooth ER region. Bar = 0.2 μm. Reproduced with permission from Fukui *et al.* (1992).

Table III

**Cytochrome P-450 Content of Liver Microsomes from Untreated and
Phenobarbital-Treated Rats (nmoles/mg protein)**

	Sato and Omura (1978)	Matsuura *et al.* (1978)		Fujii-Kuriyama *et al.* (1979)
PB treatment:	−	−	+	−
Total microsomes	1.0 (5%)[a]	0.62	1.81	—
Rough microsomes	—	0.45	0.57	0.71
Smooth microsomes	—	1.02	2.55	1.1

[a]Percent of total microsomal protein.

gomeric and membrane-spanning aggregate for the topology of microsomal P-450 has been suggested by Schwarz *et al.* (1990).

These results support the model proposed by Peterson *et al.* (1976) that P-450 molecules are clustered in the ER membranes around a small number of NADPH P-450 reductase molecules. However, whether P-450 molecules are actually clustered around reductase molecules *in vivo* in the ER membranes should be reexamined carefully by rapid-freezing, freeze-substitution technique combined with highly sensitive immunoelectron microscopy.

In marked contrast to the ER and outer nuclear envelope, the other cytoplasmic organelles, such as Golgi apparatus, plasma membranes, and mitochondria, were barely labeled with ferritin particles (Matsuura *et al.*, 1978). It is to be noted that P-450(PB) was not detected in the *trans* or *cis* regions of the Golgi apparatus as shown in Figure 10. Even the transfer vesicles containing very-low-density lipoprotein particles were not labeled with ferritin particles. This observation strongly suggests that when transfer vesicles are formed by budding from transitional elements of the ER, P-450(PB) is excluded from such regions and is not transported to the Golgi apparatus (Matsuura and Tashiro, 1979; Brands *et al.*, 1985; Yamamoto *et al.*, 1985).

Thus, P-450(PB) does not get transported out of the ER system and is indeed a resident membrane protein of ER. Its presence on outer nuclear membranes is consistent with the cytophilosophy that the ER is derived from the outer nuclear membranes or vice versa and that these two membranes are practically identical (Franke, 1974; Matsuura *et al.*, 1983).

5.   INDUCTION OF P-450 AND PROLIFERATION OF ER

The administration of drugs such as PB induces a marked increase in the ER of hepatocytes, with a parallel enhancement in the activity of drug-metabolizing enzymes (Remmer and Merker, 1965; Conney, 1967; Jones and Fawcett, 1965;

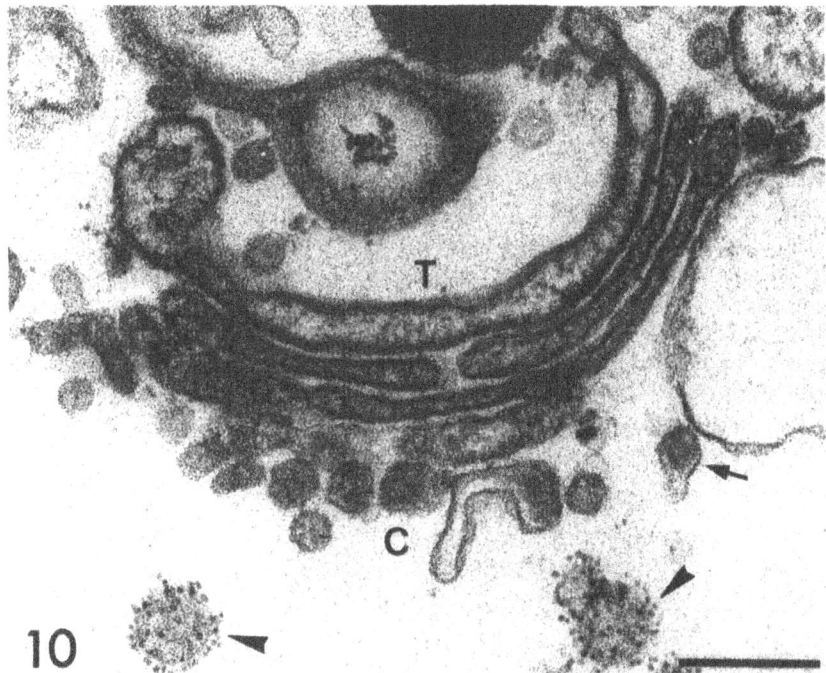

FIGURE 10. Rat liver Golgi apparatus incubated with ferritin anti-P-450(PB) antibody conjugates. Note lack of labeling of both the *cis* (C) and *trans* (T) sides of the Golgi apparatus. The transfer vesicles containing very-low-density lipoprotein (arrow) are not labeled. Note marked labeling of the microsomes (arrowheads). Bar = 0.2 μm. Reproduced with permission from Yamamoto *et al.* (1985).

Orrenius and Ericsson, 1966; Stäubli *et al.*, 1969). Stäubli *et al.* (1969) demonstrated a collinearity between the proliferation of smooth ER by PB treatment and the increase in the activity of the drug-metabolizing enzymes.

Hydrophobic drugs are usually oxygenated in liver by NADPH cytochrome P-450 reductase and P-450 isozymes (phase I metabolism), then further metabolized by phase II enzymes (e.g., glucuronidation, and sulfate glutathione, or glycine conjugation) to be converted to excretable products. Most of these phase I and II enzymes are localized in the ER, and it is quite reasonable that they are usually induced in parallel with the proliferation of smooth ER. PB treatment leads to a two- to threefold increase in the microsomal content of P-450 (Table III). When such microsomes were labeled by preembedding ferritin immunoelectron microscopy, the cytoplasmic surfaces of the smooth microsomes were almost completely covered with ferritin particles as shown in Figure 2.

It has become evident that multiple forms of P-450 exist, and more than ten

different forms have been isolated from rat liver (Gonzales, 1988; Nebert and
Gonzales, 1987). P-450IIB1 and IIB2 are the main members of the PB-inducible
P-450 family and are induced more than 30-fold by treatment with PB (Adesnik
et al., 1981; Omiecinski *et al.*, 1985; Thomas *et al.*, 1983).

Recently we have investigated the induction of P-450IIB in rat hepatocyte
ER by immunoelectron microscopy, by using monospecific anti-P-450IIB antibody, which was prepared by immunoabsorption of anti-P-450(PB) antibody 2
with liver microsomes prepared from control rats (not treated with PB). As
shown in Figure 11, the labeling density of the rough ER was practically zero in
control rats, and increased markedly after PB treatment to a saturation level of ~
0.6 particles/μm membrane, which corresponds to n 0.4 nmol P-450/mg protein.

Similar quantitative immunoelectron microscopic analysis of the induction
of P-450(PB) has been reported by Marti *et al.* (1990), using monoclonal anti-
P-450(PB) antibody. According to these authors, microsomal P-450(PB) increased from 0.059 ± to 0.014 nmol/mg protein to 0.896 ± 0.046 nmol/mg, in
untreated animals and maximally induced rats, respectively. The spectrally mea-

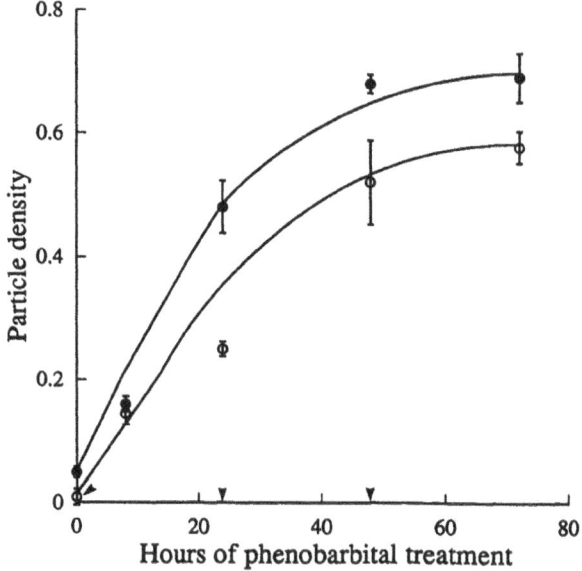

FIGURE 11. Increase in the specific labeling density of gold particles on rat hepatocyte rough ER
during PB induction. Ultrathin LR White sections were incubated with nonimmunoabsorbed or
immunoabsorbed antibodies against P-450IIB and nonimmune IgG, respectively, and the specific
binding (number of ferritin particles per micrometer of rough microsome membrane) for nonimmunoabsorbed (●) and immunoabsorbed antibodies (○) was calculated by subtracting control binding. Arrowheads indicate the time of PB injection. Average of three experiments ± SD.

sured P-450 increased from 0.447 ± 0.042 nmol/mg to 2.05 ± 0.174 nmol/mg. The amount of P-450(PB) of total spectrally measurable P-450 increased from 13.2% in untreated animals to 43.7% in maximally induced animals.

After cessation of PB treatment, the proliferated ER membranes and the induced microsomal P-450s are rapidly removed as described in the following section.

The PB induction response involves an increase in several mRNAs, including those coding for P-450IIB/IIB2, NADPH P-450 reductase, epoxide hydrolase, UDP-glucuronosyltransferase and also for other P-450s. Unfortunately, data regarding a specific PB receptor and cis-acting DNA sequence elements in this induction process are lacking. This is in marked contrast to the MC/2,3,7,8-tetrachlorodibenzoparadioxin (TCDD) induction response, where a TCDD receptor has been identified and the molecular mechanism for induction has been extensively studied as described by Nebert and Gonzales (1987), Gonzales (1988), and Fujii-Kuriyama et al. (1992).

6. TURNOVER AND DEGRADATION

The concentration of a membrane protein in the cell is determined by an equilibrium between its synthesis and degradation. Induction of microsomal P-450 could be explained by either an altered rate of synthesis, degradation, or both. The increase in the transcription of P-450 genes has been described in detail (Fujii-Kuriyama et al., 1992; Gonzales, 1988; Nebert and Gonzales, 1987). We will consider only the turnover and degradation of microsomal P-450 in hepatocytes.

Table IV shows the half-life data of P-450(PB) and some other microsomal membrane proteins (Gasser et al., 1982; Sadano and Omura, 1983; Parkinson et al., 1983; Shiraki and Guengerich, 1984). The estimation of the half-life of P-450 may be affected by several factors such as the reutilization of amino acid, the previous treatment with inducers such as PB, and the specificity of the antibody used for immunoprecipitation; thus, the data obtained must be considered with caution.

The following three conclusions, however, can be deduced from Table IV. First, there are large differences in the half lives of the microsomal proteins in the liver of untreated rats. Second, this difference in half lives becomes significantly smaller, converging to a constant value of ~ 24 h, when rats are treated with PB. Third, the heme of P-450(PB) turns over faster than its protein moiety.

As to the degradation of microsomal P-450, several possible routes can be suggested. The first is that P-450 is transported to the lysosomal compartment via the Golgi apparatus to be degraded there. We have shown, however, that P-450(PB) is not detectable in the Golgi even when P-450(PB) was maximally

Table IV
Half Lives of Microsomal Proteins

Treatment: Moieties[a]:	Gasser et al. (1982) PB(+)		Sadano and Omura (1983) PB(-)		Sadano and Omura (1983) PB(+)		Parkinson et al. (1983) Aroclor 1254-treated		Shiraki and Guengerich (1984) PB(-)		Shiraki and Guengerich (1984) PB(+)		Shiraki and Guengerich (1984) NF(+)	
	P	H	P	H	P	H	P	H	P	H	P	H	P	H
Total protein			35	20	25	—	82	41	20 ± 1	15 ± 1	29 ± 1	22 ± 2	36 ± 2	39 ± 2
NADPH P-450 reductase			35		25				29 ± 1	—	38 ± 6	—	23 ± 2	—
Cytochrome h$_5$			50	40	30				—	—	—	—	—	—
P-450 (PB)	19	12	25	15	20		37	28	19 ± 1	10 ± 2	37 ± 2	16 ± 2	20 ± 2	12 ± 6

[a]P, protein moiety; H, heme moiety.

induced and the content in the ER rapidly decreased after cessation of PB treatment, as shown in Figure 10 (Matsuura and Tashiro, 1979; Yamamoto *et al.*, 1985). Similar results have been reported by Brands *et al.* (1985) and Marti *et al.* (1990). Therefore, this cannot be a major route of degradation.

The second is that the ER membranes containing P-450 are directly segregated and degraded in the autophagosome–autolysosome system. The wide differences in the half lives of the microsomal proteins including P-450 do not support the possibility that microsomal proteins are exclusively degraded by autophagocytic processes. As pointed out above, however, this difference became significantly smaller when rats were treated with PB, suggesting active involvement of autophagic processes under such a condition.

In fact, Bolender and Weibel (1973) showed by morphometric analysis that the removal of PB-induced excess ER membranes was associated with an increase in autophagic activity, and we also showed that a decrease in the content of microsomal proteins and P-450(PB) after cessation of PB treatment parallels an increase in the number of autophagic vacuoles (Masaki *et al.*, 1984).

In a subsequent study, we prepared autophagosomes and autolysosomes almost completely free from contamination by other organelles and presented biochemical and ferritin immunoelectron microscopic (Figure 12) evidence that ER membranes heavily loaded with P-450(PB) are segregated and degraded in the autophagic vacuoles (Masaki *et al.*, 1987). Quite interestingly, only when the proteolytic processes in the autolysosomes were inhibited by administering leupeptin was the content of the autolysosomes efficiently labeled with ferritin particles (Yamamoto *et al.*, 1990b), clearly indicating that P-450(PB) is in fact degraded in the autolysosomes.

The third possibility, that microsomal P-450(PB) is degraded in the ER, is suggested by the probable presence of proteolytic enzymes within the ER (see Fra and Sitia, this volume), as indicated by the regulated degradation of 3-hydroxy-3-methylglutaryl CoA reductase (Chun *et al.*, 1990), and by the disposal of newly synthesized abnormal or unassembled T cell antigen receptor (Klausner and Sitia, 1990; Lippincott-Schwartz *et al.*, 1988). No experimental evidence, however, has been presented on ER degradation of microsomal P-450.

Molecular signals that lead P-450 molecules to rapid degradation have not been identified. Recently, however, it has been shown that phosphorylation of serine-128 of P-450IIB1 or IIB2 causes a conformational change of the reduced cytochrome P-450 to its enzymatically inactive form, P-420, then to dissociation of heme to produce apo P-420, which is rapidly degraded (Pyerin and Taniguchi, 1989; Koch and Waxman, 1989). This dissociation of heme may explain the faster turnover of the heme moiety than of the protein moiety (Sadano and Omura, 1983; Shiraki and Guengerich, 1984).

Eliasson *et al.* (1990) pointed out that phosphorylation of serine-129 in P-450IIE1 molecules is catalyzed by cAMP-dependent protein kinase, and re-

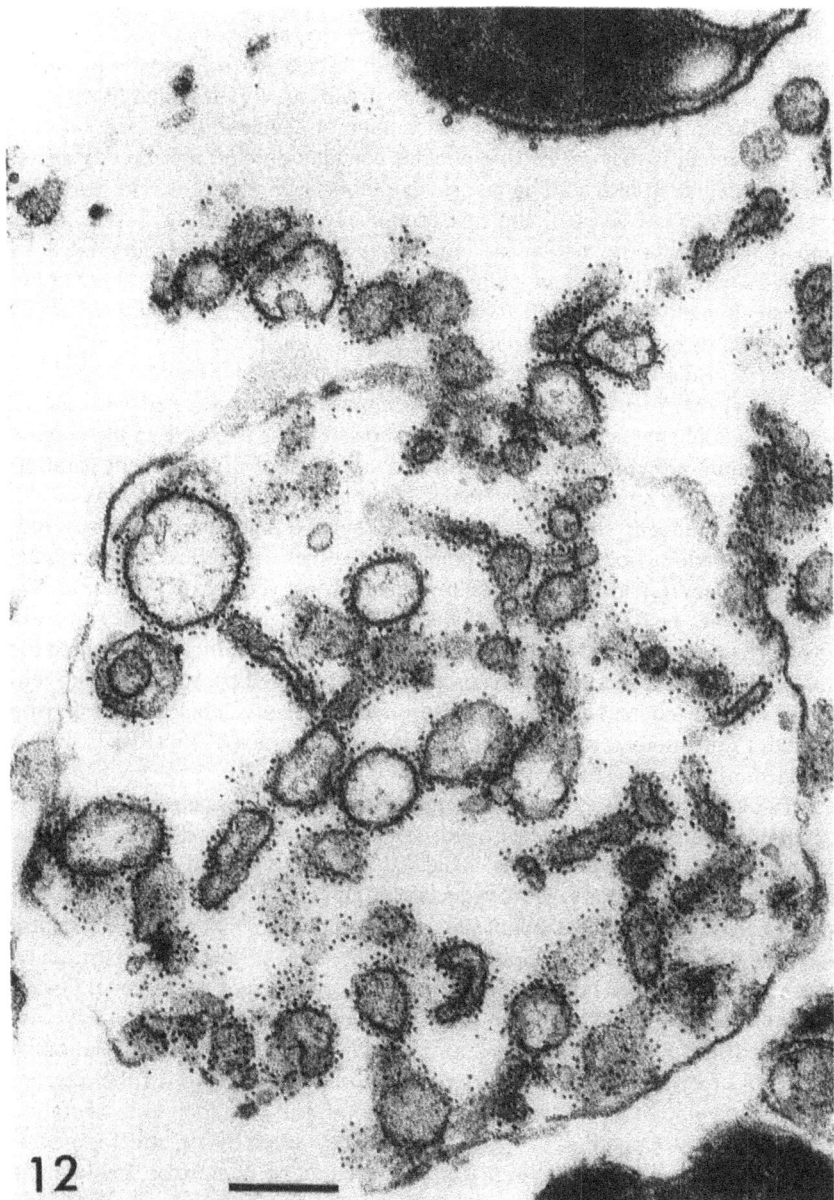

FIGURE 12. Autolysosomal fraction incubated with ferritin anti-P-450(PB) antibody conjugates. The ER structures that were engulfed within an autophagosome are markedly labeled with ferritin particles, whereas the limiting membrane of the autophagosome was barely labeled. Bar = 0.2 μm. Reproduced with permission from Masaki *et al.* (1987).

ported that glucagon or 8-bromoadenosine 3',5'-cyclic AMP causes an enhanced rate of P-450IIE1 degradation in the hepatocytes as well as phosphorylation on Ser-129. Interestingly, substrates for this enzyme, such as ethanol and imidazole, protect the enzyme from phosphorylation and therefore degradation in hepatocytes.

It has not been clarified whether the phosphorylated P-450(PB) is selectively degraded by some cytosolic or microsomal proteinase or whether portions of the ER membrane heavily loaded with phosphorylated P-450 are selectively degraded by autophagocytosis.

In this connection, it is very interesting to point out that a number of investigators have proposed that the limiting membranes surrounding the autophagosomes are derived from ER membranes (Arstila and Trump, 1968; Ericsson, 1969; Reunanen and Hirsimaki, 1983; Dunn, 1990).

As illustrated in Figure 12, however, we have shown that P-450(PB) was not detected on the limiting membranes, even after maximal induction of P-450(PB) (Masaki et al., 1987; Yamamoto et al., 1990b). The limiting membranes were labeled with antibodies against autolysosomal membrane antigens (Yamamoto et al., 1990a) and also with wheat germ agglutinin and Ricinus communis agglutinin. We suggested, therefore, that the limiting membranes are not derived from the ER membranes but from some post-ER membranes loaded with complex-type oligosaccharides. Even if the former possibility were correct, the ER membranes would have to be transformed extensively to be converted to the autophagosomal membrane, given the complete exclusion of P-450 molecules. This problem has not been settled and should be investigated further.

NOTE ADDED IN PROOF: Recently, further evidence has been reported that the N-terminus of P-450 molecules is luminally oriented (Szczesna-Skorupa and Kemper, 1993; Harada, 1993).

7. REFERENCES

Adesnik, M., Bar-Nun, S., Maschio, F., Zunich, M., Lippman, A., and Bard, E., 1981, Mechanism of induction of cytochrome P-450 by phenobarbital, J. Biol. Chem. 256:10340–10345.

Arstila, A. U., and Trump, B. F., 1968, Studies on cellular autophagocytosis. The formation of autophagic vacuoles in the liver after glucagon administration, Am. J. Pathol. 53:687–733.

Bar-Nun, S., Kreibich, G., Adesnik, M., Alterman, L., Negishi, M., and Sabatini, D. D., 1980, Synthesis and insertion of cytochrome P-450 into endoplasmic reticulum membranes, Proc. Natl. Acad. Sci. USA 77:965–969.

Bolender, R. P., and Weibel, E. R., 1973, A morphometric study of the removal of phenobarbital-induced membranes from hepatocytes after cessation of treatment, J. Cell Biol. 56:746–761.

Brands, R., Snider, M. D., Hino, Y., Park, S. S., Gelboin, H. V., and Rothman, J. E., 1985, Retention of membrane proteins by the endoplasmic reticulum, J. Cell Biol. 101:1724–1732.

Brown, C. A., and Black, S. D., 1989, Membrane topology of mammalian cytochromes P-450 from liver endoplasmic reticulum. Determination by trypsinolysis of phenobarbital-treated microsomes, *J. Biol. Chem.* **264**:4442–4449.

Chun, K. T., Bar-Nun, S., and Simoni, R. D., 1990, The regulated degradation of 3-hydroxy-3-methylglutaryl-CoA reductase requires a short-lived protein and occurs in the endoplasmic reticulum, *J. Biol. Chem.* **265**:22004–22010.

Conney, A. H., 1967, Pharmacological implications of microsomal enzyme induction, *Pharmacol. Rev.* **19**:317–366.

De Lemos-Chiarandini, C., Frey, A. B., Sabatini, D. D., and Kreibich, G., 1987, Determination of the membrane topology of the phenobarbital-inducible rat liver cytochrome P-450 isoenzyme PB-4 using site-specific antibodies, *J. Cell Biol.* **104**:209–219.

Dunn, W. A., Jr., 1990, Studies on the mechanisms of autophagy: Formation of the autophagic vacuole, *J. Cell Biol.* **110**:1923–1933.

Eliasson, E., Johansson, I., and Ingelman-Sundberg, M., 1990, Substrate-, hormone-, and cAMP-regulated cytochrome P-450 degradation, *Proc. Natl. Acad. Sci. USA* **87**:3225–3229.

Ericsson, J. L. E., 1969, Studies on induced cellular autophagy. II. Characterization of the membranes bordering autophagosomes in parenchymal liver cells, *Exp. Cell Res.* **56**:393–405.

Franke, W. W., 1974, Structure, biochemistry and functions of the nuclear envelope, *Int. Rev. Cytol. Suppl.* **4**:71–236.

Fujii-Kuriyama, Y., Negishi, M., Mikawa, R., and Tashiro, Y., 1979, Biosynthesis of cytochrome P-450 on membrane-bound ribosomes and its subsequent incorporation into rough and smooth microsomes in rat hepatocytes, *J. Cell Biol.* **81**:510–519.

Fujii-Kuriyama, Y., Mizukami, Y., Kawajiri, K., Sogawa, K., and Muramatsu, M., 1982, Primary structure of a cytochrome P-450: Coding nucleotide sequence of phenobarbital-inducible cytochrome P-450 cDNA from rat liver, *Proc. Natl. Acad. Sci. USA* **79**:2793–2797.

Fujii-Kuriyama, Y., Imataka, H., Sogawa, K., Yasumoto, K., and Kikuchi, Y., 1992, Regulation of CYP1A1 expression, *FASEB J.* **6**:706–710.

Fukui, Y., Yamamoto, A., Masaki, R., Miyauchi, K., and Tashiro, Y., 1992, Quantitative immunocytochemical analysis of the induction of cytochrome P-450IIB in rat hepatocytes, *J. Histochem. Cytochem.* **40**:73–82.

Gasser, R., Hauri, H. P., and Meyer, U. A., 1982, The turnover of cytochrome P-450b, *FEBS Lett.* **147**:239–242.

Gonzales, F. J., 1988, The molecular biology of cytochrome P-450s, *Pharmacol. Rev.* **40**:243–288.

Gonzales, F. J., Nebert, D. W., Hardwick, J. P., and Kasper, C. B., 1985, Complete cDNA and protein sequence of a pregnenolone 16-alpha-carbonitrile-induced cytochrome P-450: A representative of a new gene family, *J. Biol. Chem.* **260**:7435–7441.

Gonzales, F. J., Kimura, S., Song, B.-J., Pastewka, J., Gelboin, H. V., and Hardwick, J. P., 1986, Sequence of two related P-450 mRNAs transcriptionally increased during rat development: An R. dre. 1 sequence occupies the complete 3' untranslated region of a liver mRNA, *J. Biol. Chem.* **261**:10667–10672.

Harada, N., 1993 *Biohchem., Biophys. Res. Commun.* **156**:725–732

Imai, Y., and Sato, R., 1974, A gel-electrophoretically homogeneous preparation of cytochrome P-450 from liver microsomes of phenobarbital-pretreated rabbits, *Biochem. Biophys. Res. Commun.* **60**:8–14.

Jones, A L., and Fawcett, D. W., 1965, Hypertrophy of the agranular reticulum in hamster liver induced by phenobarbital (with a review on the functioning of this organelle in liver), *J. Histochem. Cytochem.* **14**:215–232.

Klausner, R., and Sitia, R., 1990, Protein degradation in the endoplasmic reticulum, *Cell* **62**:611–614.

Koch, J. A., and Waxman, D. J., 1989, Posttranslational modification of hepatic cytochrome P-450. Phosphorylation of phenobarbital-inducible P-450 forms PB-4 (IIB1) and PB-5(IIB2) in isolated rat hepatocytes and in vivo, *Biochemistry* **28**:3145–3152.

Li, Y. C., Wang, D. P., and Chiang, Y. L., 1990, Regulation of cholesterol 7α-hydroxylase in the liver. Cloning, sequencing, and regulation of cholesterol 7α-hydroxylase mRNA, *J. Biol. Chem.* **265**:12012–12019.

Lippincott-Schwartz, J., Bonifacino, J. S., Yuan, L. C., and Klausner, R. D., 1988, Degradation from the endoplasmic reticulum: Disposing of newly synthesized proteins, *Cell* **54**:209–220.

McClellan-Green, P. D., Negishi, M., and Goldstein, J. A., 1989, Characterization of a cDNA for rat P-450g, a highly polymorphic, male-specific cytochrome in the P-450IIC subfamily, *Biochemistry* **28**:5832–5839.

Marti, U., Hauri, H.-P., and Meyer, U. A., 1990, Induction of cytochrome P-450 by phenobarbital in rat liver visualized by monoclonal antibody immunoelectron microscopy in situ, *Eur. J. Cell Biol.* **52**:193–200.

Masaki, R., Matsuura, S., and Tashiro, Y., 1984, A biochemical and electron microscopic study of changes in the content of cytochrome P-450 in rat livers after cessation of treatment with phenobarbital, β-naphthoflavone or 3-methylcholanthrene, *Cell Struct. Funct.* **9**:53–66.

Masaki, R., Yamamoto, A., and Tashiro, Y., 1987, Cytochrome P-450 and NADPH-cytochrome P-450 reductase are degraded in the autolysosomes in rat liver, *J. Cell Biol.* **104**:1207–1215.

Matsuura, S., and Tashiro, Y., 1979, Immunoelectron-microscopic studies of endoplasmic reticulum–Golgi relationships in the intracellular transport process of lipoprotein particles in rat hepatocytes, *J. Cell Sci.* **39**:273–290.

Masuda-Mikawa R., Fujii-Kuriyama Y., Negishi M., and Tashiro Y., 1979, Purification and partial characterization of hepatic microsomal cytochrome P-450s from phenobarbital and 3-methyl-cholanthrene-treated rats. *J. Biochem.* **86**:1383–1394.

Matsuura, S., Fujii-Kuriyama, Y., and Tashiro, Y., 1978, Immunoelectron microscope localization of cytochrome P-450 on microsomes and other membrane structures of rat hepatocytes, *J. Cell Biol.* **78**:503–519.

Matsuura, S., Fujii-Kuriyama, Y., and Tashiro, Y., 1979, Quantitative immunoelectron microscopic analyses of the distribution of cytochrome P-450 molecules on rat liver microsomes, *J. Cell Sci.* **36**:413–435.

Matsuura, S., Masuda, R., Omori, K., Negishi, M., and Tashiro, Y., 1981, Distribution and induction of cytochrome P-450 in rat liver nuclear envelope, *J. Cell Biol.* **91**:212–220.

Matsuura, S., Masuda, R., Sakai, O., and Tashiro, Y., 1983, Immunoelectron microscopy of the outer membrane of rat hepatocyte nuclear envelopes in relation to the rough endoplasmic reticulum, *Cell Struct. Funct.* **8**:1–9.

Monier, S., Van Luc, P., Kreibich, G., Sabatini, D. D., and Adesnik, M., 1988, Signals for the incorporation and orientation of cytochrome P-450 in the endoplasmic reticulum membrane, *J. Cell Biol.* **107**:457–470.

Nagata, K., Matsunaga, T., Gillette, J., Gelboin, H. V., and Gonzales, F. J., 1987, Rat testosterone 7-alpha-hydroxylase: Isolation, sequence, and expression of cDNA and its developmental regulation and induction by 3-methylcholanthrene, *J. Biol. Chem.* **262**:2787–2793.

Namiki, M., Kitamura, M., Buczko, E., and Dufau, M. L., 1988, Rat testis P-450 17 alpha cDNA: The deduced amino acid sequence, expression and secondary structural configuration, *Biochem. Biophys. Res. Commun.* **157**:705–712.

Nebert, D. W., and Gonzales, F. J., 1987, P-450 genes: Structure, evolution, and regulation, *Annu. Rev. Biochem.* **56**:945–993.

Negishi, M., Fujii-Kuriyama, Y., and Tashiro, Y., 1976, Site of biosynthesis of cytochrome P-450 in hepatocytes of phenobarbital treated rats, *Biochem. Biophys. Res. Commun.* **71**:1153–1160.

Nelson, D. R., and Strobel, H. W., 1988, On the membrane topology of vertebrate cytochrome P-450 proteins, *J. Biol. Chem.* **263**:6038–6050.

Omiecinski, C. J., Walz, F. G., Jr., and Vlasuk, G. P., 1985, Phenobarbital induction of rat liver cytochromes P-450b and P-450e, *J. Biol. Chem.* **260**:3247–3250.

Orrenius, S., and Ericsson, J. L. E., 1966, Enzyme–membrane relationship in phenobarbital induction of synthesis of drug-metabolizing enzyme system and proliferation of endoplasmic membranes, *J. Cell Biol.* **28**:181–198.

Ozols, J., Heinemann, F. S., and Johnson, E. R., 1985, The complete amino acid sequence of a constitutive form of liver microsomal cytochrome P-450, *J. Biol. Chem.* **260**:5427–5434.

Parkinson, A., Thomas, P. E., Ryan, D. E., and Levin, W., 1983, The in vivo turnover of rat liver microsomal epoxide hydrolase and both the apoprotein and heme moieties of specific cytochrome P-450 isozymes, *Arch. Biochem. Biophys.* **225**:216–236.

Peterson, J., Ebel, R. E., O'Keeffe, D. H., Matsubara, T., and Estabrook, R. W., 1976, Temperature dependence of cytochrome P-450 reduction, *J. Biol. Chem.* **251**:4010–4016.

Porter, T. D., and Coon, M. J., 1991, Cytochrome P-450. Multiplicity of isoforms, substrates, and catalytic and regulatory mechanisms, *J. Biol. Chem.* **266**:13469–13472.

Pyerin, W., and Taniguchi, H., 1989, Phosphorylation of hepatic phenobarbital-inducible cytochrome P-450, *EMBO J.* **8**:3003–3010.

Remmer, H., and Merker, H. J., 1965, Effect of drugs on the formation of smooth endoplasmic reticulum and drug-metabolizing enzymes, *Ann. N.Y. Acad. Sci.* **123**:79–97.

Reunanen, H., and Hirsimaki, P., 1983, Studies on vinblastine-induced autophagocytosis in mouse liver. IV. Origin of membranes, *Histochemistry* **79**:59–67.

Sadano, H., and Omura, T., 1983, Turnover of two drug-inducible forms of microsomal cytochrome P-450 in rat liver, *J. Biochem.* **93**:1375–1383.

Sakaguchi, M., Mihara, K., and Sato, R., 1984, Signal recognition particle is required for cotranslational insertion of cytochrome P-450 into microsomal membranes, *Proc. Natl. Acad. Sci. USA* **81**:3361–3364.

Sakaguchi, M., Mihara, K., and Sato, R., 1987, A short amino-terminal segment of microsomal cytochrome P-450 functions both as an insertion signal and as a stop-transfer sequence, *EMBO J.* **6**:2425–2431.

Sakaguchi, M., Tomiyoshi, R., Kuroiwa, T., Mihara, K., and Omura, T., 1992, Functions of signal and signal-anchor sequences are determined by the balance between the hydrophobic segment and the N-terminal charge, *Proc. Natl. Acad. Sci. USA* **89**:16–19.

Sato, R., and Omura, T., 1978, *Cytochrome P-450*, Kodansha Ltd., Tokyo Academic Press, New York.

Sato, T., Sakaguchi, M., Miraha, D., and Omura, T., 1990, The amino-terminal structures that determine topological orientation of cytochrome P-450 in microsomal membrane, *EMBO J.* **9**:2391–2397.

Schwartz, D., Pirrwitz, J., Meyer, H. W., Coon, M. J., and Ruckpaul, K., 1990, Membrane topology of microsomal cytochrome P-450: Saturation transfer EPR and freeze-fracture electron microscopy studies, *Biochem. Biophys. Res. Commun.* **171**:175–181.

Shiraki, H., and Guengerich, F. P., 1984, Turnover of membrane proteins: Kinetics of induction and degradation of seven forms of rat liver microsomal cytochrome P-450, NADPH-cytochrome P-450 reductase, and epoxide hydrolase, *Arch. Biochem. Biophys.* **235**:86–96.

Sogawa, K., Gotoh, O., Kawajiri, K., and Fujii-Kuriyama, Y., 1984, Distinct organization of methylcholanthrene- and phenobarbital-inducible cytochrome P-450 genes in the rat, *Proc. Natl. Acad. Sci. USA* **81**:5066–5070.

Song, B.-J., Gelboin, H. V., Park, S.-S., Yang, C. S., and Gonzales, F. J., 1986, Complementary DNA and protein sequences of ethanol-inducible rat and human cytochrome P-450s, *J. Biol. Chem.* **261**:16689–16697.

Stäubli, W., Hess, R., and Weibel, E. R., 1969, Correlated morphometric and biochemical studies on the liver cell. II. Effect of phenobarbital on rat hepatocytes, *J. Cell Biol.* **41**:92–112.

Strom, A., Mode, A., Zaphiropoulos, P., Nilsson, A. G., Morgan, E., and Gustafsson, J. A., 1988, Cloning and pretranslational hormonal regulation of testosterone 16 alpha-hydroxylase (P-450 16 alpha) in male rat liver. *Acta Endocrino. (Copenhagen)* **118**:314–320.

Szczesna-Skorupa, E., and Kemper, B., 1989, NH_2-terminal substitutions of basic amino acids induce translocation across the microsomal membrane and glycosylation of rabbit cytochrome P-450IIC2, *J. Cell Biol.* **108**:1237–1243.

Szczesna-Skorupa, E., and Kemper, B., 1993, *J. Biol. Chem.* **268**:1757–1762.

Szczesna-Skorupa, E., Browne, N., Mead, D., and Kemper B., 1988, Positive charges at the NH_2terminus convert the membrane-anchor signal peptide of cytochrome P-450 to a secretory signal peptide, *Proc. Natl. Acad. Sci. USA* **85**:738–742.

Tarr, G. E., Black, S. D., Fujita, V. S., and Coon, M. J., 1983, Complete amino acid sequence and predicted membrane topology of phenobarbital-induced cytochrome P-450 (isozyme 2) from rabbit liver microsomes, *Proc. Natl. Acad. Sci. USA* **80**:6552–6556.

Thomas, P. E., Lu, A. Y. H., West, S. B., Ryan, D., Miwa, G. I., and Levin, W., 1978, Accessibility of cytochrome P-450 in microsomal membranes: Inhibition of metabolism by antibodies to cytochrome P-450, *Mol. Pharmacol.* **13**:819–831.

Thomas, P. E., Reik, L. M., Ryan, D. E., and Levin, W., 1983, Induction of two immunochemically related rat liver cytochrome P-450 isozymes, cytochromes P-450c and P-450d, by structurally diverse xenobiotics, *J. Biol. Chem.* **258**:4590–4598.

Vergeres, G., Winterhalter, K. H., and Richter, C., 1991, Localization of the N-terminal methionine of rat liver cytochrome P-450 in the lumen of the endoplasmic reticulum, *Biochim. Biophys. Acta* **1063**:235–241.

Welton, A. F., and Aust, S. D., 1974, The effects of 3-methylcholanthrene and phenobarbital induction of the structure of the rat liver, *Biochim. Biophys. Acta* **373**:197–210.

Welton, A. F., O'Neal, F. O., Chaney, L. C., and Aust, S. D., 1975, Multiplicity of cytochrome P-450 hemoproteins in rat liver microsomes, *J. Biol. Chem.* **250**:5631–5639.

Yamamoto, A., Masaki, R., and Tashiro, Y., 1985, Is cytochrome P-450 transported from the endoplasmic reticulum to the Golgi apparatus in rat hepatocytes? *J. Cell Biol.* **101**:1733–1740.

Yamamoto, A., Masaki, R., Fukui, Y., and Tashiro, Y., 1990a, Absence of cytochrome P-450 and presence of autolysosomal membrane antigens on the isolation membranes and autophagosomal membranes in rat hepatocytes, *J. Histochem. Cytochem.* **38**:1571–1581.

Yamamoto, A., Masaki, R., and Tashiro, Y., 1990b, Characterization of the isolation membranes and the limiting membranes of autophagosomes in rat hepatocytes by lectin cytochemistry, *J. Histochem. Cytochem.* **38**:573–580.

Yoshioka, H., Morohashi, K.-I., Sogawa, K., Miyata, T., Kawajiri, K., Hirose, T., Inayama, S., Fujii-Kuriyama, Y., and Omura, T., 1987, Structural analysis and specific expression of microsomal cytochrome P-450(M-1) mRNA in male rat livers, *J. Biol. Chem.* **262**:1706–1711.

Zaphiropoulos, P. G., Mode, A., Strom, A., Moller, C., Fernandez, C., and Gustafsson, J.-A., 1988, cDNA cloning, sequence, and regulation of a major female-specific and growth hormone-inducible rat liver cytochrome P-450 active in 15-beta-hydroxylation of steroid sulfates, *Proc. Natl. Acad. Sci. USA* **85**:4214–4217.

Chapter 14

NADH-Cytochrome b$_5$ Reductase and Cytochrome b$_5$

The Problem of Posttranslational Targeting to the Endoplasmic Reticulum

Nica Borgese, Antonello D'Arrigo,
Marcella De Silvestris, and Grazia Pietrini

1. INTRODUCTION

A large number of endoplasmic reticulum (ER) enzymes, many of which are involved in the metabolism of lipids and drugs, have a cytosolically exposed active site and only a small lumenal domain (or possibly no lumenal amino acid residues at all) so that large portions of their polypeptide chain must not be translocated across the ER membrane. The biosynthesis of this class of proteins, their mechanism of targeting to the ER and of correct insertion into the phospholipid bilayer, represents an important aspect of ER biogenesis.

Abbreviations used in this chapter: cyt b$_5$, cytochrome b$_5$; ER, endoplasmic reticulum; HMG-CoA, 3-hydroxy-3-methylglutaryl coenzyme A; OM cyt b$_5$, outer membrane cytochrome b$_5$; OMM, outer mitochondrial membrane(s); reductase, NADH-cytochrome b$_5$ reductase; SRP, signal recognition particle.

Nica Borgese, Antonello D'Arrigo, Marcella De Silvestris, and Grazia Pietrini   CNR Center for Cytopharmacology and Department of Pharmacology, University of Milan, Milan, Italy.
Subcellular Biochemistry, Volume 21: Endoplasmic Reticulum, edited by N. Borgese and J. R. Harris. Plenum Press, New York, 1993.

Table I summarizes the information available on the mechanism of insertion of ER proteins with cytoplasmically oriented active domains. Evidence for this topography is generally based on the accessibility of the active site to membrane-impermeant reagents, such as proteases or antibodies [e.g., Liscum *et al.* (1985) for 3-hydroxy-3-methylglutaryl coenzyme A (HMG-CoA) reductase; De Lemos-Chiarandini *et al.* (1987) for phenobarbital-induced cytochrome P-450; Borgese and Meldolesi (1976) for NADH-cytochrome b_5 reductase]. The mechanism of insertion of the proteins listed in column 4 of Table I was deduced from the results of cell-free translation and membrane-binding experiments, with mRNA purified from free or membrane-bound polysome fractions or with synthetic transcripts.

As can be seen from column 4, different pathways of insertion are followed by different proteins. Of the 11 proteins listed in Table I, none have cleavable signal sequences, but 5 were shown to be inserted cotranslationally into the membrane, suggesting that they are targeted to the ER via the same signal recognition particle (SRP) pathway used for translocated proteins (see Klappa *et al.* and Simon and Blobel, this volume). Indeed, for cytochrome P-450, for HMG-CoA reductase, and for sarcoplasmic reticulum Ca^{2+}-ATPase, the dependence on SRP was shown directly (Sakaguchi *et al.*, 1984; Brown and Simoni, 1984; Anderson *et al.*, 1983). Both HMG-CoA reductase and sarcoplasmic reticulum Ca^{2+}-ATPase are multispanning proteins with translocated portions, so that it would have been surprising if they had been inserted via an SRP-independent pathway. The ribophorins, which were previously thought to be involved in ribosome binding, and which both have cleavable signal sequences (Rosenfeld *et al.*, 1984), have not been included in Table I, since it has been recently demonstrated that they are part of the lumenally active oligosaccharyl transferase (Kelleher *et al.*, 1992).

The other six proteins listed in Table I are inserted into the ER membrane by an alternative pathway. In Table I, we have indicated this alternative pathway as "posttranslational," based mainly on the finding that the corresponding mRNA is recovered with free polysomes in cell fractionation experiments, in contrast with the situation for SRP-dependent proteins, whose mRNA is always found associated with bound polysomes. Although it is possible that *in vivo* some of these "posttranslationally" inserted proteins already interact in some way with the ER membrane during synthesis and that this interaction is disrupted upon homogenization, the distribution of their mRNAs with the free polysome fraction indicates that the mechanism of their insertion must be different from the SRP-dependent one. SRP-independence for the *in vitro* insertion into microsomes of cytochrome (cyt) b_5 (Anderson *et al.*, 1983) and for the α subunit of SRP receptor (Andrews *et al.*, 1989) has been shown directly. We will refer to this alternative, SRP-independent mechanism as "posttranslational" or SRP-independent throughout the chapter. If the list of Table I turns out to be represen-

Table I

Mechanism of Insertion of Mammalian ER Proteins with Cytoplasmically Oriented Active Domains

Protein	Number of residues—Swiss Prot Accession No.[a]	Position of hydrophobic stretch (or stretches)[b]	Mechanism of insertion	References
Microsomal aldehyde dehydrogenase	—	—	Posttranslational[c]	Takagi et al. (1985)
Sarcoplasmic reticulum Ca^{2+}-ATPase	1001—P04919(rb)	Eight potential transmembrane segments, of which the first close to the N-terminus (60–78)	Cotranslational[d,e]	Chyn et al. (1979), Anderson et al. (1983)
Cytochrome b_5	133—P00169(rb); P00173(r)	C-terminal (108–126)	Posttranslational[c,f]	Rachubinski et al. (1980), Okada et al. (1982), Anderson et al. (1983)
Cytochrome P-450 (phenobarbital-induced)	491—P00176(r); P12789 (rb)	N-terminal (3–20)	Cotranslational[d,e]	Bar-Nun et al. (1980), Sakaguchi et al. (1984)
Epoxide hydrolase	455—P07099(h); P07687(r)	N-terminal (5–19)	Cotranslational[d]	Ohlsson et al. (1981), Okada et al. (1982)
Heme oxygenase 1	289—P06762(r)	C-terminal (271–289)	Posttranslational[c]	Ishizawa et al. (1983)
HMG-CoA reductase	888—P04035(h)	N-terminal (1–339, contains 7 or 8 membrane-spanning regions)	Cotranslational[e]	Brown and Simoni (1984), Olender and Simoni (1992)
NADH-cytochrome b_5 reductase	300—P20070(r); P07514(b)	N-terminal (1–23)[g]	Posttranslational[c]	Borgese and Gaetani (1980, 1983), Okada et al. (1982)
NADPH-cytochrome P-450 reductase	678—P00388(r)	Close to N-terminus (28–44)	Cotranslational[d]	Gonzales and Kasper (1980), Okada et al. (1982)
α subunit SRP receptor	638—P066259(c)	N-terminal (1–22 and 64–79)	Posttranslational[f]	Andrews et al. (1989)
Stearyl CoA desaturase	358—P07308(r)	Internal (73–113)	Posttranslational[c]	Thiede and Stritmatter (1985)

[a] The letters within parentheses refer to the animal species: r, rat; rb, rabbit; b, bovine; c, canine; h, human. Where sequences are available from several species, not all have been listed.
[b] Numbers within parentheses refer to residue positions.
[c] Site of synthesis on free polysomes demonstrated.
[d] Site of synthesis on membrane-bound polysomes demonstrated.
[e] Dependence on SRP demonstrated.
[f] Independence from SRP demonstrated.
[g] N-terminal glycine is myristylated.

tative, over half of the ER proteins with cytoplasmically located active domains might use such alternative mechanism(s) for membrane insertion. However, much less attention has been directed to the understanding of these biogenetic events than to those involving the translocation of proteins into the ER lumen.

Examination of column 3 of Table I shows that the membrane-binding portion of the proteins that follow the SRP-dependent pathway is in all cases N-terminal. With the exception of the multispanners HMG-CoA reductase and sarcoplasmic reticulum Ca^{2+}-ATPase, the hydrophobic stretch is short, barely long enough to span the bilayer once as an α helix. It is thought that the N-terminus of these proteins is translocated across the ER membrane, implying that the N-terminal sequence functions both as signal- and as stop-transfer sequence (see Tashiro *et al.*, this volume). In contrast to the situation for the cotranslationally inserted enzymes, no generalizations can be drawn for the posttranslationally inserted proteins. Cyt b_5 and heme oxidase have C-terminal hydrophobic stretches, NADH-cytochrome b_5 reductase (reductase) has a myristylated N-terminal anchor, the α subunit of SRP receptor has two hydrophobic stretches, one at the extreme N-terminus and one more internal, stearyl CoA desaturase has an internal stretch; the primary structure of aldehyde dehydrogenase is not available at the time of this writing.

Two important questions must be addressed concerning the class of posttranslationally inserted proteins: first, are they truly integral membrane proteins; i.e., once inserted into the membrane, is the association maintained exclusively by interactions between hydrophobic portions of the protein and the fatty acyl chains of the interior of the bilayer? Second, are these proteins targeted to the ER, or can they bind nonspecifically to any phospholipid bilayer? The first question is not always easy to answer. The decision that a protein is "integral" is often based on its resistance to extraction by alkaline solutions or chaotropic agents, or on its capacity to associate with mild detergents. This definition is, however, purely operational, and does not give direct information on the nature of the interaction with the phospholipid bilayer. Direct evidence is provided if it can be demonstrated that the protein has transmembrane topology, although this can be quite difficult if only a few amino acid residues are exposed on one side of the membrane (see discussion below for cyt b_5). Although all of the posttranslationally inserted proteins listed in Table I appear to be tightly associated with the ER membrane, only in the case of cyt b_5 and of its reductase has the interaction of the hydrophobic stretch with the phospholipid fatty acyl side chains been shown directly by chemical or physicochemical methods (Fleming and Strittmatter, 1978; Kensil and Strittmatter, 1986; see Section 4).

The second question, regarding the extent of the specificity in the subcellular localization of posttranslationally inserted proteins, has been addressed for the α subunit of the SRP receptor and for cyt b_5 and its reductase. Whereas it has been generally accepted that the SRP receptor is an ER protein (Meyer *et al.*,

1982), it has been widely believed that cyt b_5 and its reductase are non-specifically inserted in all phospholipid bilayers. Our work on the subcellular distribution of these two proteins, which will be reviewed in the present chapter, demonstrates that both have a restricted subcellular distribution, which is difficult to reconcile with a random insertion process.

We have chosen cyt b_5 and its reductase as models for the study of post-translational insertion of ER membrane proteins, with C-terminal and N-terminal membrane anchors, respectively. In this chapter, we will review our own work and that of others on these two proteins, and then discuss possible biogenetic mechanisms involved in their localization, as well as more general implications for proteins associated with the cytosolic face of membranes.

In recent years it has become apparent that both cyt b_5 and reductase are present as different isoforms with different localization in mammalian cells. All of the work on the structure and interaction with lipids has been carried out on the well-known microsomal forms of the two enzymes. We will discuss specifically the more recently discovered isoforms in Section 3.

2. NADH-CYTOCHROME b_5 REDUCTASE AND CYTOCHROME b_5: A VERSATILE COUPLE IN FUNCTION AND INTRACELLULAR LOCATION

Both cyt b_5 and reductase belong to protein families with members in evolutionarily distant organisms.

Cyt b_5 belongs to a family of hemoproteins characterized by the so-called "cytochrome b_5 fold," consisting in a bundle of four α helices that form a heme-binding crevice, and a five-stranded mixed β-pleated sheet at the bottom of the crevice (Mathews *et al.*, 1972; Guiard and Lederer, 1979). In addition to cyt b_5, the family includes the mammalian outer mitochondrial membrane (OMM) isoform of cyt b_5 (Lederer *et al.*, 1983), and the hemebinding domains of baker's yeast flavocytochrome b_2, of liver sulfite oxidase (Guiard and Lederer, 1979), and of higher plant and fungal nitrate reductases (Campbell and Kinghorn, 1990).

NADH-cytochrome b_5 reductase is a member of a recently discovered fla-voenzyme family of dehydrogenases-electron transferases, which at present comprises as members: the dinucleotide-binding domains of plant and fungal nitrate reductases, mammalian NADPH-cytochrome P-450 reductase, mammalian brain nitric oxide synthase, bacterial sulfite reductase, *Bacillus mega-terium* cytochrome P-450$_{BM-3}$, and plant ferredoxin-NADP$^+$ reductase (Karplus *et al.*, 1991; Hyde *et al.*, 1991; Bredt *et al.*, 1991). The three-dimensional structure of the latter enzyme has been determined by X-ray diffraction, and shown to comprise a FAD binding domain with an antiparallel β-barrel core and an NADP$^+$-binding domain with α/β structure (Karplus *et al.*, 1991). NADH-cyt b$_5$ reductase is most closely related to the FAD/NADH-binding moiety of plant NADH-nitrate reductases, with which it has nearly 50% sequence identity.

Proteins of both families have diverse intracellular localizations, e.g., cytosolic (as for NADH-nitrate reductase), peripherally associated with a membrane (as for ferredoxin-NADP$^+$ reductase), or integrated in the membrane by a short hydrophobic stretch, with the hydrophilic, active domain exposed to the cytoplasm (as reductase and cyt b$_5$). Members of both families are often found as domains of larger multicenter redox enzymes. Thus, baker's yeast flavocytochrome b$_2$ contains a heme-binding domain of the cyt b$_5$ family fused to an FMN-binding domain (Xia and Mathews, 1990). Higher plant NADH-nitrate reductase is composed of a cyt b$_5$-like heme-binding domain, a flavoprotein domain closely related to reductase, and a molybdenum-pterin binding domain homologous to that of mammalian sulfite oxidase of the mitochondrial intermembrane space (Campbell and Kinghorn, 1990). These observations suggest that both cyt b$_5$- and reductase-like proteins arose early in evolution and that they were used as building blocks for the construction of multicenter redox enzymes by gene fusion events.

In mammalian cells (with the exception of erythrocytes; see Section 3), cyt b$_5$ and reductase have been converted to membrane-bound enzymes, with their active domains exposed at the cytosol. The localization of this electron transport chain on membranes concentrates the components to a two-dimensional space, and is thought to promote a discrete orientation of the active domains, which would facilitate interaction of the enzymes with each other (Enoch *et al.*, 1977). On the ER, cyt b$_5$ is known to donate its electrons to a variety of acceptors involved in diverse aspects of lipid metabolism, such as delta^5, delta^6, and delta^9 desaturation of fatty acids (Strittmatter *et al.*, 1974; Okayasu *et al.*, 1977; Lee *et al.*, 1977), desaturation of plasmalogen precursor alkyl and acyl side chains (Paltauf *et al.*, 1974), prostaglandin synthesis (Strittmatter *et al.*, 1982), and cholesterol biosynthesis (Fukushima *et al.*, 1981).

On the ER membrane, cyt b$_5$ is also involved in processes for which reductase is not required. Thus, it can accept electrons from an alternative reductase, NADPH-cyt P-450 reductase (Dailey and Strittmatter, 1980), and it participates in drug metabolism, interacting with some (but not all) forms of cyt P-450 (Aoyama *et al.*, 1990). In the latter function, it is thought to form a ternary complex with cyt P-450 and NADPH-cyt P-450 reductase (Tamburini and Schenkman, 1987).

Thus, while reductase has only one acceptor on the ER membrane, cyt b$_5$ is promiscuous, with multiple acceptors and also more than one donor. In all of these interactions, the same negatively charged surface area surrounding the exposed heme edge of cyt b$_5$ appears to be involved, with formation of complementary charge pairs between carboxylate groups of that region and appropriately spaced amino groups at the surface of the various acceptors or donors (Dailey and Strittmatter, 1979, 1980; Tamburini *et al.*, 1985; Strittmatter *et al.*, 1992).

An important question is whether cyt b_5 and reductase form relatively stable complexes within the ER membrane or whether they are distributed randomly and interact by random collision dependent on the translational movement of the membrane anchors within the phospholipid bilayer. In the liver ER, cyt b_5 is present in approximately tenfold molar excess over reductase (Strittmatter *et al.*, 1972), thus the majority of the cytochrome could not be contained in such a complex, while all of the reductase could theoretically be complexed with cyt b_5. The existence of such a complex could have implications for the mechanism of targeting of the reductase to the membrane and of its retention in the ER. The problem was investigated by Rogers and Strittmatter nearly 20 years ago (Rogers and Strittmatter, 1974), who, on the basis of kinetic analysis of the reduction of excess exogenous cyt b_5 bound to microsomes or of endogenous cyt b_5 in microsomes in which reductase had been partially inactivated, concluded that no complexes were detectable and that interactions between the two proteins must be very rapid compared with their rates of diffusion within the membrane.

3. ISOFORMS OF CYTOCHROME b_5 AND OF NADH-CYTOCHROME b_5 REDUCTASE AND THEIR SUBCELLULAR DISTRIBUTION

While cyt b_5 and reductase are present in most mammalian cells as membrane-bound proteins, both enzymes exist also in a soluble form in erythrocytes. In the erythrocyte cytoplasm, soluble cyt b_5 reduces methemoglobin, and the reductase–cyt b_5 system constitutes the most important enzymatic system for the maintenance of hemoglobin in the reduced state (Hultquist and Passon, 1971). Erythrocyte reductase deficiency is the most common cause of hereditary methemoglobinemia in man (Scott, 1960).

In addition to these soluble isoforms and to the "classical" microsomal enzymes, novel membrane-bound isoforms of cyt b_5 and of the reductase have been discovered, the biogenesis and significance of which will be discussed below.

3.1. Isoforms of Cytochrome b_5: Soluble, Microsomal, and Outer Mitochondrial Membrane Forms

Initial comparison of the primary structures of the soluble and microsomal forms of bovine cyt b_5 demonstrated that the soluble isoform was identical to the N-terminal, cytoplasmic catalytic domain of the microsomal enzyme (residues

1–97) (Slaughter *et al.*, 1982). This suggested that the two isoforms were products of the same gene. The authors also proposed that the soluble form might be
derived from the microsomal enzyme by posttranslational proteolytic processing
during erythrocyte maturation, resulting in the release of the catalytic domain
from the hydrophobic C-terminal membrane anchor (see Figure 1 and Section 4
for more details on the membrane anchor of cyt b_5). However, when sequence
comparison was carried out in other species (rabbit, man, and pig), it was found
that the C-terminal residue of the soluble form had no counterpart in the microsomal enzyme (Schafer and Hultquist, 1983; Kimura *et al.*, 1984). These data
suggested that soluble and microsomal forms are generated from separate
mRNAs. Indeed, as anticipated, Giordano and Steggles (1991) have isolated a
cDNA from human reticulocytes that codes for soluble cyt b_5 and that has a 24
nucleotide (nt) insertion between codon 96 and 97 of the corresponding liver
transcript. The inserted sequence starts with two in-frame codons (of which the
second one is different from the corresponding codon of the liver transcript)
followed by a stop codon. Since the rest of the sequence of this cDNA is identical
to the liver transcript, the data strongly suggest that the mRNAs coding for the

A

Glu-**Ser-Asn-Ser-Ser-Trp-Trp-Thr-Asn-Trp-Val-Ile-Pro-Ala-Ile-Ser-Ala-**
    -30   -20

                              +   -   -
Leu-Val-Val-Ala-Leu-Met-Tyr-Arg-Leu-Tyr-Met-Ala-Glu-Asp-COOH
         -10   -1

B

                             +
Myr-<u>Gly-Ala-Gln-Leu-Ser-Thr</u>-Leu-Ser-Arg-**Val-Val-Leu-Ser-Pro-**
       1   10

                                +
Val-Trp-Phe-Val-Tyr-Ser-Leu-Phe-Met-Lys-
                     2 0

FIGURE 1. Primary structures of membrane anchors of rat liver cytochrome b_5 (panel A) and
NADH-cytochrome b_5 reductase (panel B). Residues of cyt b_5 anchor are numbered from the
C-terminus with negative numbers. In both panels, residues in boldface are part of a stretch of
uncharged amino acids known to interact with the phospholipid bilayer; the charge of basic or acidic
residues flanking these stretches are shown. The N-terminal glycine of reductase is myristylated
(*Myr*). The six residues that constitute the signal for myristylation are underlined (panel B).

microsomal and soluble forms of cyt b_5 are generated from the same gene by tissue-specific alternative splicing.

It has been known for many years that the outer mitochondrial membranes (OMM) contain spectrally detectable cyt b_5 (Parsons *et al.*, 1968; Fukushima *et al.*, 1972), and it was believed that microsomal cyt b_5 also accounted for the outer mitochondrial form. Indeed, anti-cyt b_5 antibodies recognized the cytochrome on OMM in immunoelectron microscopy (Fowler *et al.*, 1976) and enzyme inhibition experiments (Borgese and Meldolesi, 1976). Enzyme inhibition experiments also indicated that the catalytic domain of cyt b_5 on OMM was exposed to the cytoplasm (Borgese and Meldolesi, 1976). However, when Ito (1980a) purified a tryptic hemopeptide with b_5 spectral characteristics from mitochondria, he concluded that it was not microsomal cyt b_5. Subsequently, the tryptic hemopeptide of this protein was sequenced (Lederer *et al.*, 1983) and found to have $\sim 60\%$ identity with the hemopeptide of the microsomal protein. Since antibodies raised against the mitochondrial hemoeptide were more effective in inhibiting cyt b_5-dependent reduction of cytochrome c in the mitochondrial than in the microsomal fraction, it was concluded that newly characterized hemopeptide was derived from an OMM isoform of cyt b_5 (OM cyt b_5) (Ito, 1980b). Since the sequence difference between microsomal and OM cyt b_5 are distributed throughout the hemopeptides, the results of Lederer *et al.* (1983) also indicated that these two isoforms are products of different genes.

3.2. Subcellular Distribution of Cytochrome b_5 Isoforms in Liver Cells

An interesting question opened by the discovery of OM cyt b_5 was whether any microsomal cyt b_5 was on the OMM at all. Indeed, the two isoforms would be expected to cross-react with polyclonal antibodies, thus casting doubt on all previous immunological studies (Fowler *et al.*, 1976; Borgese and Meldolesi, 1976; Ito, 1980b). An exclusion of microsomal cyt b_5 from OMM would be in contrast with the widely held belief that cyt b_5 can insert into *any* phospholipid bilayer, thanks to its C-terminal insertion sequence (Benazko *et al.*, 1982; Anderson *et al.*, 1983).

D'Arrigo *et al.* (1993) have recently investigated this problem, using anti-peptide antibodies monospecific for each of the two cyt b_5 isoforms. They first used the anti-OM cyt b_5 antibody to identify the holocytochrome as a 23-kDa polypeptide, bound to the OMM in an alkali- and urea-resistant fashion. They then used the two antibodies to probe Western blots of well-characterized subcellular fractions (Figure 2A, B). Quantitative analysis of the blots and comparison with values for marker enzymes in the fractions revealed that OM cyt b_5 was not present on ER membranes, while microsomal cyt b_5 was present on OMM at extremely low concentrations, less than 5% its concentration on ER membranes. The analysis also showed that most (if not all) of the spectrally detectable cyt b_5

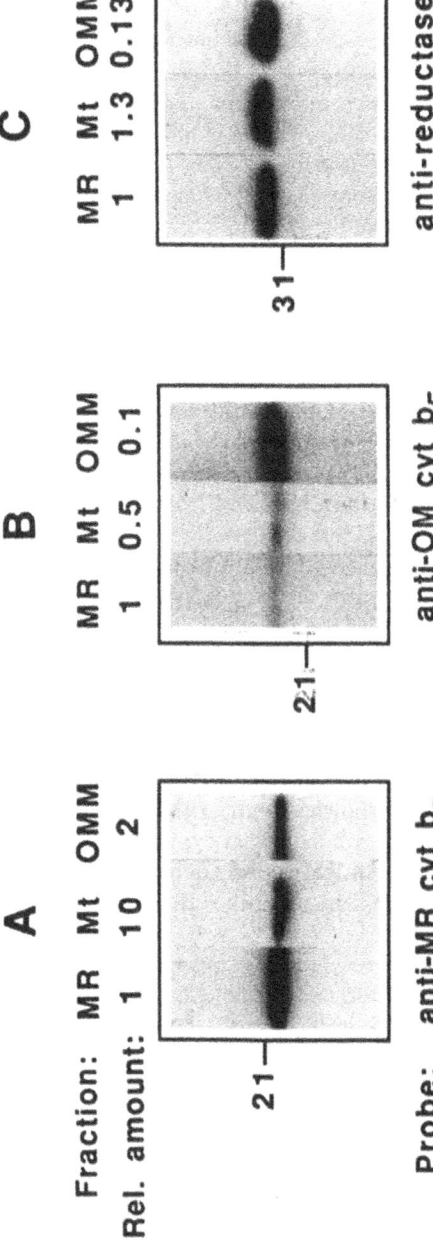

FIGURE 2. Subcellular distributions of rat liver microsomal cyt b$_5$ (MR cyt b$_5$), outer mitochondrial membrane cyt b$_5$ (OM cyt b$_5$), and NADH-cyt b$_5$ reductase (reductase). Microsomal (MR), mitochondrial (Mt), and outer mitochondrial membrane (OMM) fractions were probed in Western blots with antipeptide antibodies against MR cyt b$_5$ (A), OM cyt b$_5$ (B), and with antibodies raised against purified reductase (C). The relative amount of cell fraction per lane (Rel. amount) is expressed on a protein basis and is normalized to the microsomal fraction. The numbers on the left of the panels indicate M_r ($\times 10^{-3}$) and position of standards.

on the surface of mitochondria is due to the OM-specific form, which, however, is expressed at lower concentration on its target membrane than is its microsomal counterpart on the ER membrane (see Figure 4). These data suggest that novel posttranslational targeting mechanisms result in the exclusion of microsomal cyt b_5 from OMM and direct it to the ER. In agreement with our conclusion, Mitoma and Ito (1992) have recently shown that microsomal cyt b_5, overexpressed in COS cells, codistributes with microsomal markers in cell fractionation experiments. (Possible targeting pathways for microsomal cyt b_5 will be discussed in Section 5.) As far as OM cyt b_5 is concerned, it is likely that it shares a common pathway with other OMM proteins. These are known to be delivered to their target membrane via a receptor-mediated mechanism (Soellner et al., 1989, 1990). The proteins analyzed so far do not have cleavable mitochondrial localization signals (Steger et al., 1990); however, for their insertion they do use a receptor (MOM 19) involved also in the import of matrix-directed proteins, which carry positively charged cleavable pre-sequences (Soellner et al., 1989, 1990). Once OM cyt b_5 has been cloned, it will be interesting to see whether it competes for transport with matrix-directed mitochondrial precursor proteins.

3.3. Myristylated and Nonmyristylated Isoforms of NADH-Cytochrome b_5 Reductase

The membrane anchor of microsomal reductase is N-myristylated (Ozols et al., 1984; see Figure 1). The first seven residues of the primary translation product (which contains the initiator Met) constitute the signal for myristylation, a cotranslational modification catalyzed by a cytoplasmic enzyme (for review see Gordon et al., 1991). The myristylation consensus sequence is followed by a group of three residues, of which the third one is basic, and a subsequent stretch of 14 uncharged amino acids.

As in the case of cyt b_5, also for the reductase, amino acid sequence analysis revealed that the soluble form was identical to the cytoplasmic catalytic domain of the membrane-bound myristylated enzyme, and it was suggested that the soluble enzyme was generated by posttranslational proteolysis during erythrocyte maturation (Yubisui et al., 1986, 1987). Recently, however, Pietrini et al. (1992) have shown that soluble reductase is encoded in a separate mRNA, which is generated from the reductase gene by an alternative promoter mechanism (Figure 3A,B). The first exon of the reductase gene contains the 5' noncoding sequence, the initiator AUG, and only six codons which specify the myristylation consensus. This exon is preceded by a housekeeping promoter, is expressed ubiquitously, and is spliced to the third exon of the gene. A specific erythroid mRNA is generated by initiation of transcription from exon 2, which is preceded by an erythroid-specific promoter. Thus, the codons specifying the myristylation con-

A

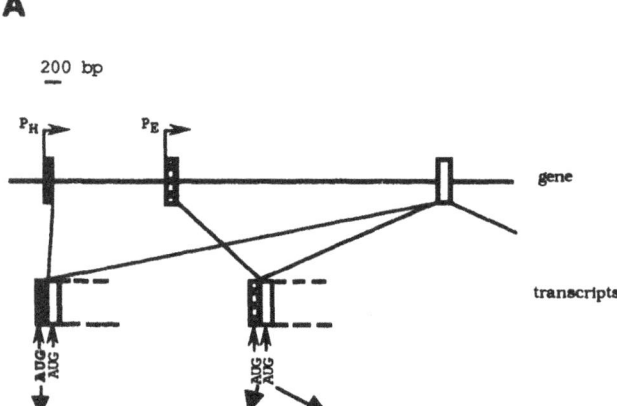

B

L   MET Gly Ala Gln Leu Ser Thr
                                                Leu Ser Arg <u>Val</u>

R   MET Leu Gly Pro Leu Leu Trp Thr Ala Ser Leu Pro Val

L&R   <u>Val Leu Ser Pro Val Trp Phe Val Tyr Ser Leu Phe MET</u> Lys.....

FIGURE 3. Organization of 5′ portion of reductase gene: generation of three protein products by a combination of transcriptional and translational mechanisms. (A) In the top line, which represents the gene, the symbols indicate: P_H, housekeeping promoter; P_E, erythroid promoter; filled rectangle, exon encoding myristylation consensus sequence; checkered rectangle, exon encoding reticulocyte-specific hydrophobic sequence; open rectangle, first common exon. In the middle line, representing the transcripts, the first AUG on the left (in boldface) indicates a strong initiation codon, while the other AUGs (in normal lettering) indicate weak initiation codons. The third line depicts the protein products differing at the N-termini: protein with N-terminal filled rectangle: myristylated version of the reductase; protein with N-terminal checkered rectangle: reticulocyte product with hydrophobic, nonmyristylated N-terminus; product without N-terminal rectangle: soluble reductase. (B) Comparison of N-terminal sequences of products encoded in reductase mRNAs cloned from liver (L) or reticulocytes (R). The liver mRNA, expressed ubiquitously, codes for 7 N-terminal amino acids, which constitute the myristylation consensus sequence (top line). The extended reticulocyte form has, in the place of the myristylation consensus, 13 uncharged residues. The soluble product starts from the downstream MET present in the common portion of the two mRNAs. Reproduced from Pietrini *et al.* (1992) with permission of the Rockefeller University Press.

sensus can be excluded or included in the transcript in a tissue-specific manner. The erythroid mRNA was found to be bifunctional (Figure 3A,B). It generates two polypeptides: a minor product, which is an N-terminally extended form of the reductase, and which starts from a weak initiation codon (Kozak, 1991), and a major product, which begins from a downstream AUG. The generation of these two products can be explained by the scanning model for initiation of translation (Kozak, 1991). The first weak initiation codon causes "leaky" scanning of small ribosomal subunits, so that a large proportion of them bypass it and initiate translation at the downstream AUG. This downstream AUG is in the common portion of the two reductase transcripts and its use excludes from the polypeptide product the entire membrane anchor, with generation of the soluble form of the reductase. It is not used in the ubiquitous transcript, which codes for myristylated reductase, because the first AUG which precedes the codons specifying the myristylation consensus is in a strong context for initiation and does not permit "leaky" scanning of the small ribosomal subunit.

The unexpected finding of the study of Pietrini *et al.* (1992) was the existence of the hitherto undescribed third reductase isoform (the N-terminally extended erythroid polypeptide). This isoform has at its N-terminus 12 uncharged reticulocyte-specific residues in addition to 17 residues of the membrane anchor of myristylated reductase (residues 7–23 of the myristylated form; Figure 3B). Thus, it has a very hydrophobic, nonmyristylated N-terminal region, and, indeed, the authors found that it interacts with microsomes *in vitro*. The functional significance of this novel reductase isoform is as yet unclear. However, since it is specifically expressed in erythroid cells and since membrane-bound reductase is present on the plasma membrane of erythrocytes (Choury *et al.*, 1981; Borgese *et al.*, 1982), it is tempting to speculate that the N-terminal anchor of this third isoform has specific targeting information that differs from that of the anchor of the ubiquitously expressed, myristylated enzyme.

3.4. Subcellular Distribution of Reductase in Liver Cells

In liver cells, reductase is present on OMM as well as on ER membranes (Sottocasa *et al.*, 1967). In contrast to the situation for cyt b_5, the OMM and microsomal forms of the reductase appear to be the same protein. They have the same apparent molecular weight, generate identical peptide maps, and are immunologically indistinguishable (Kuwahara *et al.*, 1978; Meldolesi *et al.*, 1980); they also have the same amino acid composition and are both myristylated (Borgese and Longhi, 1990). Moreover, only one reductase transcript was detected in rat liver, and results of Southern blotting of rat genomic DNA were consistent with the presence of a single reductase gene (Pietrini *et al.*, 1988).

To investigate the subcellular distribution of reductase in rat liver, Borgese and Pietrini (1986) probed Western blots of well-characterized subcellular fractions with polyclonal monospecific antireductase antibodies (Figure 2C). They found that the reductase is most concentrated in OMM, followed by ER membranes, and that it is present in low concentration or absent in other membranes. The high concentration of reductase in OMM had been overlooked in previous studies in which enzyme assays were used to estimate reductase concentration. When reductase is assayed by the rotenone-insensitive NADH-cytochrome c reductase assay, which measures the reduction of cytochrome c via endogenous cyt b_5, the concentration of reductase in OMM is underestimated relative to microsomes, because of the low concentration of OM cyt b_5. The higher concentration of reductase in OMM might be at least in part explained by the slow rate of degradation of mitochondrial reductase compared with its microsomal counterpart (Borgese *et al.*, 1980).

Thus, in the case of reductase, it appears that the same protein inserts into two biogenetically unrelated membranes. However, it cannot at present be excluded that minor differences exist between the mitochondrial and ER forms, which escaped the analyses carried out so far. Other enzymes, i.e., phospholipid synthesizing enzymes in yeast (Kuchler *et al.*, 1986), long-chain acyl CoA synthetase (Miyazawa *et al.*, 1985), and aldehyde dehydrogenase II (Horton and Barrett, 1975), have been reported to be distributed between OM and ER membranes. However, formal proof that the same protein molecule is involved is lacking in all of these cases.

3.5. Comparison between the Subcellular Localizations of Cytochrome b_5 and Reductase Isoforms in Rat Liver

Figure 4 summarizes the quantitative information on the intracellular distribution of cyt b_5 isoforms and reductase reviewed in the previous sections. The figure represents the following points: (1) two different gene products account for the cyt b_5 of OMM and ER (closed squares and open triangles, respectively, in panel A), while reductase activity in the two organelles is due to of the same protein (circles in panel B); (2) cyt b_5 is much more concentrated in ER membranes than OM cyt b_5 in OMM; the reverse is true for reductase, which is more concentrated in OMM (see legend to Figure 4); (3) cyt b_5 is in excess over reductase in the ER, while the reverse is true for OMM. Possible molecular mechanisms whereby this steady-state situation is achieved will be discussed in the following sections.

What is the function of the OMM reductase–cyt b_5 system? The microsomal acceptors of cyt b_5 have not been found on OMM, but Ito *et al.* (1981) showed that OM cyt b_5 is involved in ascorbate regeneration from semidehydroascorbate. Why this system has been placed on the surface of mitochondria rather than on the ER is an open question.

A

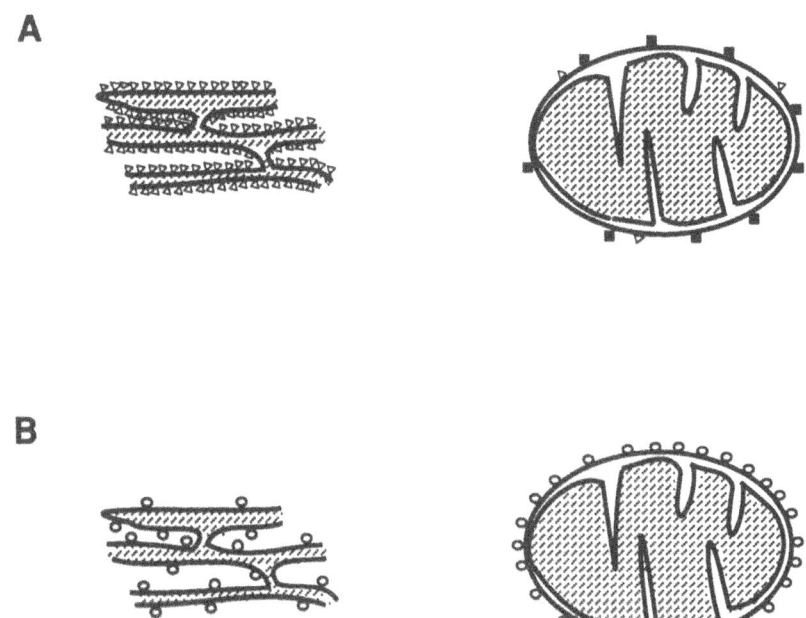

B

FIGURE 4. Cartoon representation of relative average densities of cytochromes b_5 (panel A) and NADH-cytochrome b_5 reductase (panel B). Microsomal (MR) cyt b_5 (open triangles in panel A) is concentrated on the ER membrane, where it is present in $\sim$ tenfold molar excess over reductase (open circles in panel B). Hardly any MR cyt b_5 is present on outer mitochondrial membranes, where nearly all of the measurable cyt b_5 is contributed by outer membrane (OM) cyt b_5 (closed squares in panel A). OM cyt b_5 is, however, not an abundant protein of the outer mitochondrial membrane, where it is present at $\sim$ threefold lower concentration than reductase. The latter is more concentrated on OMM than on ER membranes. Although the cartoon shows a homogeneous distribution of the enzymes on the plane of the membrane, microheterogeneity cannot be excluded. Drawn from data in Borgese and Pietrini (1986) and D'Arrigo *et al.* (1993).

4. INTERACTIONS WITH MEMBRANES

What are the molecular mechanisms that result in the steady-state distribution of cyt b_5 isoforms and of reductase depicted in Figure 4? As stated in the Introduction, the synthesis both of microsomal cyt b_5 and of reductase occurs on ribosomes that are recovered in the free polysome fraction (Rachubinski *et al.*, 1980; Borgese and Gaetani, 1980), implying that *in vivo* they are delivered to their target membranes posttranslationally. For cyt b_5 this is not surprising, since its membrane anchor is at the C-terminus, which becomes available only after release of the polypeptide from the ribosome. For reductase it is less expected and implies that the N-terminal anchor is incapable of interacting tightly with the membrane as long as it is part of a nascent peptide. Although *N*-myristylation is a cotranslational event (for review see Gordon *et al.*, 1991), it is possible that it occurs late in translation, and this could explain the delayed interaction of the anchor with the membrane. Here, we will first discuss what is known on the topology of the membrane anchors of microsomal cyt b_5 and reductase within the phospholipid bilayer, and then turn to the results of *in vitro* binding experiments using natural and artificial membranes.

4.1. Membrane Anchors

4.1.1. Microsomal cyt b_5

The membrane-binding segment of cyt b_5 contains 23 uncharged residues and a short polar region at the extreme C-terminus (Figure 1). The initial 8-residue stretch of the anchor (position -30 to -23) is rich in polar residues, while the following 15-residue stretch (from -22 to -8) is definitely hydrophobic. The sequence of the entire membrane-binding segment is well conserved in all species in which it has been determined (Ozols, 1989). This segment is required for *in vitro* interaction of cyt b_5 with microsomal membranes (Strittmatter *et al.*, 1974) as well as with liposomes (Dailey and Strittmatter, 1978). Moreover, a chimeric protein, consisting of *Escherichia coli* β-galactosidase coupled to the cyt b_5 membrane anchor, spontaneously associates with biological membranes (George *et al.*, 1991). Fluorescence studies have demonstrated that the Trp in position -25 (Trp 109) is in direct contact with the fatty acyl chains of the bilayer (Fleming and Strittmatter, 1978; Tennyson and Holloway, 1986). Circular dichroism (Dailey and Strittmatter, 1978) and Fourier-transform infrared spectroscopy (Holloway and Buchheit, 1990) indicate that the membrane anchor is predominantly α-helical. The short C-terminal polar region is important for the stabilization of the membrane association. Indeed, deletion of the C-terminal residues, although not abolishing the *in vitro* association of cyt b_5 with microsomes or liposomes, results in a "loose" type of binding, i.e., the bound cytochrome can exchange with added liposomes or microsomes (Dailey and Strittmatter, 1981a) (see Section 4.2).

A question that still remains open is whether the membrane anchor of cyt b_5 spans the entire membrane or is restricted to the outer leaflet, with the C-terminus looping back to the cytoplasmic face of the bilayer (hairpin conformation). In the first case, insertion of cyt b_5 into its target membrane would also involve the translocation of the C-terminal polar segment. To study this question, Strittmatter's group has used a model system, in which the nonpolar tryptic peptide (containing the membrane anchor) or cyt b_5 holoenzyme are incorporated into preformed dimyristoylphosphatidylcholine liposomes, in a manner thought to be identical to the *in vivo* mode of insertion. Using membrane-impermeant reagents, they demonstrated that the C-terminal residues are exposed on the outside of the vesicles (Dailey and Strittmatter, 1981b; Arinc *et al.*, 1987). Physical studies on the same model system were also consistent with the idea that the membrane anchor does not penetrate beyond the bilayer midplane (Chester *et al.*, 1992). On the basis of these results, this group has concluded that cyt b_5 in the ER membrane has a hairpin conformation. Takagaki *et al.* (1983b), using a different experimental approach, reached the opposite conclusion. These authors reconstituted cyt b_5 by detergent dialysis into dipalmitoylphosphatidylcholine liposomes containing asymmetrically distributed radioactive photoactivatable phospholipids, and after cross-linking determined the extent of radioactive labeling of peptide fragments obtained by chemical cleavage. The results indicated that, when cyt b_5 is "tightly" bound to the liposomes in a nonexchangeable form (see Section 4.2), the hydrophobic portion traverses the entire bilayer. Gogol and Engelman (1984), on the basis of neutron scattering studies, also concluded that the membrane anchor penetrates deeply into phospholipid bilayers. The discrepancy between these results from different laboratories might be caused by the different model systems utilized. In this respect, it is important to consider the data on cyt b_5 bound to ER membranes *in vivo*. Ozols (1989) investigated the situation in intact microsomes and found that the six C-terminal amino acids of endogenous cyt b_5 are removed by trypsin, bringing support for the hairpin model. However, in this study, control experiments to demonstrate that the microsomal vesicles were sealed were not carried out. Recently, Vergeres and Waskell (1992) studied the subcellular distribution of wild-type and mutant rat cyt b_5 expressed in yeast. Mutation of Pro-115 (position -19 in Figure 1)—which, according to the model of Strittmatter's group, should be responsible for the hairpin conformation (Dailey and Strittmatter, 1978)—to Ala had no effect on the association of cyt b_5 with microsomes, suggesting that the membrane anchor is a transmembrane α-helix. In conclusion, although it appears likely that cyt b_5 is a transmembrane protein, there is, surprisingly, no general consensus on this question.

4.1.2. Microsomal Reductase

As explained in Section 3.3, the membrane anchor of reductase consists of a myristylation consensus sequence (the first 7 residues of the primary translation product) and a downstream stretch of 14 uncharged amino acids (Figure 1). As in the case of cyt b_5, the membrane anchor is required for interaction of reductase with microsomes or liposomes (Spatz and Strittmatter, 1973), and the sequence is well conserved in the three species in which it has been determined (Ozols *et al.*, 1984; Pietrini *et al.*, 1988; Tomatsu *et al.*, 1989).

Much less work has been done on the structure of the reductase membrane anchor than on that of cyt b_5. Kensil and Strittmatter (1986) found that the isolated reductase nonpolar peptide could be reconstituted into phospholipid vesicles with an increase in fluorescence quantum yield of Trp-16 consistent with this amino acid being in a hydrophobic environment. They also used fluorescence energy transfer analysis to estimate a distance of ~ 2 nm for this Trp residue from the outer surface of the bilayer. These data are consistent with either a hairpin or a transmembrane topology for the reductase membrane anchor.

4.2. *In Vitro* Binding to Natural and Artificial Membranes

Both *in vitro* synthesized microsomal cyt b_5 and reductase have been observed to bind posttranslationally to added microsomes (Okada *et al.*, 1982; Borgese and Gaetani, 1983; Pietrini *et al.*, 1992). The binding of cyt b_5 was shown to occur also to trypsin-treated microsomes (Benazko *et al.*, 1982), to be alkali-resistant, and to not require SRP (Anderson *et al.*, 1983). However, the biological meaning of these observations and their consequent utility for establishing a model system for the study of the mechanisms of localization of the two proteins are doubtful. This is because cyt b_5 and reductase will bind to any phospholipid bilayer *in vitro*.

Early experiments showed that both microsomal cyt b_5 and reductase could associate with preformed phosphatidylcholine liposomes, where they were then active in electron transport reactions (Robinson and Tanford, 1975; Rogers and Strittmatter, 1975; Strittmatter and Rogers, 1975). Cyt b_5 was also found to bind to a variety of biological membranes, some of which do not normally contain the cytochrome (Remacle, 1978). In the meantime it was observed that both proteins, bound to preformed liposomes, could be transferred between vesicles (Roseman *et al.*, 1977; Enoch *et al.*, 1977). While this phenomenon was initially thought to reflect the situation *in vivo*, it was subsequently discovered that the endogenous proteins, bound *in vivo*, do not in fact exchange with added artificial or natural membrane vesicles (Enoch *et al.*, 1979; Poengsen and Ullrich, 1980). Enoch *et al.* (1979) distinguished between "loose" and "tight" binding for cyt b_5. In both types of binding the membrane anchor interacts with the phospholipid fatty acyl chains (Freire *et al.*, 1983); however, in the "loose" conformation the cytochrome rapidly exchanges between vesicles, and the C-terminus is sensitive to carboxypeptidase Y. In contrast, the "tightly" bound form is nonexchangeable and carboxypeptidase Y-resistant and is therefore thought to reflect the physiological mode of binding. Takagaki *et al.* (1983a,b) provided evidence that in the "loose" form the membrane anchor is in a hairpin conformation, while it spans the entire membrane in the "tight" conformation (see Section 4.1). Enoch *et al.* found that the "tight" conformation was present in: (1) endogenous cyt b_5 in microsomes; (2) cyt b_5 bound to microsomes *in vitro;* (3) cyt b_5 incorporated into egg phosphatidylcholine vesicles of dimyristoylphosphatidylcholine. Preformed liposomes made with other types of phospholipids, including mammalian microsomal lipids, were found to support only the "loose" type of binding. Thus, although the significance of the "tight" binding to dimyristoylphosphatidylcholine liposomes is not fully understood, it appears that cyt b_5 is not able to tightly insert into protein-free preformed bilayers composed of physiological mixtures of phospholipids.

Conditions for the "tight" binding of reductase have not been reported. It should be mentioned that also NADPH-cyt P-450 reductase, which *in vivo* is inserted into the ER membrane cotranslationally, can interact with liposomes, binding in a "loose" but functional form (Poengsen and Ullrich, 1980).

In conclusion, probably many proteins with a short hydrophobic sequence at one extremity can interact with phospholipid bilayers, but it is not clear that this reflects what is occurring *in vivo*. To begin to study the molecular basis of the interaction of posttranslationally inserted proteins, it will be necessary to establish cell-free systems that distinguish between physiologically relevant and irrelevant binding.

5.   CONCLUSIONS AND IMPLICATIONS

The evidence reviewed in this chapter supports the idea that posttranslational targeting mechanisms control the subcellular localization of cyt b_5 and reductase isoforms, and possibly of a large number of other ER enzymes. While the forms directed to the OMM most likely share a common targeting pathway with other OMM proteins (Soellner *et al.*, 1989, 1990; Steger *et al.*, 1990), the presumptive posttranslational targeting pathways to the cytoplasmic face of the ER remain to be discovered. Future research will be directed at uncovering on the one hand the signals and, on the other hand, the features of the membranes involved in these as yet uninvestigated targeting pathways. The availability of cDNA clones for cyt b_5 and reductase, and the possibility of expressing wild-type, mutant, or chimeric forms of the enzymes *in vivo* and *in vitro*, should be very helpful in the search for putative targeting signals. Using this approach, Mitoma and Ito (1992) have recently reported that in COS cells, the C-terminal ten amino acid residues of microsomal cyt b_5 are necessary for its targeting to the ER. On the other hand, Vergeres and Waskell (1992) found that in yeast, a mutant rat microsomal cyt b_5, in which Ala-131 and Glu-132 (positions -2 and -3 of Figure 1A) were changed to lysines, remained normally membrane associated. Further work with mutants is required to nail down the targeting sequence of microsomal cyt b_5.

Concerning the characteristics of the target membranes involved in recognition, the evidence presented in the previous section indicates that, in the case of microsomal cyt b_5 and of reductase, lipids alone are not sufficient for physiologically relevant binding of these enzymes. Thus, although the lipid composition of different membranes may contribute to determining the localization pattern of the two proteins, it is reasonable to search for specific protein receptors involved in targeting. There are two ways in which such protein receptors could work (Figure 5). A "stoichiometric" receptor would work by binding with high affinity to the cytoplasmic domain, the membrane anchor, or both domains of the posttranslationally targeted protein, forming a stable complex with it. Figure 5A shows a model in which the binding of a protein (P) to the membrane would be caused by two interactions: binding of its cytoplasmic domain to a receptor (R) and interaction of a short hydrophobic portion (e.g., a fatty acid) with the interior of the bilayer. Clearly, there are many other ways in which a membrane protein could interact with a stoichiometric receptor. A "catalytic" receptor would act by somehow facilitating insertion of the posttranslationally targeted protein, which would then dissociate from the receptor and remain stably integrated in the bilayer (Figure 5B). We believe that both for microsomal cyt b_5 and for reductase, a catalytic mechanism is more likely to be involved, because: (1) in the case of cyt b_5, the work of Strittmatter's group showed that a vast excess of cyt b_5 could be bound to microsomes in a physiologically relevant manner (Strittmatter *et al.*, 1972; Enoch *et al.*, 1979); (2) similarly, for reductase, we have observed that the at least tenfold *in vivo* overexpressed protein remains entirely bound to membranes (G. Pietrini, unpublished observations). If "stoichiometric" receptors were involved, one would have to postulate that they are present in vast excess over the proteins they bind.

It is possible that these "catalytic" receptors handle groups of proteins carrying related topogenic signals, in the same way that the SRP-dependent system handles all signal-sequence-carrying translocated proteins. In this respect, it is of interest that the prototype "nontransmembrane" tyrosine phosphatase, PTP-1B, has been recently reported to be localized to the ER, and the information for this localization has been found to reside in a hydrophobic C-terminal region (Frangioni *et al.*, 1992). It is tempting to speculate that PTP-1B and microsomal cyt b_5 share a common targeting mechanism. A plausible candidate that could be involved is a recently described microsomal protein that uses ATP to facilitate membrane insertion of signal-sequence-bearing precursors, both in the SRP-dependent pathway and, for certain precursors, in a pathway that bypasses SRP (Klappa *et al.*, 1991).

The question as to whether posttranslationally inserted proteins use a catalytic or stoichiometric mechanism is of relevance also for other organelles. For instance, specific small GTP-binding proteins are targeted to the cytosolic face of distinct subcellular compartments, where they are thought to regulate intercompartmental vesicular traffic (for review see Pfeffer, 1992). GTP-binding proteins of the rab family are isoprenylated at their C-terminus, and the hypervariable

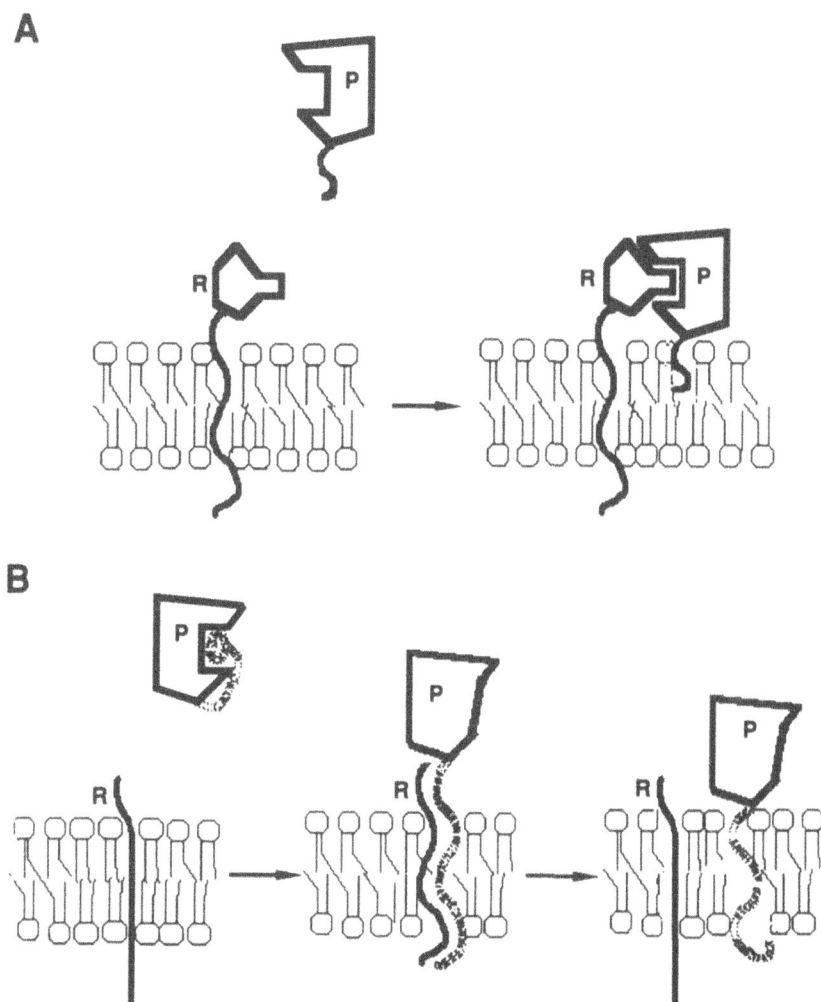

FIGURE 5. Two models for the posttranslational insertion of proteins into membranes. (A) Stoichiometric receptor: newly synthesized protein (P) binds to target membrane via direct interaction with the lipid bilayer as well as with a specific protein receptor (R). (B) Newly synthesized membrane protein (P) does not interact with phospholipid bilayers, because the membrane-binding segment (fuzzy gray curled line) is sequestered. Recognition by a membrane receptor (R; shown as black transmembrane line) triggers a conformational change of the protein, which enables the membrane-binding segment to interact with lipids. Once this interaction has occurred, the protein dissociates from its receptor and remains permanently integrated in the bilayer.

C-terminal domain has been shown to confer targeting specificity to members of the family (Chavrier *et al.*, 1991). Pfeffer (1992) has proposed a model in which each rab would have a receptor on its target membrane, which would act catalytically. On the other hand, the finding that overexpression of rab proteins led to their cytoplasmic accumulation was consistent with the presence of a stoichiometric, saturable receptor (Gorvel *et al.*, 1991). A second class of small GTP-binding proteins, the ARF proteins, has members that are N-myristylated. Also for these proteins, a catalytic receptor has been suggested (Serafini *et al.*, 1991). At variance with the ER enzymes discussed in this chapter, the small GTP-binding proteins come on and off their target membranes in a GTP hydrolysis-dependent cycle.

While there is reason to believe that many posttranslationally inserted proteins are targeted by a catalytic mechanism, there is also evidence for the presence of stoichiometric receptors in some cases. For example, the myristylated tyrosine kinases p56[lck] and p59[fyn] are in a complex with the T cell receptor-associated polypeptides (Rudd *et al.*, 1988; Veillette *et al.*, 1988; Beyers *et al.*, 1992) tyrosine kinases may not be integral membrane proteins.

In conclusion, there probably are a number of different posttranslational targeting mechanisms, related to the different characteristics of the inserted proteins, such as position of the membrane anchor, presence or absence of lipid modification, tightness of the interactions with the phospholipid bilayer. The molecular description of these mechanisms is an important goal of future research on membrane biogenesis and membrane traffic.

ACKNOWLEDGMENTS. We thank Drs. F. Clementi, A. Vitale, and A. Ceriotti for critically reading the manuscript. Unpublished work of the authors reported herein was partially supported by CNR grants: Progetto finalizzato Ingegneria Genetica, Progetto finalizzato Biotecnologie e Biostrumentazione, and Progetto Speciale PS0408.

6. REFERENCES

Anderson, D. J., Mostov, K. E., and Blobel, G., 1983, Mechanisms of integration of de novo synthesized polypeptides into membranes: Signal recognition particle is required for integration into microsomal membranes of calcium ATPase and of lens MP 26 but not of cytochrome b_5, *Proc. Natl. Acad. Sci. USA* **80:**7249–7253.

Andrews, D. W., Lauffer, L., Walter, P., and Lingappa, V. R., 1989, Evidence for a two-step mechanism involved in assembly of functional signal recognition particle receptor, *J. Cell Biol.* **108:**797–810.

Aoyama, T., Nagata, K., Yamazoe, Y., Kato, R., Matsunaga, E., Gelboin, H. V., and Gonzales, G. J., 1990, Cytochrome b_5 potentiation of cytochrome P-450 catalytic activity demonstrated by a vaccinia virus-mediated *in situ* reconstitution system, *Proc. Natl. Acad. Sci. USA* **87**:5425–5429.

Arinc, E., Rzepecki, L. M., and Strittmatter, P., 1987, Topography of the C-terminus of cytochrome b_5 tightly bound to dimyristoylphosphatidylcholine vesicles, *J. Biol. Chem.* **262**:15563–15567.

Bar-Nun, S., Kreibich, G., Adesnik, M., Alterman, L., Negishi, M., and Sabatini, D. D., 1980, Synthesis and insertion of cytochrome P-450 into endoplasmic reticulum membranes, *Proc. Natl. Acad. Sci. USA* **77**:965–969.

Benazko, P., Brehn, S., Pfeil, W., and Rapoport, T. A., 1982, Different modes of interaction of the signal sequence of carp preproinsulin and of the insertion sequence of rabbit cytochrome b_5, *Eur. J. Biochem.* **123**:121–126.

Beyers, A. D., Spruyt, L. L., and Williams, A. F., 1992, Molecular associations between the T-lymphocyte antigen receptor complex and the surface antigens CD2, CD4, or CD8 and CD5, *Proc. Natl. Acad. Sci. USA* **89**:2945–2949.

Borgese, N., and Gaetani, S., 1980, Site of synthesis of rat liver NADH-cytochrome b_5 reductase, an integral membrane protein, *FEBS Lett.* **112**:216–220.

Borgese, N., and Gaetani, S., 1983, *In vitro* synthesis and posttranslational insertion into microsomes of the integral membrane protein, NADH-cytochrome b_5 oxidoreductase, *EMBO J.* **2**:1263–1269.

Borgese, N., and Longhi, R., 1990, Both the outer mitochondrial membrane and the microsomal forms of cytochrome b_5 reductase contain covalently bound myristic acid. Quantitative analysis on the polyvinylidene difluoride-immobilized proteins, *Biochem. J.* **266**:341–347.

Borgese, N., and Meldolesi, J., 1976, Immunological similarity of the NADH-cytochrome c electron transport system in microsomes, Golgi complex, and mitochondrial outer membranes of rat liver cells, *FEBS Lett.* **63**:231–234.

Borgese, N., and Pietrini, G., 1986, Distribution of the integral membrane protein NADH-cytochrome b_5 reductase in rat liver cells, studied with a quantitative radioimmunoblotting assay, *Biochem. J.* **239**:393–403.

Borgese, N., Pietrini, G., and Meldolesi, J., 1980, Localization and biosynthesis of NADH-cytochrome b_5 reductase, an integral membrane protein, in rat liver cells. III. Evidence for the independent insertion and turnover of the enzyme in various subcellular compartments, *J. Cell Biol.* **86**:38–45.

Borgese, N., Macconi, D., Parola, L., and Pietrini, G., 1982, Rat erythrocyte NADH-cytochrome b_5 reductase. Quantitation and comparison between the membrane-bound and soluble forms using an antibody against the rat liver enzyme, *J. Biol. Chem.* **257**:13854–13861.

Bredt, D. S., Hwang, P. M., Glatt, C. E., Lowenstein, C., Reed, R. R., and Snyder, S. H., 1991, Cloned and expressed nitric oxide synthase structurally resembles cytochrome P-450 reductase, *Nature* **351**:714–718.

Brown, D. A., and Simoni, R. D., 1984, Biogenesis of 3-hydroxy-3-methylglutaryl-coenzyme A reductase, an integral glycoprotein of the endoplasmic reticulum, *Proc. Natl. Acad. Sci. USA* **81**:1674–1678.

Campbell, W. H., and Kinghorn, J. R., 1990, Functional domains of assimilatory nitrate reductases and nitrite reductases, *Trends Biochem. Sci.* **15**:315–319.

Chavrier, P., Gorvel, J.-P., Stelzer, R., Simons, K., Gruenberg, J., and Zerial, M., 1991, Hypervariable C-terminal domain of rab proteins acts as a targeting signal, *Nature* **353**:769–772.

Chester, D. W., Skita, V., Young, H. S., Mavromoustakos, T., and Strittmatter, P., 1992, Bilayer structure and physical dynamics of the cytochrome b_5 dimyristoylphosphatidylcholine interactions, *Biophys. J.* **61**:1224–1243.

Choury, D., Leroux, A., and Kaplan, J.-C., 1981, Membrane-bound cytochrome b_5 reductase(MetHb reductase) in human erythrocytes. Study in normal and methemoglobinemic patients, *J. Clin. Invest.* **67**:149–155.

Chyn, T., Martonosi, A. N., Morimoto, T., and Sabatini, D. D., 1979, *In vitro* synthesis of the Ca^{2+} transport ATPase by ribosomes bound to sarcoplasmic reticulum membranes, *Proc. Natl. Acad. Sci. USA* **76**:1241–1245.

Dailey, H. A., and Strittmatter, P., 1978, Structural and functional properties of the membrane binding segment of cytochrome b_5, *J. Biol. Chem.* **253**:8203–8209.

Dailey, H. A., and Strittmatter, P., 1979, Modification and identification of cytochrome b_5 carboxyl groups involved in protein–protein interaction with cytochrome b_5 reductase, *J. Biol. Chem.* **254**:5388–5396.

Dailey, H. A., and Strittmatter, P., 1980, Characterization of the interaction of amphipathic cytochrome b_5 with stearyl coenzyme A desaturase and NADPH-cytochrome P-450 reductase, *J. Biol. Chem.* **255**:5184–5189.

Dailey, H. A., and Strittmatter, P., 1981a, The role of COOH-terminal anionic residues in binding cytochrome b_5 to phospholipid vesicles and biological membranes, *J. Biol. Chem.* **256**:1677–1680.

Dailey, H. A., and Strittmatter, P., 1981b, Orientation of the carboxyl and NH_2 termini of the membrane-binding segment of cytochrome b_5 on the same side of phospholipid bilayers, *J. Biol. Chem.* **256**:3951–3955.

D'Arrigo, A., Manera, E., Longhi, R., and Borgese, N., 1992, The specific subcellular localization of 2 isoforms of cytochrome b_5 suggests novel targeting pathways, *J. Biol. Chem.* **268**:2802–2808.

De Lemos-Chiarandini, C., Frey, A. B., Sabatini, D. D., and Kreibich, G., 1987, Determination of the membrane topology of the phenobarbital-inducible rat liver cytochrome P-450 isoenzyme PB-4 using site-specific antibodies, *J. Cell Biol.* **104**:209–219.

Enoch, H. G., Fleming, P. J., and Strittmatter, P., 1977, Cytochrome b_5 and cytochrome b_5 reductase–phospholipid vesicles. Intervesicle protein transfer and orientation factors in protein–protein interactions, *J. Biol. Chem.* **252**:5656–5660.

Enoch, H. G., Fleming, P. J., and Strittmatter, P., 1979, The binding of cytochrome b_5 to phospholipid vesicles and biological membranes. Effect of orientation on intermembrane transfer and digestion by carboxypeptidase Y, *J. Biol. Chem.* **254**:6483–6488.

Fleming, P. J., and Strittmatter, P., 1978, The nonpolar peptide segment of cytochrome b_5—binding to phospholipid vesicles and identification of the fluorescent tryptophanyl residue, *J. Biol. Chem.* **253**:8198–8202.

Fowler, S., Remacle, J., Trouet, A., Beaufay, H., Berthet, J., Wibo, M., and Hauser, P., 1976, Analytical study of microsomes and isolated subcellular membranes from rat liver. V. Immunological localization of cytochrome b_5 by electron microscopy: Methodology and application to various subcellular fractions, *J. Cell Biol.* **71**:535–550.

Frangioni, J. V., Beahm, P. H., Shifrin, V., Jost, C. A., and Neel, B. G., 1992, The nontransmembrane tyrosine phosphatase PTP-1B localizes to the endoplasmic reticulum via its 35 amino acid C-terminal sequence, *Cell* **68**:545–560.

Freire, E., Markello, T., Rigell, C., and Holloway, P. W., 1983, Calorimetric and fluorescence characterization of interactions between cytochrome b_5 and phosphatidylcholine bilayers, *Biochemistry* **22**:1675–1680.

Fukushima, H., Grinstead, G. F., and Gaylor, J. L., 1981, Total enzymic synthesis of cholesterol from lanosterol. Cytochrome b_5-dependence of 4-methyl sterol oxidase, *J. Biol. Chem.* **256**:822–826.

Fukushima, L., Ito, A., Omura, T., and Sato, R., 1972, Occurrence of different types of cytochrome b_5-like hemoprotein in liver mitochondria and their intramitochondrial localization, *J. Biochem.* **71**:447–461.

George, S. K., Xu, Y.-H., Benson, L. A., Pratsch, L., Peters, R., and Ihler, G. M., 1991, Cytochrome b_5 and a recombinant protein containing the cytochrome b_5 hydrophobic domain spontaneously associate with the plasma membranes of cells, *Biochim. Biophys. Acta* **1066**:131–143.

Giordano, S. J., and Steggles, A. W., 1991, The human liver and reticulocyte cytochrome b₅ mRNAs are products from a single gene, *Biochem. Biophys. Res. Commun.* **178**:38–44.

Gogol, E. P., and Engelman, D. M., 1984, Neutron scattering shows that cytochrome b₅ penetrates deeply into the lipid bilayer, *Biophys. J.* **46**:491–495.

Gonzales, F. J., and Kasper, C. B., 1980, Phenobarbital induction of NADPH-cytochrome c (P-450) oxidoreductase messenger ribonucleic acid, *Biochemistry* **19**:1790–1796.

Gordon, J., Duronio, R. J., Rudnick, D. A., Adams, S. P., and Gokel, G. W., 1991, Protein N-myristoylation, *J. Biol. Chem.* **266**:8647–8650.

Gorvel, J.-P., Chavrier, P., Zerial, M., and Gruenberg, J., 1991, rab5 controls early endosome fusion *in vitro*, *Cell* **64**:915–925.

Guiard, B., and Lederer, F., 1979, The "cytochrome b₅ fold": Structure of a novel protein super-family, *J. Mol. Biol.* **135**:639–650.

Holloway, P. W., and Buchheit, C., 1990, Topography of the membrane-binding domain of cyto-chrome b₅ in lipids by Fourier-transform infrared spectroscopy, *Biochemistry* **29**:9631–9637.

Horton, A. A., and Barrett, M. C., 1975, The subcellular localization of aldehyde dehydrogenase in rat liver, *Arch. Biochem. Biophys.* **167**:426–436.

Hultquist, D. E., and Passon, P. G., 1971, Catalysis of methemoglobin reduction by erythrocyte cytochrome b₅ and cytochrome b₅ reductase, *Nature* **229**:252–254.

Hyde, G. E., Crawford, N. M., and Campbell, W. H., 1991, The sequence of squash NADH:nitrate reductase and its relationship to the sequences of other flavoprotein oxidoreductases. A family of flavoprotein pyridine nucleotide cytochrome reductases, *J. Biol. Chem.* **266**:23542–23547.

Ishizawa, S., Yoshida, T., and Kikuchi, G., 1983, Induction of heme oxygenase in rat liver. Increase of the specific mRNA by treatment with various chemicals and immunological identity of the enzymes in various tissues as well as the induced enzymes, *J. Biol. Chem.* **258**:4220–4225.

Ito, A., 1980a, Cytochrome b₅-like hemoprotein of outer mitochondrial membrane; OM cytochrome b. I. Purification of OM cytochrome b from rat liver mitochondria and comparison of its molecular properties with those of cytochrome b₅, *J. Biochem.* **87**:63–71.

Ito, A., 1980b, Cytochrome b₅-like hemoprotein of outer mitochondrial membrane; OM cytochrome b. II. Contribution of OM cytochrome b to rotenone-insensitive NADH-cytochrome c reductase activity, *J. Biochem.* **87**:73–80.

Ito, A., Hajashi, S. T., and Yoshida, T., 1981, Participation of a cytochrome b₅-like hemoprotein of outer mitochondrial membrane (OM-cytochrome b) in NADH-semidehydroascorbic acid reduc-tase activity of rat liver, *Biochem. Biophys. Res. Commun.* **101**:591–598.

Karplus, P. A., Daniels, M. J., and Herriott, J. R., 1991, Atomic structure of ferredoxin-NADP⁺ reductase: Prototype for a structurally novel flavoenzyme family, *Science* **251**:60–66.

Kelleher, D. J., Kreibich, G., and Gilmore, R., 1992, Oligosaccharyltransferase activity is associ-ated with a protein complex composed of ribophorins I and II and a 48 kDa protein, *Cell* **69**:55–65.

Kensil, C. R., and Strittmatter, P., 1986, Binding and fluorescence properties of the membrane domain of NADH-cytochrome b₅ reductase—Determination of the depth of trp-16 in the bilay-er, *J. Biol. Chem.* **261**:7316–7321.

Kimura, S., Abe, K., and Sugita, Y., 1984, Differences in C-terminal amino acid sequences between erythrocyte and liver cytochrome b₅ isolated from pig and human, *FEBS Lett.* **169**:143–146.

Klappa, P., Mayinger, P., Pipkorn, R., Zimmermann, M., and Zimmermann, R., 1991, A microso-mal protein is involved in ATP-dependent transport of presecretory proteins into mammalian microsomes, *EMBO J.* **10**:2795–2803.

Kozak, M., 1991, An analysis of vertebrate mRNA sequences: Intimations of translational control, *J. Cell Biol.* **115**:887–904.

Kuchler, K., Daum, G., and Paltauf, F., 1986, Subcellular and submitochondrial localization of phospholipid-synthesizing enzymes in *Saccharomyces cerevisiae, J. Bacteriol.* 165:901–910.

Kuwahara, S., Okaka, Y., and Omura, T., 1978, Evidence for molecular identity of microsomal and mitochondrial NADH-cytochrome b_5 reductases of rat liver, *J. Biochem.* 83:1049–1046.

Lederer, G., Ghrir, R., Guiard, B., Cortial, S., and Ito, A., 1983, Two homologous cytochrome b_5 in a single cell, *Eur. J. Biochem.* 132:95–102.

Lee, T.-C., Baker, R. C., Stephens, N., and Snyder, R., 1977, Evidence for participation of cytochrome b_5 in microsomal delta-6 desaturation of fatty acids, *Biochim. Biophys. Acta* 489:25–31.

Liscum, A., Finer-Moore, J., Stroud, R. M., Luskey, K. L., Brown, M. S., and Goldstein, J. L., 1985, Domain structure of 3-hydroxy-3-methylglutaryl coenzyme A reductase, a glycoprotein of the endoplasmic reticulum, *J. Biol. Chem.* 260:522–530.

Mathews, F. S., Levine, M., and Argos, P., 1972, Three-dimensional Fourier synthesis of calf liver cytochrome b_5 at 2.8 Å resolution, *J. Mol. Biol.* 64:449–464.

Meldolesi, J., Corte, G., Pietrini, G., and Borgese, N., 1980, Localization and biosynthesis of NADH-cytochrome b_5 reductase, an integral membrane protein, in rat liver cells. II. Evidence that a single enzyme accounts for the activity in its various subcellular locations, *J. Cell Biol.* 85:231–237.

Meyer, D. I., Louvard, D., and Dobberstein, B., 1982, Characterization of molecules involved in protein translocation using a specific antibody, *J. Cell Biol.* 92:579–583.

Mitoma, J.-Y., and Ito, A., 1992, The carboxy-terminal 10 amino acid residues of cytochrome b_5 are necessary for its targeting to the endoplasmic reticulum, *EMBO J.* 11:4197–4204.

Miyazawa, S., Hashimoto, T., and Yokota, S., 1985, Identity of long-chain acyl-coenzyme A synthetase of microsomes, mitochondria, and peroxisomes in rat liver, *J. Biochem.* 98:723–733.

Ohlsson, R. I., Lane, C. D., and Guengerich, F. P., 1981, Synthesis and insertion, both *in vivo* and *in vitro*, of rat-liver cytochrome P-450 and epoxide hydratase into *Xenopus laevis* membranes, *Eur. J. Biochem.* 115:367–373.

Okada, Y., Frey, A. B., Guenthner, T. M., Oesch, F., Sabatini, D. D., and Kreibich, G., 1982, Studies on the biosynthesis of microsomal membrane proteins. Site of synthesis and mode of insertion of cytochrome b_5, cytochrome b_5 reductase, cytochrome P-450 reductase, and epoxide hydrolase, *Eur. J. Biochem.* 122:393–402.

Okayasu, T., Ono, T., Shinojima, K., and Imai, Y., 1977, Involvement of cytochrome b_5 in oxidative desaturation of linoleic acid to τ-linolenic acid in rat liver microsomes, *Lipids* 12:267–271.

Olender, E. H., and Simoni, R. D., 1992, The intracellular targeting and membrane topology of 3-hydroxy-3-methylglutaryl-CoA reductase, *J. Biol. Chem.* 267:4223–4235.

Ozols, J., 1989, Structure of cytochrome b_5 and its topology in microsomal membranes, *Biochim. Biophys. Acta* 997:121–130.

Ozols, J., Carr, S. A., and Strittmatter, P., 1984, Identification of the NH_2-terminal blocking group of NADH-cytochrome b_5 reductase as myristic acid and the complete amino acid sequence of the membrane-binding domain, *J. Biol. Chem.* 259:13349–13354.

Paltauf, F., Prough, R. A., Siler Masters, B. S., and Johnston, J. M., 1974, Evidence for the participation of cytochrome b_5 in plasmalogen biosynthesis, *J. Biol. Chem.* 249:2661–2662.

Parsons, D. F., Williams, G. R., Thompson, W., Wilson, D., and Chance, B., 1968, Improvements in the procedure for purification of mitochondrial outer and inner membrane. Comparison of the outer membrane with smooth endoplasmic reticulum, in: *Mitochondrial Structure and Compartmentation* (E. Quagliarello, S. Papa, E. C. Slater, and J. M. Tager, eds.), pp. 29–70, Adriatica Editrice, Bari.

Pfeffer, S. R., 1992, GTP-binding proteins in intracellular transport, *Trends Cell Biol.* **2**:41–46.

Pietrini, G., Carrera, P., and Borgese, N., 1988, Two transcripts encode rat cytochrome b_5 reductase, *Proc. Natl. Acad. Sci. USA* **85**:7246–7250.

Pietrini, G., Aggujaro, D., Carrera, P., Malyzsko, J., Vitale, A., and Borgese, N., 1992, A single mRNA, transcribed from an alternative, erythroid-specific promoter, codes for 2 non-myristylated forms of NADH-cytochrome b_5 reductase, *J. Cell Biol.* **117**:975–986.

Poengsen, J., and Ullrich, V., 1980, Transfer of cytochrome b_5 and NADPH-cytochrome c reductase between membranes, *Biochim. Biophys. Acta* **596**:248–263.

Rachubinski, R. A., Verma, D. P. S., and Bergeron, J. J. M., 1980, Synthesis of rat liver microsomal cytochrome b_5 by free polysomes, *J. Cell Biol.* **84**:705–716.

Remacle, J., 1978, Binding of cytochrome b_5 to membranes of isolated subcellular organelles from rat liver, *J. Cell Biol.* **79**:291–313.

Robinson, E. C., and Tanford, C., 1975, The binding of deoxycholate, Triton X-100, sodium dodecylsulfate, and phosphatidylcholine vesicles to cytochrome b_5, *Biochemistry* **14**:369–378.

Rogers, M. J., and Strittmatter, P., 1974, Evidence for random distribution and translational movement of cytochrome b_5 in endoplasmic reticulum, *J. Biol. Chem.* **249**:895–900.

Rogers, P., and Strittmatter, P., 1975, The interaction of NADH-cytochrome b_5 reductase and cytochrome b_5 bound to egg lecithin liposomes, *J. Biol. Chem.* **250**:5713–5718.

Roseman, M. A., Holloway, P., Calabro, M. A., and Thompson, T. E., 1977, Exchange of cytochrome b_5 between phospholipid vesicles, *J. Biol. Chem.* **252**:4842–4849.

Rosenfeld, M. G., Marcantonio, E. E., Hakimi, J., Ort, V. M., Atkinson, P. H., Sabatini, D. D., and Kreibich, G., 1984, Biosynthesis and processing of ribophorins in the endoplasmic reticulum, *J. Cell Biol.* **99**:1076–1082.

Rudd, C. E., Trevillyans, J. M., Dasgupta, J. D., Wong, L. L., and Schlossman, S. F., 1988, The CD4 receptor is complexed in detergent lysates to a protein-tyrosine kinase (pp 58) from human T lymphocytes, *Proc. Natl. Acad. Sci. USA* **85**:5190–5194.

Sakaguchi, M., Mihara, K., and Sato, R., 1984, Signal recognition particle is required for co-translational insertion of cytochrome P-450 into microsomal membranes, *Proc. Natl. Acad. Sci. USA* **81**:3361–3364.

Schafer, D. A., and Hultquist, D. E., 1983, Purification and structural studies of rabbit erythrocyte cytochrome b_5, *Biochem. Biophys. Res. Commun.* **115**:807–813.

Scott, E. M., 1960, The relation of diaphorase of human erythrocytes to inheritance of methemoglobinemia, *J. Clin. Invest.* **39**:1176–1179.

Serafini, T., Orci, L., Amherdt, M., Brunner, M., Kahn, R. A., and Rothman, J. E., 1991, ADP-ribosylation factor is a subunit of the coat of Golgi-derived COP-coated vesicles: A novel role for a GTP-binding protein, *Cell* **67**:239–253.

Slaughter, S. R., Williams, C. H., Jr., and Hultquist, D. E., 1982, Demonstration that bovine erythrocyte cytochrome b_5 is the hydrophilic segment of microsomal cytochrome b_5, *Biochim. Biophys. Acta* **705**:228–237.

Soellner, T., Griffiths, G., Pfaller, R., Pfanner, N., and Neupert, W., 1989, MOM 19, an import receptor for mitochondrial precursor proteins, *Cell* **59**:1061–1070.

Soellner, T., Pfaller, R., Griffiths, G., Pfanner, N., and Neupert, W., 1990, A mitochondrial import receptor for the ADP/ATP carrier, *Cell* **62**:107–115.

Sottocasa, G. L., Kuylenstierna, B., Ernster, L., and Bergstrand, A., 1967, An electron transport system associated with the outer membrane of liver mitochondria, *J. Cell Biol.* **32**:415–438.

Spatz, L., and Strittmatter, P., 1973, A form of reduced nicotinamide adenine dinucleotide-cytochrome b₅ reductase containing both the catalytic site and an additional hydrophobic membrane-binding segment, *J. Biol. Chem.* **248**:793–799.

Steger, H. F., Soellner, T., Kiebler, M., Dietmeier, K. A., Pfaller, R., Trülzsch, K. S., Tropschug, M., Neupert, W., and Phanner, N., 1990, Import of ADP/ATP carrier into mitochondria: Two receptors act in parallel, *J. Cell Biol.* **111**:2353–2363.

Strittmatter, P., and Rogers, M. J., 1975, Apparent dependence of interactions between cytochrome b₅ and cytochrome b₅ reductase upon translational diffusion in dimyristoyl lecithin liposomes, *Proc. Natl. Acad. Sci. USA* **72**:2658–2661.

Strittmatter, P., Rogers, M. J., and Spatz, L., 1972, The binding of cytochrome b₅ to liver microsomes, *J. Biol. Chem.* **247**:7188–7194.

Strittmatter, P., Spatz, L., Corcoran, D., Rogers, M. J., Setlow, B., and Redline, R., 1974, Purification and properties of rat liver microsomal stearyl-coenzyme A desaturase, *Proc. Natl. Acad. Sci. USA* **71**:4565–4569.

Strittmatter, P., Machuga, E. T., and Roth, G. J., 1982, Reduced pyridine nucleotides and cytochrome b₅ as electron donors for prostaglandin synthetase reconstituted in dimyristyl phosphatidylcholine vesicles, *J. Biol. Chem.* **257**:11883–11886.

Strittmatter, P., Kittler, J. M., Coghill, J. E., and Ozols, J., 1992, Characterization of lysyl residues of NADH-cytochrome b₅ reductase implicated in charge-pairing with active-site carboxyl residues of cytochrome b₅ by site-directed mutagenesis of an expression vector for the flavoprotein, *J. Biol. Chem.* **267**:2519–2533.

Takagaki, Y., Radhakrishnan, R., Wirtz, K. W. A., and Khorana, H. G., 1983a, The membrane-embedded segment of cytochrome b₅ as studied by cross-linking with photoactivatable phospholipids I. The transferable form, *J. Biol. Chem.* **258**:9128–9135.

Takagaki, Y., Radhakrishnan, R., Wirtz, K. W. A., and Khorana, H. G., 1983b, The membrane-embedded segment of cytochrome b₅ as studied by cross-linking with photoactivatable phospholipids II. The nontransferable form, *J. Biol. Chem.* **258**:9136–9142.

Takagi, Y., Ito, A., and Omura, T., 1985, Biogenesis of microsomal aldehyde dehydrogenase in rat liver, *J. Biochem.* **98**:1647–1652.

Tamburini, P. P., and Schenkman, J. B., 1987, Purification to homogeneity and enzymological characterization of a functional covalent complex composed of cytochromes P-450 isozyme 2 and b₅ from rabbit liver, *Proc. Natl. Acad. Sci. USA* **84**:11–15.

Tamburini, P. P., White, R. E., and Schenkman, J. B., 1985, Chemical characterization of protein–protein interactions between cytochrome P-450 and cytochrome b₅, *J. Biol. Chem.* **260**:4007–4015.

Tennyson, J., and Holloway, P. W., 1986, Fluorescence studies of cytochrome b₅ topography—Incorporation of cytochrome b₅ into brominated phosphatidylcholine vesicles by deoxycholate, *J. Biol. Chem.* **261**:14196–14200.

Thiede, M. A., and Strittmatter, P., 1985, The induction and characterization of rat liver stearyl-CoA desaturase mRNA, *J. Biol. Chem.* **260**:14459–14463.

Tomatsu, S., Kobayashi, Y., Fukumaki, Y., Yubisui, Y., Orii, T., and Sakaki, Y., 1989, The organization and the complete nucleotide sequence of the human NADH-cytochrome b₅ reductase gene, *Gene* **80**:353–361.

Veillette, A., Bookman, M. A., Horak, E. M., and Bolen, J. B., 1988, The CD4 and CD8 T cell surface antigens are associated with the internal membrane tyrosine-protein kinase p56[lck], *Cell* **55**:301–308.

Vergeres, G., and Waskell, L., 1992, Expression of cytochrome b₅ in yeast and characterization of mutants of the membrane-anchoring domain, *J. Biol. Chem.* **267**:12583–12591.

Xia, Z.-X., and Mathews, F. S., 1990, Molecular structure of flavocytochrome b_2 at 2.4 Å resolution, *J. Mol. Biol.* **212**:837–863.

Yubisui, T., Miyata, T., Iwanaga, S., Tamura, M., and Takeshita, M., 1986, Complete amino acid sequence of NADH-cytochrome b_5 reductase purified from human erythrocytes, *J. Biochem.* **99**:407–422.

Yubisui, T., Naitoh, Y., Zenno, S., Tamura, M., Takeshita, M., and Sakaki, Y., 1987, Molecular cloning of cDNAs of human liver and placenta NADH-cytochrome b_5 reductase, *Proc. Natl. Acad. Sci. USA* **84**:3609–3613.

Chapter 15

Motility and Construction of the Endoplasmic Reticulum in Living Cells

Christopher Lee and Lan Bo Chen

1. INTRODUCTION

Since the discovery of the endoplasmic reticulum (ER) (Porter *et al.*, 1945), investigation of the biochemistry of this organelle has revealed its fundamental importance in protein synthesis, metabolism of lipids, and other functions (for review see Palade, 1956; Fawcett, 1981). However, one of the most characteristic features of the ER, its intricate tubuloreticular structure, has until recently received little attention; both its cell biological basis and function have remained relatively obscure. Recently, however, the development of vital staining techniques for the ER has created a new opportunity for observing it in live cells, and for scrutinizing the process of its assembly. A broad variety of approaches, including *in vitro* study of cytoskeletal motility, and investigation of ER motility and cytoskeletal interactions in living cells, have elucidated the basic mechanisms whereby the structure of ER is formed.

Christopher Lee and Lan Bo Chen   Dana-Farber Cancer Institute and Harvard Medical School, Boston, Massachusetts 02115. *Present address of C.L.*: Department of Cell Biology, Stanford University Medical Center, Stanford, California 94305.
Subcellular Biochemistry, Volume 21: Endoplasmic Reticulum, edited by N. Borgese and J. R. Harris. Plenum Press, New York, 1993.

2. ER STRUCTURE *IN VIVO*

An important early step in this research was the development of methods for visualizing ER in whole, living cells, and the transition from electron to light microscopy that it dictated. Although electron microscopy has played a tremendously important role in the description of the ER, its usual requirement that samples be prepared as thin sections made it difficult to perceive the large-scale ER organization of whole cells. By this technique the ER typically appears as a highly complex, chaotic assembly of fragmented tubules and occasional cisternae (Fawcett, 1981). In contrast, examination of whole ER structure either by whole-mount EM or light microscopy revealed a relatively simple reticulum of interconnecting tubules (Terasaki *et al.,* 1984) (Figure 1), reminiscent of the earliest descriptions of the ER (Porter *et al.,* 1945). Consequently, recently discovered methods for visualizing the ER by fluorescence microscopy have greatly facilitated the study of its structure, by providing simple means for observing the ER and its development in whole and even live cells (Terasaki *et al.,* 1984, 1986; Lee and Chen, 1988; Terasaki, 1989). Such techniques were first employed to study the effects of disruption of microtubules or filamentous actin on ER, and showed that microtubules appeared to be essential for the long-term integrity of the ER network (Terasaki *et al.,* 1986; Louvard *et al.,* 1982). These results were supported by a variety of biochemical and morphological evidence for interaction of microtubules with membranes of the ER (for review see Lee and Chen, 1988). In this atmosphere of circumstantial evidence, microtubules seemed a logical starting point for studying the cell biological basis of ER structure, and attention turned to determining if and how they might be involved.

Fortunately, examination of ER structure *in vivo* suggested a simple physical mechanism whereby its complex reticular architecture could be formed, which further suggested a specific role for microtubules (Lee and Chen, 1988). Earlier observations had noted an unusual feature of the geometry of ER networks; nearly all of the junctions between ER tubules join three tubules (as opposed to four, five, etc.) (Lee and Chen, 1988). Based on this observation, it was proposed that ER might form through an iterative process of branching and intersection, since both tubule branching and intersection give rise to triple junctions (Lee and Chen, 1988). As an initial test of this hypothesis, a high-sensitivity videocamera was used to observe live epithelial cells (CV-1) stained with the ER-specific dye $DiOC_6(3)$, to see if such movements actually occurred. The resulting recordings of ER behavior in living cells showed it to be a highly dynamic structure, and demonstrated that tubule branching and intersection were indeed a characteristic motility of interphase ER networks (Figure 2). These movements were observed to produce the typical features of ER structure— linear tubules, polygonal reticulum, and triple junctions—consistent with the hypothesis that they were the basis of ER construction. In addition, other forms

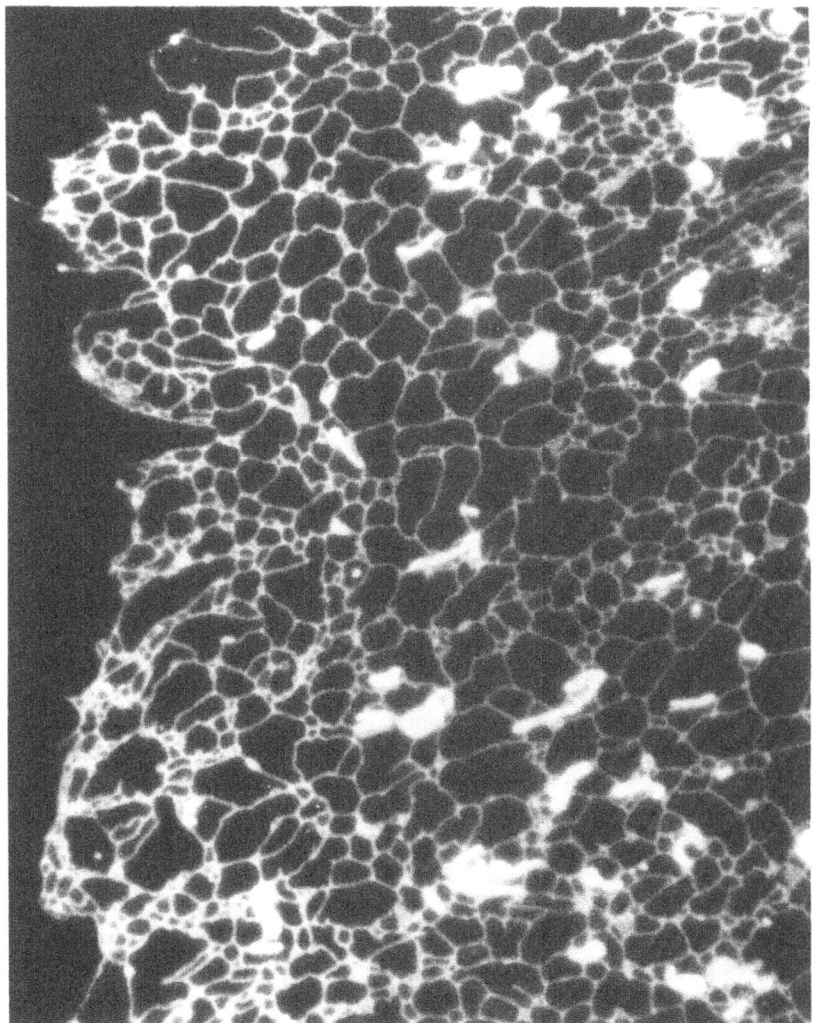

FIGURE 1. Endoplasmic reticulum in an African monkey kidney epithelial CV-1 cell stained with the fluorescent dye $DiOC_6(3)$. From Lee and Chen (1988).

of ER motility were observed that had the complementary effects of either rearranging or disassembling the reticular structure, and it was proposed that the combination of these motility effects might be responsible for determining the structure of ER in living cells. The nature of these movements suggested the involvement of the cytoskeleton, and in light of the extensive evidence of ER–microtubule interactions, it was suggested that microtubule-associated motility

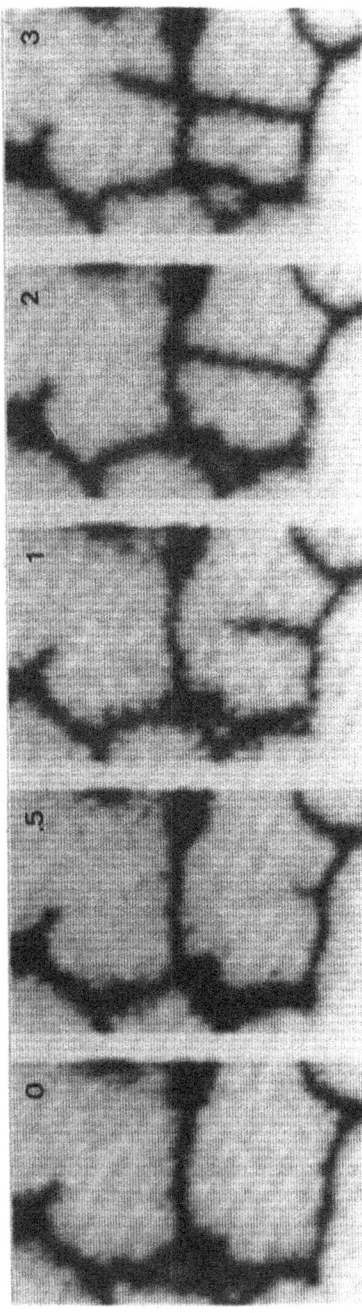

FIGURE 2. Tubule branching of the endoplasmic reticulum in a living CV-1 cell stained with DiOC$_6$(3). These images were recorded real-time on videotape; subsequently, individual frames were cut and averaged by image processor. The times for each photograph are given in seconds. From Lee and Chen (1988).

might be the basis of ER motility, and thus of ER construction (Lee and Chen, 1988).

3. *IN VITRO* MODELS OF ER CONSTRUCTION

At the same time, *in vitro* studies of microtubule motility produced a strikingly similar result: ER-like tubuloreticular networks were observed to form in purified cellular extracts through microtubule-associated motility (Dabora and Sheetz, 1988; Vale and Hotani, 1988) by a process of repeated branching and intersection. In one case, a crude extract of microsomal membranes was mixed with the cytosolic fraction under conditions promoting microtubule polymerization; upon addition of ATP, microsomal membranes were observed to move along individual microtubules, often extending to form tubules, which further branched and intersected to form elaborate polygonal networks reminiscent of ER (Dabora and Sheetz, 1988). Electron microscopic examination of the membranes in these structures revealed ribosomes bound to their surfaces, indicating that at least some of the membranes composing the network originated from ER. The branching motility observed in this system appeared similar to that reported *in vivo*, and seemed to operate through pulling of the leading edge of the membrane branch along a single microtubule. Evaluation of the biochemical requirements of this process showed that it was ATP-dependent, and similar in its vanadate and *N*-ethylmaleimide sensitivities to the microtubule-motility-generating protein kinesin (Dabora and Sheetz, 1988; Vale *et al.*, 1985). These results suggested that tubule branching motility observed *in vivo* might operate through binding of the ER membrane to a motile protein such as kinesin or dynein, which would then pull the ER tubule along a microtubule substrate.

In a second set of *in vitro* experiments, Vale and Hotani (1988) observed a similar phenomenon while assaying microtubule motility on a glass surface coated with the motile protein kinesin. Coincidentally, membrane microsomes were found to copurify with the kinesin preparation, and in the motility assay appeared to bind to moving microtubules, which pulled the membrane out into tubular structures. Over time, these tubules branched and intersected to form a polygonal network similar in appearance to the ER (Vale and Hotani, 1988). Although this work was similar to that of Dabora and Sheetz (1988) in that it showed that branching and intersection of membranes could generate network structures, it differed in two important respects. First, the minute quantities of membrane present in their preparations prevented their identification. Second, the mechanism by which microtubules were observed to generate branching motility was different. While Dabora and Sheetz observed movement of membrane branches along stationary microtubules, Vale and Hotani found membrane adhesion to motile microtubules. In the latter case, it appeared that the membranes bound directly to the microtubule, while the former suggested binding to a motile intermediary, which would then cause it to move along a microtubule substrate. Although both models indicated microtubules as the basis of the branching motility, their apparent disagreement over the mechanism by which this motion is generated made it difficult to draw definite conclusions about this process.

4. ER CONSTRUCTION IN LIVING CELLS

To resolve such uncertainties, in further *in vivo* studies we have investigated the mechanism and cytoskeletal requirements of ER motility and construction in living cells, by taking advantage of the microtubule-disrupting effects of nocodazole on ER structure (Lee *et al.*, 1989). Prolonged treatment of cultured cells with nocodazole causes the ER network to collapse and aggregate in the perinuclear region; however, upon transfer to drug-free medium the ER re-forms its normal structure within 15–20 min (Terasaki *et al.*, 1986). This experimental system permitted direct observation of the development of the ER, by staining cells with $DiOC_6(3)$ and recording its reconstruction after transfer to drug-free medium, with a high-sensitivity videocamera (Figure 3). These results confirmed that the ER's reticular configuration was formed through a process of tubule branching and intersection. More importantly, it was possible to test the role of different elements of the cytoskeleton, namely microtubules, filamentous actin, and intermediate filaments, in the motility and construction of the ER. First, drugs specifically disrupting these cytoskeletal structures were tested for their effect on ER construction. Drugs inhibiting microtubule polymerization completely blocked ER construction; in contrast, disruption of filamentous actin by cytochalasin B, and of vimentin intermediate filaments by cycloheximide, had no effect on ER construction. Furthermore, reductions of the density of microtubule polymerization by transfer to recovery medium containing various low concentrations of nocodazole produced correlated decreases in the density of ER construction; ER tubules only extended where microtubules were present, and were exactly aligned with them.

Intriguingly, hyperstabilization of microtubules by transferring the cells to recover in 20 μM taxol induced a high density of abnormally short microtubules, but dramatically inhibited construction of the ER (Lee *et al.*, 1989). Specifically, the ER failed to extend into regions where microtubules were present but were so short that they did not intersect. This apparent requirement for a contiguous microtubule network indicated that ER was constructed by branching along the network of microtubules, and is consistent with the model of Dabora and Sheetz (1988), but not with that of Vale and Hotani (1988). In the former model, the ER forms on an already existing network of microtubules; thus, the ER can only

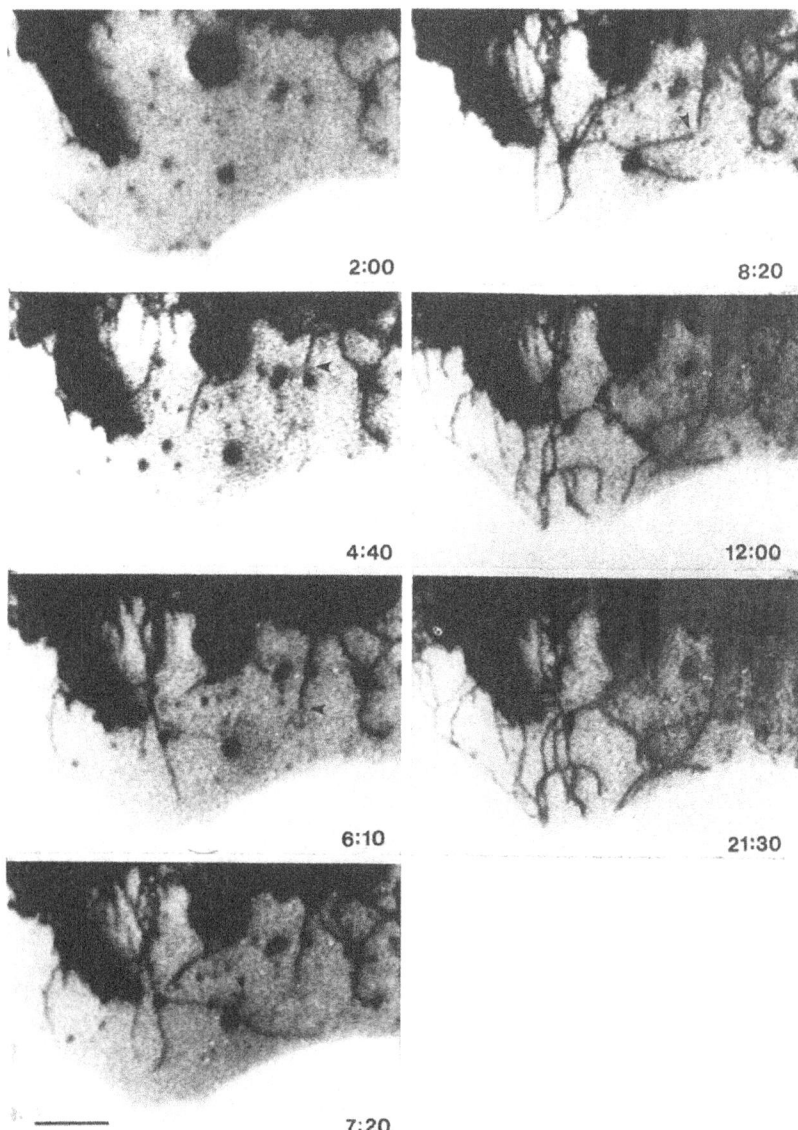

FIGURE 3. Construction of the endoplasmic reticulum during nocodazole recovery in a living CV-1 cell stained with DiOC$_6$(3). The time (min:s) after transfer to drug-free medium is indicated on each photograph. The endoplasmic reticulum network was formed by first extending long, unbranched tubules away from the perinuclear region. These tubules then formed branches that intersected to produce a reticulum. From Lee *et al.* (1989).

form a network structure in regions where the microtubules form a network. In the experiments of the latter group the membrane networks *did* extend into regions of nonintersecting microtubules, since the membrane branches were gradually built up through the individual action of passing microtubules (Vale and Hotani, 1988). The failure of the ER in living cells to extend into regions of nonintersecting microtubules thus favors the model of Dabora and Sheetz, and indicates the involvement of microtubules as a dynamic framework on which the ER is assembled through branching and intersection of ER tubules, along a primarily stationary microtubule network.

5. FUTURE PROSPECTS

It seems likely that current research on ER motility and structure will build on these results primarily by focusing on characterization of the molecules involved. First, further studies of the cytoskeletal bases of ER motility are needed to determine the involvement of actin filaments and microtubules in different species. The reports described above have focused exclusively on mammalian cells; in this case, microtubules appear to be the only element of the cytoskeleton important for ER motility (Lee and Chen, 1988; Dabora and Sheetz, 1988; Vale and Hotani, 1988; Lee *et al.*, 1989). However, studies of characean alga have shown movement of ER along actin filaments (Kachar and Reese, 1988), and it would be interesting to see what other species employ actin in the generation of ER motility, and to test the role of microtubules in such cases. At a more refined level, it would be of great interest to identify the proteins involved in microtubule-driven ER motility in mammalian cells, and it is likely that further studies of these phenomena *in vitro* will provide the means to characterize the molecular motor(s) of this process, MAPs, and the ER membrane protein(s) involved. Such work could be important not only in providing a complete description of the process whereby the ER is constructed, but also as an essential opening into questions of regulation—of the mechanisms by which the organization and distribution of the ER is specified and controlled in different cell types. Specifically, understanding of how cells establish and alter ER organization could be quite important in elucidating the mechanisms controlling the extraordinary structural changes accompanying cell division: the breakdown of the nuclear membrane, the partition of organelle membranes during cytokinesis, and the subsequent reassembly of interphase organelle structure. Understanding the molecules and interactions involved in regulating ER assembly could provide an important first glimpse of the molecular basis of such processes.

Furthermore, these investigations of ER construction suggest an interesting paradigm in which cytoskeletal motility acts as the mechanism for establishing specific organelle structures and intracellular localization. Microtubules, in par-

ticular, appear to be involved in the distribution of not only ER, but also mito-chondria, lysosomes, and the Golgi apparatus. All of these organelles have been observed in some cases to form tubular networks similar to the ER (Johnson *et al.*, 1980; Swanson *et al.*, 1987; Lin and Queally, 1982), and to undergo microtubule-associated motility (Ho *et al.*, 1987; Phaire-Washington *et al.*, 1980; Summerhayes *et al.*, 1983). Mitochondria have been directly observed to form networks through branching and intersection (Lee and Chen, unpublished results), and this motility has also been found to be microtubule-dependent (Summerhayes *et al.*, 1983). In such cases, live staining techniques can be especially useful for elucidating how organelle structures and distributions are formed, by directly observing their development. Furthermore, nocodazole re-covery may be useful for studying the role of microtubules in the organization of these structures. In cells treated overnight with nocodazole, all phase-contrast visible membrane organelles are aggregated in the region around the nucleus; however, within 2–4 h after transfer to drug-free medium, the intracellular distribution of organelles appears normal (Lee and Chen, unpublished results). This system may therefore be useful for studying the process through which other organelles are organized into their normal distribution, and for elucidating both the similarities and differences of microtubule involvement in the structure of many different organelles.

6. REFERENCES

Dabora, S. L., and Sheetz, M. P., 1988, The microtubule-dependent formation of a tubulovesicular network with characteristics of the ER from cultured cell extracts, *Cell* **54:**27–35.

Fawcett, D. W., 1981, The endoplasmic reticulum, in: *The Cell*, Saunders, Philadelphia.

Ho, W. C., Allan, V. J., van Meer, G., and Kreis, T., 1987, Interaction of the Golgi apparatus with microtubules, *J. Cell Biol.* **105:**262a.

Johnson, L. V., Walsh, M. L., and Chen, L. B., 1980, Localization of the mitochondria in living cells with rhodamine 123, *Proc. Natl. Acad. Sci. USA* **77:**990–994.

Kachar, B., and Reese, T. S., 1988, The mechanism of cytoplasmic streaming in characean algal cells: Sliding of endoplasmic reticulum along actin filaments, *J. Cell Biol.* **106:**1545–1552.

Lee, C., and Chen, L. B., 1988, Dynamic behavior of endoplasmic reticulum in living cells, *Cell* **54:**37–46.

Lee, C., Ferguson, M. K., and Chen, L. B., 1989, Construction of the endoplasmic reticulum, *J. Cell Biol.* **109:**2045–2055.

Lin, J. J., and Queally, S. A., 1982, A monoclonal antibody that recognizes Golgi-associated protein of cultured fibroblast cells, *J. Cell Biol.* **92:**108–112.

Louvard, D., Reggio, H., and Warren, G., 1982, Antibodies to the Golgi complex and endoplasmic reticulum, *J. Cell Biol.* **92:**92–107.

Palade, G. E., 1956, The endoplasmic reticulum, *J. Biophys. Biochem. Cytol.* **2:**85–98.

Phaire-Washington, L., Silverstein, S. C., and Wang, E., 1980, Phorbol myristate acetate stimulates microtubule and 10-nm filament extension and lysosome redistribution in mouse macrophages, *J. Cell Biol.* **86:**641–655.

Porter, K. R., Claude, A., and Fullam, E. F., 1945, A study of tissue culture cells by electron microscopy: Methods and preliminary observations, *J. Exp. Med.* **81**:233–241.

Summerhayes, I. C., Wong, D., and Chen, L. B., 1983, Effect of microtubules and intermediate filaments on mitochondrial distribution, *J. Cell Sci.* **61**:87–105.

Swanson, J., Bushnell, A., and Silverstein, S. C., 1987, Tubular lysosome morphology and distribution within macrophages depend on the integrity of cytoplasmic microtubules, *Proc. Natl. Acad. Sci. USA* **84**:1921–1925.

Terasaki, M., 1989, Fluorescent labeling of endoplasmic reticulum, *Methods Cell Biol.* **29**:125–135.

Terasaki, M., Song, J., Wong, J. R., Weiss, M. J., and Chen, L. B., 1984, Localization of endoplasmic reticulum in living and glutaraldehyde-fixed cells with fluorescent dyes, *Cell* **38**:101–108.

Terasaki, M., Chen, L. B., and Fujiwara, K., 1986, Microtubules and the endoplasmic reticulum are highly interdependent structures, *J. Cell Biol.* **103**:1557–1568.

Vale, R. D., and Hotani, H., 1988, Formation of membrane networks *in vitro* by kinesin-driven microtubule movement, *J. Cell Biol.* **107**:2233–2241.

Vale, R. D., Reese, T. S., and Sheetz, M. P., 1985, Identification of a novel force-producing protein, kinesin, involved in microtubule-based motility, *Cell* **43**:39–50.

Index

The manufacturer's authorised representative in the EU is Springer
Nature Customer Service Centre GmbH, Europaplatz 3, 69115 Heidelberg,
Germany. If you have any concerns regarding our products, please
contact ProductSafety@springernature.com

Printed and bound by CPI Group (UK) Ltd, Croydon, CR0 4YY
24/04/2026
02096348-0001